SINGULAR PERTURBATION METHODS IN CONTROL: ANALYSIS AND DESIGN

SINGULAR PERTURBATION METHODS IN CONTROL: ANALYSIS AND DESIGN

PETAR V. KOKOTOVIĆ

Co-ordinated Science Laboratory
University of Illinois
Urbana, USA

HASSAN K. KHALIL

Department of Electrical Engineering and Systems Science
Michigan State University
East Lansing, USA

JOHN O'REILLY

Department of Electronic and Electrical Engineering
University of Strathclyde
Glasgow, UK

1986

ACADEMIC PRESS
Harcourt Brace Jovanovich, Publishers

London • Orlando • San Diego • New York • Boston
Austin • Tokyo • Sydney • Toronto

ACADEMIC PRESS INC. (LONDON) LTD
24/28 Oval Road, London NW1 7DX

United States Edition Published by
ACADEMIC PRESS INC.
Orlando, Florida 32887

British Library Cataloguing in Publication Data
Kokotović, Petar
Singular perturbation methods in control:
analysis and design
ISSN 0076-5392
1. automatic control—Mathematical models
I. Title II. Khalil, Hassan K.
III. O'Reilly, John IV. Series
629.8'312 TJ213
ISBN 0-12-417635-6

PREFACE

In this book, control theory merges with singular perturbation techniques to form a two-time-scale methodology for the modeling, analysis and design of control systems. Its goal is to simplify the software and hardware implementation of control algorithms, while improving their robustness properties.

The text is intended for control engineers and graduate students who seek an introduction to singular perturbation methods in control. At the same time, the book aims to provide research workers with sketches of significant current developments and hints of important future problems in the areas of robust, adaptive, stochastic and nonlinear control. No previous knowledge of singular perturbation techniques is assumed.

Ever since Prandtl's work at the beginning of this century, singular perturbation techniques have been a traditional tool of fluid dynamics. Their use spread to other areas of mathematical physics and engineering, where the same terminology of "boundary layers" and "inner" and "outer" matched asymptotic expansions continued to be used. In control systems, boundary layers are a characteristic of system two-time-scale behavior. They appear as initial and terminal "*fast transients*" of state trajectories and represent the "*high-frequency*" parts of the system response. High-frequency and low-frequency models of electrical circuits, which have had a long history of their own, are naturally incorporated in the new two-time-scale methodology.

In the control literature, the singular perturbation approach of Tikhonov (1952) and Vasil'eva (1963) was first applied to optimal control and regulator design by Kokotović and Sannuti (1968) and Sannuti and Kokotović (1969a, b) and, more specifically, to flight-path optimization by Kelley and Edelbaum (1970) and Kelley (1970a, b, 1971a, b). Applications to broader classes of control problems followed at an increasing rate, as evidenced by more than 500 references† surveyed by Kokotović (1984) and Saksena, O'Reilly and Kokotović (1984). For control engineers, singular perturbations are a means of taking into account neglected high-frequency phenomena and considering them in a separate fast time-scale. This is achieved by treating a change in the dynamic order of a system of differential equations as a parameter perturbation, which, being more abrupt than a regular perturbation, is called a singular perturbation. The practical advantages of such a "parameterization" of changes in model order are significant, because the order of every real dynamic system is higher than that of the model used to represent the system.

Chapter 1 shows that the parameterization of the model order using a perturbation parameter ε to multiply the derivatives of some of the state variables is common to most physical systems with small time constants, inertias and other physical quantities which can be expressed as multiples of ε. Neglecting these quantities means setting $\varepsilon = 0$, thus eliminating some of the derivatives from the model and hence reducing its order. It is further shown how a broader class of dynamic systems can be brought to this "standard" singular perturbation form. A geometric property of this form is that it possess a slow manifold which is an equilibrium manifold for the fast phenomena. Examples in Chapter 1 include models of DC motors, PID controllers, aircraft, voltage regulators and adaptive systems. Their study is pursued in subsequent chapters.

Chapter 2 is dedicated to properties of linear time-invariant systems that exhibit two-time-scale behavior because of the presence of both a group of small eigenvalues and a group of large eigenvalues. These systems can be transformed into a slow and a fast part, each of which can be separately analyzed for stability, controllability and other system properties. Readers familiar with the elements of linear systems theory (Kailath, 1980; Chen, 1984) but less accustomed to some of the nonlinear topics in Chapter 1 may choose Chapter 2 as their point of departure.

Chapter 3 deals with linear feedback control design for linear time-invariant systems. For state-feedback control, design proceeds by way of an exact decomposition of the state-feedback control problem, be it an eigenvalue assignment problem or an optimal linear regulator problem, into separate slow and fast problems. Advantage is thereby taken of the singularly perturbed nature of the problem to design a well-conditioned composite feedback control, the sum of slow and fast controls, which solves the original ill-conditioned control problem to within a specified order-of-ε accuracy. Conditions are also provided for overcoming the fact that, unlike state feedback, static output-feedback design based upon a slow

† A selection from these references is reprinted in a volume published by IEEE Press, edited by Kokotović and Khalil (1986).

model is nonrobust in the sense that it may not stabilize the original system. The only prerequisite for Chapter 3 is a basic knowledge of linear time-invariant control theory such as that contained in Kwakernaak and Sivan (1972) or Anderson and Moore (1971).

Singularly perturbed systems with white-noise inputs are more complex because in the limit their fast variables also behave as white-noise processes. In optimization problems this leads to unbounded functionals. Chapter 4 shows that a two-time-scale near-optimal design of filters and regulators is possible with major savings in on-line as well as off-line computations. Reading Chapter 4 requires a familiarity with linear least-squares estimation and linear–quadratic–Gaussian optimal control at the level presented in Kwakernaak and Sivan (1972).

The remaining three chapters deal with time-varying and nonlinear problems, where the benefits of a separation of time scales are even more pronounced. A perturbation approach is often the most effective, and sometimes the only way to solve such problems.

Time-varying systems in Chapter 5 differ from their time-invariant counterparts in Chapter 2 owing to time variations of the coefficients as a potential new cause of multiple-time-scale behavior. It is shown, however, that if the parameter variations are slow compared with the fast dynamics, then the time-scale phenomena remain qualitatively the same as in time-invariant systems. The methods of Chapter 5 take advantage of this structure by introducing a "frozen" parameter treatment of the fast system and showing when this conceptually appealing approximation is valid. Some familiarity with linear time-varying systems, as presented in, for example, Chen (1984) or Miller and Michel (1982), is assumed. Results of Chapter 5 are used in both Chapters 6 and 7, which, although closely related, can be studied independently of each other.

Control problems in Chapter 6 are of the trajectory optimization type. They are restricted to finite time intervals, and hence must take into account the fast phenomena at both ends of the trajectory. Initial and end conditions for these "boundary layer" phenomena must be properly "matched". Chapter 6 circumvents the matching difficulty by an explicit use of the Hamiltonian property of optimization problems. Thanks to this property the initial and end layers can be separated by an exact transformation. Chapter 6 assumes a knowledge of a standard course on optimal control such as Bryson and Ho (1975) or Athans and Falb (1966).

Nonlinear stability and stabilization problems in Chapter 7 are greatly simplified by exploitation of the two-time-scale system property. A Lyapunov function for a nonlinear singularly perturbed system is constructed via a two-time-scale stability analysis, which is in many respects analogous to the linear analysis of Chapter 5 because the slow states appear as slowly varying parameters in the fast system. A two-stage state feedback design, the so-called composite control design, is used to obtain stabilizing and near-optimal feedback controllers. Unlike the finite-time near-optimal control of Chapter 6, the feedback nature of the near-optimal composite control is required for stabilization over an infinite time interval. For Chapter 7, an acquaintance with the elements of nonlinear systems analysis such as are contained in Vidyasagar (1978) or Millar and Michel (1982) is assumed.

The chapters of the book may be studied sequentially or in a number of other ways. For example, readers interested in stochastic control would concentrate on Chapter 4 after familiarizing themselves with the contents of Chapter 3 on linear feedback control. Chapter 5 on time-varying systems could be read immediately after Chapter 2. Other possibilities are that Chapter 6 on optimal control and Chapter 7 on nonlinear systems could be proceeded to immediately after Chapter 1 and Chapter 5.

May 1986

P. V. Kokotović
H. K. Khalil
J. O'Reilly

CONTENTS

ACKNOWLEDGEMENTS

From its initiation almost two decades ago, the research in singular perturbation methods for control system analysis and design described in these pages has been encouraged and fostered by M. E. Van Valkenburg, J. B. Cruz, Jr., W. R. Perkins and other friends at the Coordinated Science Laboratory and the Electrical and Computer Engineering Department, University of Illinois at Urbana. The Department of Electrical Engineering and Systems Science, Michigan State University, East Lansing and the Department of Electronic and Electrical Engineering, University of Strathclyde, Glasgow have also been highly supportive of this effort.

While responsibility for any errors or shortcomings in this book rests with the authors, its inspiration and content owe much to many colleagues, friends, co-workers and supporters. We are particularly thankful for the contributions of K. W. Chang, J. H. Chow, Z. Gajić, A. H. Haddad, F. Hoppensteadt, P. A. Ioannou, S. H. Javid, R. E. O'Malley, Jr., G. Peponides, A. Saberi, P. Sannuti, V. I. Utkin, A. B. Vasil'eva, R. R. Wilde, R. A. Yackel and K. K. D. Young. More specialized or advanced contributions, helpful in the preparation of this text but going beyond its scope, are acknowledged in the Notes and References.

Grant support from the following sources is gratefully acknowledged: for P. V. Kokotović from the National Science Foundation Grant ECS 83-11851 and the Joint Services Electronics Program under Contract N00014-84-C-0149; for H. K. Khalil under the National Science Foundation Grants ECS 82-05337 and ECS 84-10649; and for J. O'Reilly from the Carnegie Trust for the Universities of Scotland. During the Spring Semester of 1984, the support of P. V. Kokotović by the Keating-Crawford Chair of Engineering, University of Notre Dame, Indiana, provided an excellent environment for much of the joint writing.

Finally, thanks are due to Rose Harris and Dixie Murphy for their unceasing efforts in typing the many drafts that went to make the major portion of the book, and also to Bob MacFarlane for the artwork, all of whom are with the Coordinated Science Laboratory of the University of Illinois.

1 TIME-SCALE MODELING

1.1 Introduction

When a control engineer uses singular perturbation methods to solve problems in his field, the first problem he faces is one of modeling, that is, how to mathematically describe the system to be controlled. Modeling for control is parsimonious and implicit. It is parsimonious, because the model should not be more detailed than that required by the specific control task. It is implicit, because the extent of the necessary detail is not known before the control task is accomplished. Typical control tasks are optimal regulation, tracking and guidance. Since these tasks are to be accomplished in the inevitable presence of unknown disturbances, parameter variations and other uncertainties, the control system must possess a sufficient degree of robustness or insensitivity to the extraneous effects.

How do singular perturbation techniques respond to this challenge? Their key contribution, from which all other benefits follow, is at the level of modeling. Control engineers have been simplifying their models long before they were told that what they were doing was a singular perturbation. As our bibliography shows, they became aware of the new tool about fifteen years ago and have been increasingly interested in it ever since.

For the control engineer, singular perturbations legitimize *ad hoc* simplifications of dynamic models. One of them is the neglect of "small" time constants, masses, capacitances, and similar "parasitic" parameters which increase the dynamic order of the model. However, a design based on a simplified model may result in a system far from its desired performance, or even in an unstable system. If this happens, the control engineer needs a tool that will help him to improve his oversimplified design. What is required is to treat the simplified design as a first step, which captures the dominant phenomena; then the disregarded phenomena, if important, are treated in the second step.

It turns out that asymptotic expansions into reduced ("outer") and boundary-layer ("inner") series, which are the main characteristics of singular perturbation techniques, coincide with the outlined design stages. Because most control systems are dynamic, the decomposition into stages is dictated by a separation of time scales. Typically, the reduced model represents the slowest (average) phenomena, which in most applications are dominant. Boundary layer (and sublayer) models evolve in faster time scales and represent deviations from the predicted slow behavior. The goal of the second, third, and later, design stages is to make the boundary layers and sublayers asymptotically stable, so that the deviations rapidly decay. The separation of time scales also eliminates stiffness difficulties and prepares for a more efficient hardware and software implementation of the controller.

In this chapter, some of the basic concepts of singular perturbation asymptotics and time-scale modeling are introduced by way of illustrative examples. So as to develop a more intuitive feel for time-scale phenomena, technical details and proofs are postponed to subsequent chapters. Since the rest of the book is design-oriented, some of the modeling examples, developed in this chapter, will later be used to illustrate various analysis and design results.

1.2 The Standard Singular Perturbation Model

The singular perturbation model of finite-dimensional dynamic systems, extensively studied in the mathematical literature by Tikhonov (1948, 1952), Levinson (1950), Vasil'eva (1963), Wasow (1965), Hoppensteadt (1967, 1971), O'Malley (1971), etc., was also the first model to be used in control and systems theory. This model is in the explicit state-variable form in which the derivatives of some of the states are multiplied by a small positive scalar ε, that is,

$$\dot{x} = f(x, z, \varepsilon, t), \quad x(t_0) = x^0, \quad x \in R^n, \tag{2.1}$$

$$\varepsilon \dot{z} = g(x, z, \varepsilon, t), \quad z(t_0) = z^0, \quad z \in R^m \tag{2.2}$$

where a dot denotes a derivative with respect to time t, and f and g are assumed to be sufficiently many times continuously differentiable functions of their arguments x, z, ε, t. The scalar ε represents all the small parameters to be neglected. In most applications, having a single parameter is not a restriction. For example, if T_1 and T_2 are small time constants of the same order of magnitude, $O(T_1) = O(T_2)$, then one of them can be taken as ε and the other expressed as its multiple, say $T_1 = \varepsilon$, $T_2 = \alpha\varepsilon$, where $\alpha = T_2/T_1$ is a known constant. We postpone the discussion of important scaling issues and the meaning of the O-symbol until Section 1.7. In this section, our goal is to introduce ε as a modeling tool.

In control and systems theory, the model (2.1), (2.2) is a step towards "reduced-order modeling", a common engineering task. The order reduction is converted into a parameter perturbation, called "singular". When we set $\varepsilon = 0$, the dimension of the state space of (2.1), (2.2) reduces from $n + m$ to n because the differential equation (2.2) degenerates into the algebraic or transcendental equation

$$0 = g(\bar{x}, \bar{z}, 0, t), \tag{2.3}$$

where the bar is used to indicate that the variables belong to a system with $\varepsilon = 0$. We will say that the model (2.1), (2.2) is in *standard form* if and only if the following crucial assumption concerning (2.3) is satisfied.

Assumption 2.1

In a domain of interest (2.3) has $k \geqq 1$ distinct ("isolated") real roots

$$\bar{z} = \bar{\phi}_i(\bar{x}, t), \quad i = 1, 2, \ldots, k. \tag{2.4}$$

This assumption ensures that a well-defined n-dimensional reduced model will correspond to each root (2.4). To obtain the ith reduced model, we substitute (2.4) into (2.1),

$$\dot{\bar{x}} = f(\bar{x}, \bar{\phi}_i(\bar{x}, t), 0, t), \quad \bar{x}(t_0) = x^0, \tag{2.5}$$

and keep the same initial condition for the state variable $\bar{x}(t)$ as for $x(t)$. In the sequel, we shall drop the subscript i and rewrite (2.5) more compactly as

$$\dot{\bar{x}} = \bar{f}(\bar{x}, t), \quad \bar{x}(t_0) = x^0. \tag{2.6}$$

This model is sometimes called a *quasi-steady-state model*, because z, whose velocity $\dot{z} = g/\varepsilon$ can be large when ε is small, may rapidly converge to a root of (2.3), which is the quasi-steady-state form of (2.2). We will discuss this two-time-scale property of (2.1), (2.2) in the next section.

The convenience of using a parameter to achieve order reduction also has a drawback: it is not always clear how to pick the parameters to be considered as small. Fortunately, in many applications our knowledge of physical processes and components of the system sets us on the right track. Let us illustrate this by examples. In all four examples the control input u is a given differentiable function of time, $u = u(t)$.

Example 2.1

A common model for most DC motors, shown in Fig. 1.1, consists of a mechanical torque equation and an equation for the electrical transient in the armature circuit, namely,

$$J\dot{\omega} = ki, \tag{2.7}$$

$$L\dot{i} = -k\omega - Ri + u, \tag{2.8}$$

where i, u, R and L are the armature current, voltage, resistance and inductance respectively, J is the moment of inertia, ω is the angular speed,

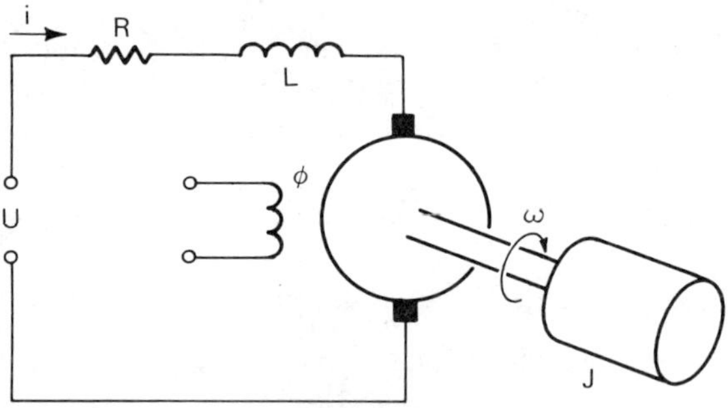

Fig. 1.1. Armature controlled DC motor.

and ki and $k\omega$ are respectively the torque and the back e.m.f. developed with constant excitation flux ϕ. In practically all well designed motors, L is small and can play the role of our parameter ε. This means that $\omega = x$, $i = z$ and the model (2.7), (2.8) is in the standard form (2.1), (2.2) whenever $R \neq 0$. Neglecting L, we solve

$$0 = -k\bar{\omega} - R\bar{i} + u$$

to obtain

$$\bar{i} = \frac{u - k\bar{\omega}}{R}, \tag{2.9}$$

which is the only root (2.4), and substitute it in (2.7). The resulting model

$$J\dot{\bar{\omega}} = -\frac{k^2}{R}\bar{\omega} + \frac{k}{R}u \tag{2.10}$$

is the commonly used first-order model of the DC motor in the form of (2.6). We shall return to (2.7), (2.8) for a discussion on scaling in Section 1.7.

Example 2.2

In the feedback system of Fig. 1.2(a) with a high-gain amplifier K and a nonlinear block $N(z)$, the choice of ε is not as obvious. However, any

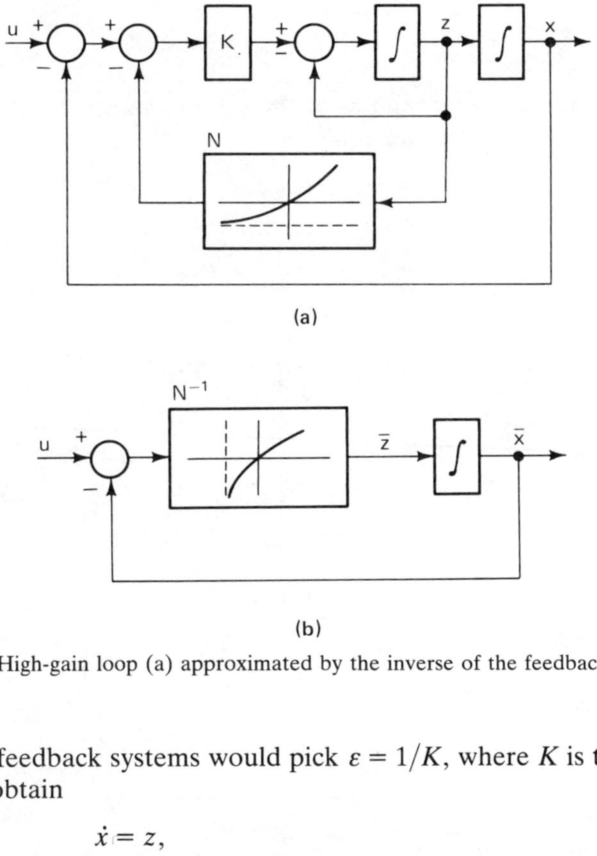

Fig. 1.2. High-gain loop (a) approximated by the inverse of the feedback path (b).

student of feedback systems would pick $\varepsilon = 1/K$, where K is the amplifier gain, and obtain

$$\dot{x} = z, \tag{2.11}$$

$$\varepsilon \dot{z} = -x - \varepsilon z - N(z) + u. \tag{2.12}$$

To be specific, let $N(z) = e^z - 1$ and obtain from (2.3) and (2.4) applied to (2.12),

$$0 = -\bar{x} - 0 - e^{\bar{z}} + 1 + u,$$

or

$$\bar{z} = N^{-1}(u - \bar{x}) = \ln(1 + u - \bar{x}). \tag{2.13}$$

Hence Assumption 2.1 is satisfied whenever $u - \bar{x} > -1$. Then, the reduced model is

$$\dot{\bar{x}} = \ln(1 + u - \bar{x}). \tag{2.14}$$

This model is represented by the block diagram in Fig. 1.2(b) in which the loop with infinite gain $\varepsilon = 0$ is replaced by the inverse of the operator in the feedback path.

Example 2.3

Using capacitor voltages v_1 and v_2 as the state variables, the *RC* circuit in Fig. 1.3(a) is described by the equations

$$R_1 C_1 \dot{v}_1 = -v_1 + v_2, \tag{2.15}$$

$$R_2 C_2 \dot{v}_2 = \frac{R_2}{R_1} v_1 - \left(1 + \frac{R_2}{R_1}\right) v_2 + u. \tag{2.16}$$

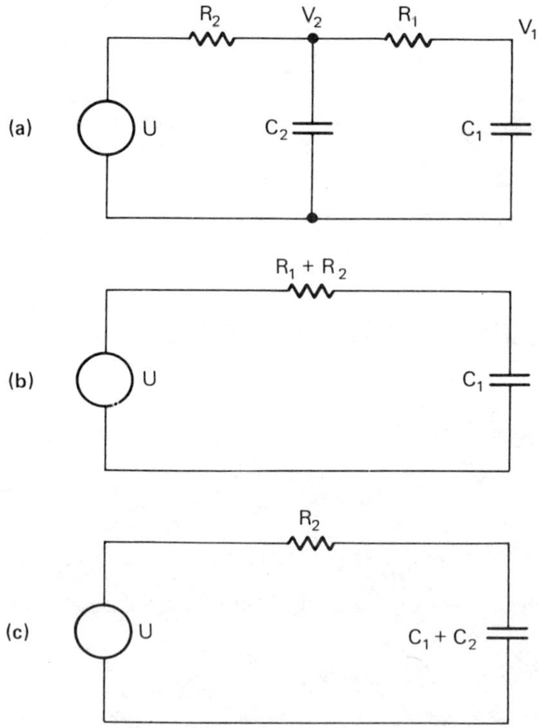

Fig. 1.3. Low-frequency "equivalents" of the circuit (a) when C_2 is small (b) and when R_1 is small (c).

Suppose first that the capacitor C_2 is small; that is, let $C_2 = \varepsilon$. Then $v_1 = x$, $v_2 = z$, and the root (2.4) obtained after letting $\varepsilon = 0$ in (2.16) is

$$\bar{v}_2 = \frac{R_2}{R_1 + R_2}\bar{v}_1 + \frac{R_1}{R_1 + R_2}u, \tag{2.17}$$

which represents an ideal voltage divider formed by the series connection of R_1 and R_2. When (2.17) is substituted into (2.15), the reduced model of the RC circuit is

$$(R_1 + R_2)C_1\dot{\bar{v}}_1 = -\bar{v}_1 + u \tag{2.18}$$

and represents the circuit with C_2 removed, as in Fig. 1.3(b), a common "low-frequency equivalent" of the original circuit in Fig. 1.3(a).

In all the examples considered thus far, the original models are in the standard form (2.1), (2.2) because Assumption 2.1 is obviously satisfied. To avoid a misleading conclusion that this is always the case, let us reconsider the RC circuit in Fig. 1.3, but this time assuming that the small parameter is the resistance R_1.

Example 2.4

Multiplying (2.16) by R_1/R_2 and letting $R_1 = \varepsilon$, (2.15) and (2.16) become

$$\varepsilon\dot{v}_1 = \frac{1}{C_1}[-v_1 + v_2], \tag{2.19}$$

$$\varepsilon\dot{v}_2 = \frac{1}{C_2}\left[v_1 - \left(1 + \frac{\varepsilon}{R_2}\right)v_2 + \frac{\varepsilon}{R_2}u\right]. \tag{2.20}$$

If this model were in the form (2.1), (2.2), both v_1 and v_2 would be considered as z-variables and (2.3) would be

$$0 = -\bar{v}_1 + \bar{v}_2. \tag{2.21}$$

However, Assumption 2.1 would then be violated because the roots of (2.3), in this case $\bar{v}_1 = \bar{v}_2$, are not distinct. The question remains as to whether the model of this RC network can be simplified by setting $\varepsilon = 0$, that is, by neglecting the small parasitic resistance R_1. For a simple answer we multiply (2.19) by C_1, (2.20) by C_2, add them together and obtain an equation without ε, that is,

$$C_1\dot{v}_1 + C_2\dot{v}_2 = -\frac{1}{R_2}v_2 + \frac{1}{R_2}u. \tag{2.22}$$

This suggests that instead of v_1 we can use

$$x = \frac{C_1 v_1 + C_2 v_2}{C_1 + C_2} \tag{2.23}$$

as a new voltage variable, which, along with $v_2 = z$, transforms (2.19), (2.20) into

$$\dot{x} = \frac{1}{R_2(C_1 + C_2)}[-z + u], \tag{2.24}$$

$$\varepsilon\dot{z} = \left(\frac{1}{C_1} + \frac{1}{C_2}\right)x - \left(\frac{1}{C_1} + \frac{1}{C_2} + \frac{\varepsilon}{R_2 C_2}\right)z + \frac{\varepsilon}{R_2 C_2}u. \tag{2.25}$$

Now (2.3) becomes

$$0 = \left(\frac{1}{C_1} + \frac{1}{C_2}\right)\bar{x} - \left(\frac{1}{C_1} + \frac{1}{C_2}\right)\bar{z} \tag{2.26}$$

and it satisfies Assumption 2.1. The substitution of $\bar{z} = \bar{x}$ into (2.24) results in the reduced model

$$\dot{\bar{x}} = \frac{1}{R_2(C_1 + C_2)}[-\bar{x} + u] \tag{2.27}$$

describing the circuit in Fig. 1.3(c). Every electrical engineer would propose this circuit as a "low-frequency equivalent" of the circuit in Fig. 1.3(a) when $R_1 = \varepsilon$ is small.

Most of the singular perturbation literature assumes that the model (2.1), (2.2) is in the standard form; that is, it satisfies Assumption 2.1. The importance of Example 2.4 is that it points out the dependence of Assumption 2.1 on the choice of state variables. In most applications a goal of modeling is to remain close to the original "physical" variables. This was possible in our Examples 2.1, 2.2, 2.3, but not in Example 2.4, where a new voltage variable (2.23) had to be introduced. However, few engineers, accustomed to the simplified "equivalent" circuit in Fig. 1.3(c), would question the "physicalness" of this new variable. On the contrary, physical properties of the circuit in Fig. 1.3(a) are more clearly displayed by the standard form (2.24), (2.25).

For a simple RC circuit we have been able to find a way to convert a nonstandard model into the standard form. A systematic approach to nonstandard models will be presented in Section 1.6.

1.3 Time-Scale Properties of the Standard Model

Singular perturbations cause a multi-time-scale behavior of dynamic systems characterized by the presence of both slow and fast transients in the system response to external stimuli. Loosely speaking, the slow response, or the "quasi-steady-state", is approximated by the reduced model (2.6), while the discrepancy between the response of the reduced model (2.6) and that of the full model (2.1), (2.2) is the fast transient. To see this, let us return to (2.1)–(2.6) and examine the variable z which has been excluded from the reduced model (2.6) and substituted by its "quasi-steady-state" $\bar{z}$. By contrast with the original variable z, starting at t_0 from a prescribed z^0, the quasi-steady-state $\bar{z}$ is not free to start from z^0, and there may be a large discrepancy between its initial value

$$\bar{z}(t_0) = \bar{\phi}(\bar{x}(t_0), t_0) \tag{3.1}$$

and the prescribed initial condition z^0. Thus, $\bar{z}$ cannot be a uniform approximation of z. The best we can expect is that the approximation

$$z = \bar{z}(t) + O(\varepsilon) \tag{3.2}$$

will hold on an interval excluding t_0, that is, for $t \in [t_1, T]$ where $t_1 > t_0$. However, we can constrain the quasi-steady-state $\bar{x}$ to start from the prescribed initial condition x^0, and hence the approximation of x by $\bar{x}$ may be uniform. In other words,

$$x = \bar{x}(t) + O(\varepsilon) \tag{3.3}$$

may hold on an interval including t_0, that is, for all t in the interval $[t_0, T]$ on which $\bar{x}(t)$ exists.

The approximation (3.2) establishes that during an initial ("boundary layer") interval $[t_0, t_1]$ the original variable z approaches $\bar{z}$ and then, during $[t_1, T]$, remains close to $\bar{z}$. Let us remember that the speed of z can be large, $\dot{z} = g/\varepsilon$. In fact, having set ε equal to zero in (2.2), we have made the transient of z instantaneous whenever $g \neq 0$. Will z escape to infinity during this transient or converge to its quasi-steady-state $\bar{z}$?

To answer this question let us analyze $\varepsilon\dot{z}$, which may remain finite, even when ε tends to zero and $\dot{z}$ tends to infinity. We set

$$\varepsilon\frac{dz}{dt} = \frac{dz}{d\tau}, \quad \text{hence } \frac{d\tau}{dt} = \frac{1}{\varepsilon}, \tag{3.4}$$

and use $\tau = 0$ as the initial value at $t = t_0$. The new time variable

$$\tau = \frac{t - t_0}{\varepsilon}, \quad \tau = 0 \text{ at } t = t_0, \tag{3.5}$$

is "stretched", that is, if ε tends to zero, τ tends to infinity even for fixed t only slightly larger than t_0. On the other hand, while z and τ almost instantaneously change, x remains very near its initial value x^0. To describe the behavior of z as a function of τ we use the so-called "boundary layer correction" $\hat{z} = z - \bar{z}$ satisfying the "boundary layer system"

$$\frac{d\hat{z}}{d\tau} = g(x^0, \hat{z}(\tau) + \bar{z}(t_0), 0, t_0) \tag{3.6}$$

with the initial condition $z^0 - \bar{z}(t_0)$, and x^0, t_0 fixed parameters. The solution $\hat{z}(\tau)$ of this initial value problem is used as a "boundary layer" correction of (3.2) for a possibly uniform approximation of z:

$$z = \bar{z}(t) + \hat{z}(\tau) + O(\varepsilon). \tag{3.7}$$

Clearly, $\bar{z}(t)$ is the slow transient of z, and $\hat{z}(\tau)$ is the fast transient of z. For the corrected approximation (3.7) to converge, after a short period, to the slow approximation (3.2), the correction term $\hat{z}(\tau)$ must decay as $\tau \to \infty$ to an $O(\varepsilon)$ quantity. Note that in the slow time scale t this decay is rapid since

$$\frac{d\hat{z}(\tau)}{dt} = \frac{d\hat{z}(\tau)}{d\tau}\frac{d\tau}{dt} = \frac{1}{\varepsilon}\frac{d\hat{z}(\tau)}{d\tau}.$$

The stability properties of the boundary layer system (3.6), which are crucial for the approximations (3.2), (3.3) and (3.7) to hold, are now stated as two separate assumptions.

Assumption 3.1

The equilibrium $\hat{z}(\tau) = 0$ of (3.6) is asymptotically stable uniformly in x^0 and t_0, and $z^0 - \bar{z}(t_0)$ belongs to its domain of attraction, so $\hat{z}(\tau)$ exists for $\tau \geqq 0$.

If this assumption is satisfied, then

$$\lim_{\tau \to \infty} \hat{z}(\tau) = 0 \tag{3.8}$$

uniformly in x^0, t_0; that is, z will come close to its quasi-steady-state $\bar{z}$ at some time $t_1 > t_0$. To ensure that z stays close to $\bar{z}$, we think as if any instant $t \in [t_1, T]$ can be the initial instant, and make the following assumption about the linearization of (3.6).

Assumption 3.2

The eigenvalues of $\partial g/\partial z$ evaluated, for $\varepsilon = 0$, along $\bar{x}(t), \bar{z}(t)$, have real parts smaller than a fixed negative number, i.e.

$$\mathrm{Re}\,\lambda\left\{\frac{\partial g}{\partial z}\right\} \leqq -c < 0. \tag{3.9}$$

Both assumptions describe a strong stability property of the boundary layer system (3.6). If z^0 is assumed to be sufficiently close to $\bar{z}(t_0)$, then Assumption 3.2 encompasses Assumption 3.1. We also note from (3.9) that the nonsingularity of $\partial g/\partial z$ implies that the root $\bar{z}(t)$ is distinct as required by Assumption 2.1. These assumptions are common in much of the singular perturbation literature (e.g. Tikhonov, 1948, 1952; Levinson, 1950; Vasil'eva, 1963; Hoppensteadt, 1971). These references contain the proof and refinements of the following fundamental theorem.

Theorem 3.1

If Assumptions 3.1 and 3.2 are satisfied, then the approximation (3.3), (3.7) is valid for all $t \in [t_0, T]$, and there exists $t_1 \geqslant t_0$ such that (3.2) is valid for all $t \in [t_1, T]$.

We will refer to this result as "Tikhonov's theorem", although the form in which it is stated here is due to Vasil'eva and other authors. We do not give the proof of this theorem. Instead, we provide independent proofs of its special cases as they are needed in subsequent chapters of this book. This approach is more in tune with control applications, in which requirements (3.8) and (3.9) are to be satisfied by controller design. Let us only point out that the proof of Theorem 3.1 makes use of two-time-scale asymptotic expansions of x, z, having some terms defined in t-scale and others in τ-scale. We illustrate this technique by substituting only the first terms

$$x = \bar{x}(t) + \hat{x}(\tau), \quad z = \bar{z}(t) + \hat{z}(\tau) \tag{3.10}$$

into (2.1), (2.2). Then, expressing the derivatives of x and z with respect to both t and τ, using (3.4), the system (2.1), (2.2) is rewritten as

$$\dot{\bar{x}}(t) + \frac{\mathrm{d}\hat{x}(\tau)}{\mathrm{d}\tau}\frac{\mathrm{d}\tau}{\mathrm{d}t} = f(\bar{x} + \hat{x}, \bar{z} + \hat{z}, \varepsilon, t), \tag{3.11}$$

$$\varepsilon\dot{\bar{z}}(t) + \varepsilon\frac{\mathrm{d}\hat{z}(\tau)}{\mathrm{d}\tau}\frac{\mathrm{d}\tau}{\mathrm{d}t} = g(\bar{x} + \hat{x}, \bar{z} + \hat{z}, \varepsilon, t). \tag{3.12}$$

Now, requiring that $\bar{x}(t)$ satisfy the reduced model (2.5), we obtain from (3.11)

$$\frac{\mathrm{d}\hat{x}(\tau)}{\mathrm{d}\tau} = \varepsilon[f(\bar{x} + \hat{x}, \bar{z} + \hat{z}, \varepsilon, t) - f(\bar{x}, \bar{z}, 0, t)] \tag{3.13}$$

and hence $\mathrm{d}\hat{x}(\tau)/\mathrm{d}\tau \to 0$ as $\varepsilon \to 0$. Therefore, $\hat{x}(\tau) = \text{const}$ for $\varepsilon = 0$, and, since $\bar{x}(t_0) = x^0$, this constant is zero; that is, $\hat{x}(\tau) = 0$ for $\varepsilon = 0$. This shows that if the bracketed quantity on the right-hand side of (3.13) remains bounded, then $x(t, \varepsilon) \to \bar{x}(t)$ as $\varepsilon \to 0$. That the error is not larger than $O(\varepsilon)$ would follow from a more detailed analysis of (3.13). Using the fact that $\hat{x}(\tau) = 0$, and requiring that $\bar{z}(t)$ satisfy (2.3), we obtain from (3.12)

$$\frac{\mathrm{d}\hat{z}(\tau)}{\mathrm{d}\tau} = g(\bar{x}(t), \hat{z}(\tau) + \bar{z}(t), \varepsilon, t) - \varepsilon\dot{\bar{z}}, \quad \hat{z}(0) = z^0 - \bar{z}(t_0), \tag{3.14}$$

where $\dot{\bar{z}}$ is determined by differentiating (2.4) with respect to t. We note that (3.14) is a more general form of the boundary layer system (3.6). If in (3.14) we substitute $t = t_0 + \varepsilon\tau$, using (3.5), and then let $\varepsilon \to 0$, we obtain (3.6) as the limiting case of (3.14). For this reasoning to make sense, the equilibrium $\hat{z}(\tau) = 0$ of (3.6) is required to possess the stability properties spelled out in Assumptions 3.1 and 3.2. General asymptotic expansion procedures use two power series in ε to represent $x(t, \varepsilon)$. The coefficients of the first series are functions of t, and those of the second are functions of τ. Two analogous series are used to represent $z(t, \varepsilon)$. After these series are substituted into (2.1), (2.2), equations similar to (3.11), (3.12) are obtained and matched, term by term, to calculate the coefficients of the series. We will comment on such higher-order approximations in subsequent chapters in connection with specific applications. Here, various aspects of Theorem 3.1 are illustrated by examples.

Example 3.1

To apply Theorem 3.1 to the high-gain feedback system in Fig. 1.2(a) we examine the boundary layer system (3.6) corresponding to (2.12), which describes the fast transient in the inner feedback loop of Fig. 1.2(a); namely, using (2.13) and $u(t_0) = u^0$,

$$\frac{\mathrm{d}\hat{z}}{\mathrm{d}\tau} = -x^0 - \mathrm{e}^{\hat{z}(\tau) + \bar{z}(t_0)} + 1 + u^0 = (-x^0 + 1 + u^0)(1 - \mathrm{e}^{\hat{z}(\tau)}). \tag{3.15}$$

Since $-x^0 + 1 + u^0 > 0$ by Assumption 2.1, as in Example 2.2, the equilibrium $\hat{z} = 0$ of (3.15) is uniformly asymptotically stable for all $\hat{z}(0)$ and

Assumption 3.1 is satisfied. Also, for $\varepsilon = 0$, $z = \bar{z}$, we have

$$\frac{\partial g}{\partial z} = -e^{\bar{z}} < 0, \tag{3.16}$$

and hence Assumption 3.2 is satisfied for all $\bar{z}$. We conclude that Theorem 3.1 holds whenever $u - \bar{x} > -1$, which is also a necessary and sufficient condition for the existence of the reduced model (2.15). While this restriction is imposed on the two-time-scale approximation, the solution of the exact system (2.11), (2.12) for $\varepsilon > 0$ is not restricted and it exists even when $u - x < -1$.

Example 3.2

In this example there are three possible but only two valid reduced models, determined by the initial condition for z. Consider the system

$$\dot{x} = \frac{x^2 t}{z}, \quad x(t_0) = x^0 = 1, \quad t_0 = 0, \tag{3.17}$$

$$\varepsilon \dot{z} = -(z + xt)(z - 2)(z - 4), \quad z(t_0) = z^0. \tag{3.18}$$

In this case (2.3) is

$$0 = -(\bar{z} + \bar{x}t)(\bar{z} - 2)(\bar{z} - 4) \tag{3.19}$$

and has three distinct roots

$$\bar{z} = -\bar{x}t, \quad \bar{z} = 2, \quad \bar{z} = 4; \tag{3.20}$$

that is, there can be three reduced models. Analyzing the boundary layer system

$$\frac{d\hat{z}}{d\tau} = -(\bar{z} + \hat{z} + \bar{x}t)(\bar{z} + \hat{z} - 2)(\bar{z} + \hat{z} - 4), \tag{3.21}$$

we can readily check that Assumptions 3.1 and 3.2 hold for $\bar{z} = -\bar{x}t$ if $z^0 < 2$, and for $\bar{z} = 4$ if $z^0 > 2$. Both assumptions are violated by $\bar{z} = 2$, which is an unstable equilibrium of (3.21). Hence, there are only two valid reduced models

$$\dot{\bar{x}} = \begin{cases} -\bar{x}, & \text{if } z^0 < 2, \\ \tfrac{1}{4}\bar{x}^2 t, & \text{if } z^0 > 2. \end{cases} \tag{3.22}$$

Note that the solution $\bar{x} = (1 - \frac{1}{8}t^2)^{-1}$ of $\dot{\bar{x}} = \frac{1}{4}\bar{x}^2 t$ escapes to infinity at $t = 2\sqrt{2}$. However, Theorem 3.1 still holds for $t \in [t_0, T]$ with $T < 2\sqrt{2}$.

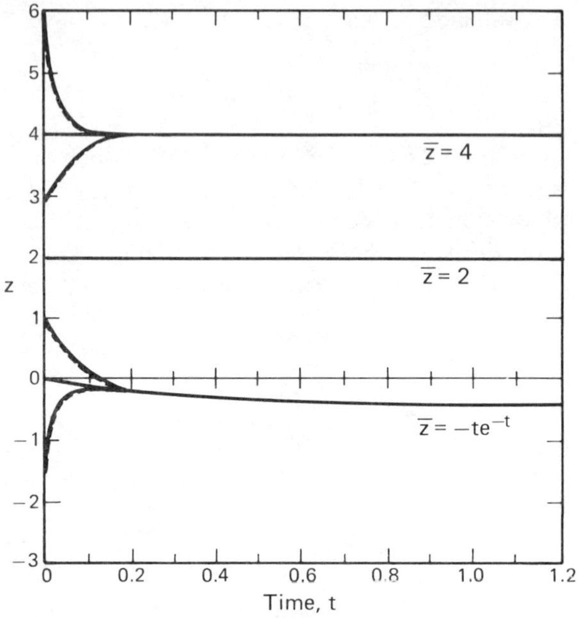

Fig. 1.4. Exact (solid) and approximate (dotted) solutions of (3.17), (3.18).

This is illustrated by simulation results in Fig. 1.4 for four different values of z^0, two for each reduced model. The approximate (dotted) and the exact (solid) trajectories for z are virtually indistinguishable, although ε is relatively large at $\varepsilon = 0.3$. The trajectories in Fig. 1.4 clearly exhibit a two-time-scale behavior. They start with a fast transient of $z(t, \varepsilon)$ from z^0 to $\bar{z}(t)$, which is the so-called boundary layer, or the "inner" part of the solution $z(t, \varepsilon)$. After the decay of this transient, they remain close to $\bar{z}(t)$, which is the slow ("outer") part of the solution, also called the "quasi-steady-state" of $z(t, \varepsilon)$. In general, the "quasi-steady-state" varies with time, as does $\bar{z}(t) = -te^{-t}$. Exceptionally, as in the case $\bar{z} = 4$, the "quasi-steady-state" can be a true steady state of z. For a fuller understanding of the general boundary layer system (3.14) it is useful to see the meaning of the term $\varepsilon\dot{\bar{z}}(t)$ for $\bar{z}(t) = -te^{-t}$ in this example.

While the preceding two examples treat boundary layers as corrective terms to the reduced slow solution, which is the more relevant part of the exact solution, in many applications both fast and slow parts are important. In such applications, the two-time-scale approximation (3.3), (3.7), validated by Theorem 3.1, is a tool for separate analysis and design of the slow and fast subsystems, the main topic of this book. The following case study formulates one of the classical control designs, the PID regulator, as a two-time-scale problem. This study is a precursor of methods to be developed in Chapters 2 and 3, and will be reconsidered there in more detail.

Case Study 3.1 Two-Time-Scale PID Control

A classical feedback loop appearing in most single-variable process control problems is shown in Fig. 1.5, with the transfer function of the PID block defined by

$$u(s) = k_P\left(1 + k_D s + k_I \frac{1}{s}\right)e(s), \quad e(s) = r(s) - y(s), \tag{3.23}$$

where k_P, k_D and k_I are the customary tuning parameters, d is an unknown disturbance, y is the output to be regulated to the desired set-point r, and

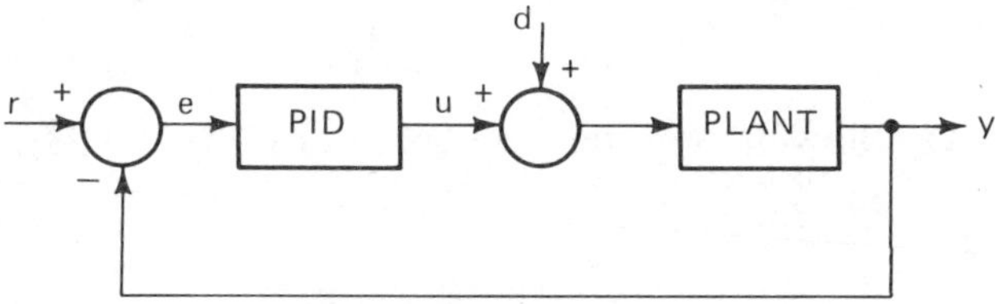

Fig. 1.5. PID control system.

s is the Laplace transform complex variable. In this example we shall treat r and d as step inputs, that is, as constants for $t > 0$. The presence of the integral term $1/s$ in (3.23) will then guarantee that the regulation error $e(t) \to 0$ as $t \to \infty$, provided that the feedback loop is asymptotically stable. The property that for constant disturbances the integral control action reduces the regulation error to zero is true for all plants, linear or nonlinear, for which the loop remains asymptotically stable. The task of the proportional gain k_P is to speed up the process, and that of the differential term k_D is to stabilize it. A classical design approach is to first choose k_P and k_D for the desired transient performance and then to insert a small integral gain k_I to slowly regulate the error to zero. In this way, the integral term contributes to the steady-state accuracy of the regulation, without much interference with the loop stability properties. For the sake of argument, assume that the plant transfer function is

$$y(s) = \frac{k_0}{s^2}[u(s) + d(s)], \tag{3.24}$$

where k_0 is the gain of the plant and is assumed to be one, $k_0 = 1$, without loss of generality. This plant can be thought of as a motor driving a heavy inertial object, or as a two-reservoir system. In the first case the disturbance d can be a load torque, while in the second case it can be a constant outflow or inflow.

To model this feedback loop as a singularly perturbed system, we need to mathematically express the fact that the integral action is much slower than the other two actions. This will be so if its gain k_I is $O(\varepsilon)$; that is, if $k_I = \varepsilon \bar{k}_I$, where $\bar{k}_I$ is $O(1)$. So as to simplify notation, we also introduce $k_1 = k_P$, $k_2 = k_P k_D$, $k_3 = k_P \bar{k}_I$ and rewrite the controller transfer function (3.23) as

$$u(s) = \left(k_1 + k_2 s + \varepsilon k_3 \frac{1}{s}\right) e(s). \tag{3.25}$$

Choosing the state variables in the fast time scale $\tau = t/\varepsilon$,

$$e_1 = e, \quad e_2 = \frac{de}{d\tau}, \quad e_3 = \varepsilon \int_0^\tau e \, d\sigma + e_3(0), \tag{3.26}$$

we obtain the following state representation of (3.24), (3.25):

$$\frac{de_1}{d\tau} = e_2, \tag{3.27}$$

$$\frac{de_2}{d\tau} = -k_1 e_1 - k_2 e_2 - k_3 e_3 - d, \tag{3.28}$$

$$\frac{de_3}{d\tau} = \varepsilon e_1. \tag{3.29}$$

Equation (3.29) shows the reason for denoting the time variable as the fast time scale τ. Namely, when $\varepsilon = 0$ the integral action e_3 is constant, which is the characteristic of the slow variable's behavior in the fast time scale; see (3.13) and the discussion following it. To observe the slow transient caused by the integral action, we rewrite the system (3.27)–(3.29) in the slow time scale $t = \varepsilon\tau$. In doing so, we also identify $e_1 = z_1$, $e_2 = z_2$ as the fast variables and $e_3 = x$ as the slow variable, and remind the reader that a dot denotes the derivative with respect to the slow time variable t. The system (3.27)–(3.29) is thus brought to the standard form (2.1), (2.2), namely,

$$\dot{x} = z_1, \tag{3.30}$$

$$\varepsilon \dot{z}_1 = z_2, \tag{3.31}$$

$$\varepsilon \dot{z}_2 = -k_1 z_1 - k_2 z_2 - k_3 x - d. \tag{3.32}$$

To satisfy Assumption 2.1 we need $k_1 \neq 0$, while to satisfy Assumptions 3.1 and 3.2 we must choose k_1 and k_2 in the PD part of the design such that the matrix in (3.31), (3.32) be Hurwitz, that is,

$$\mathrm{Re}\,\lambda \begin{bmatrix} 0 & 1 \\ -k_1 & -k_2 \end{bmatrix} < 0. \tag{3.33}$$

For the I-part of the design the reduced model of (3.30)–(3.32) is obtained with $\bar{z}_2 = 0$, $\bar{z}_1 = -(k_3\bar{x} + d)/k_1$; that is,

$$\dot{\bar{x}} = -\frac{k_3}{k_1}\bar{x} - \frac{d}{k_1} = -\bar{k}_{\mathrm{I}}\bar{x} - \frac{d}{k_{\mathrm{P}}}. \tag{3.34}$$

Equation (3.34) shows that the slow transient due to integral control is adjusted by the choice of the tuning parameter $\bar{k}_{\mathrm{I}}$, while its extent is proportional to d and inversely proportional to $k_{\mathrm{P}}\bar{k}_{\mathrm{I}}$. In this case, the boundary layer system (3.6) is

$$\frac{d\hat{z}_1}{d\tau} = \hat{z}_2, \quad \frac{d\hat{z}_2}{d\tau} = -k_1\left(\hat{z}_1 - \frac{k_3}{k_1}\bar{x} - \frac{d}{k_1}\right) - k_2\hat{z}_2 - k_3\bar{x} - d$$
$$= -k_1\hat{z}_1 - k_2\hat{z}_2; \tag{3.35}$$

that is, it is a homogeneous system with the matrix (3.33), used for the PD part of the design. The fact that the disturbance term does not appear in (3.35), but appears in (3.34), clearly indicates the roles of the PD and I actions of the controller (3.25).

1.4 Slow and Fast Manifolds

It is recalled that, as functions of time, the solutions $x(t, \varepsilon)$, $z(t, \varepsilon)$ of the singularly perturbed system (2.1), (2.2) consist of a fast boundary layer and a slow quasi-steady-state. Of the two components $x(t, \varepsilon)$ and $z(t, \varepsilon)$, the layer is significant only in $z(t, \varepsilon)$, while $x(t, \varepsilon)$ is predominantly slow since its layer is not larger than $O(\varepsilon)$. Our aim in this section is to give a geometric view of the two-time-scale behavior of $x(t, \varepsilon)$, $z(t, \varepsilon)$ as trajectories in R^{n+m}. We consider a simpler form of the system (2.1), (2.2), namely,

$$\dot{x} = f(x, z), \quad x \in R^n, \tag{4.1}$$

$$\varepsilon\dot{z} = g(x, z), \quad z \in R^m, \tag{4.2}$$

where the dependence of f and g on ε and t is suppressed. Insofar as ε is concerned, this is done only to simplify notation, while the elimination of the dependence of f and g on t is made to allow us to use the concept of *invariant manifolds*.

In the $(n+m)$-dimensional state space of x and z, an n-dimensional manifold M_ε, depending on the scalar parameter ε, can be defined by the expression

$$M_\varepsilon : z = \phi(x, \varepsilon); \quad x \in R^n, z \in R^m; \tag{4.3}$$

where it is assumed that ϕ is a sufficiently many times continuously differentiable function of x and ε. This expression reduces the dimension of the state space from $n+m$ to n, by restricting the state to remain on M_ε. For example, if in R^3 we have $n = 2$ and $m = 1$, then M_ε will be a surface defined by one scalar equation (4.3), while for $n = 1$ and $m = 2$ the two scalar equations (4.3) will define a curve. For M_ε to be an *invariant manifold* of (4.1), (4.2) the expression (4.3) must hold for all $t > t^*$ if it holds for $t = t^*$; that is,

$$z(t^*, \varepsilon) = \phi(x(t^*, \varepsilon), \varepsilon) \Rightarrow z(t, \varepsilon) = \phi(x(t, \varepsilon), \varepsilon) \quad \forall t \geq t^*. \tag{4.4}$$

Differentiating this expression with respect to t, we obtain

$$\dot{z} = \frac{\mathrm{d}}{\mathrm{d}t} \phi(x(t, \varepsilon), \varepsilon) = \frac{\partial \phi}{\partial x} \dot{x}. \tag{4.5}$$

Multiplying (4.5) by ε, and substituting $\dot{x}$ and $\dot{z}$ from (4.1), (4.2) and z from (4.3), we obtain a *manifold condition*

$$\varepsilon \frac{\partial \phi}{\partial x} f(x, \phi(x, \varepsilon)) = g(x, \phi(x, \varepsilon)), \tag{4.6}$$

which $\phi(x, \varepsilon)$ must satisfy for all x of interest and all $\varepsilon \in [0, \varepsilon^*]$, where ε^* is a positive constant. To begin with, let us analyze M_ε for $\varepsilon = 0$, that is, M_0 defined by (4.6) at $\varepsilon = 0$,

$$M_0 : z = \phi(x, 0), \quad 0 = g(x, \phi(x, 0)). \tag{4.7}$$

In (4.7) we recognize the expressions (2.3), (2.4) obtained by formally neglecting ε in (2.1), (2.2). There can be several functions $\phi(x, 0)$ satisfying (4.7), and hence several manifolds M_0. By Assumption 2.1, they have no points in common, that is, they do not intersect each other and are nontangential in the domain of interest. For a further interpretation of M_0 we rewrite (4.1), (4.2) in the fast time scale τ, defined by (3.4), (3.5). From

$$\frac{\mathrm{d}x}{\mathrm{d}\tau} = \varepsilon f(x, z), \quad \frac{\mathrm{d}z}{\mathrm{d}\tau} = g(x, z), \tag{4.8}$$

we conclude that at $\varepsilon = 0$ both $\mathrm{d}x/\mathrm{d}\tau$ and $\mathrm{d}z/\mathrm{d}\tau$ are zero for $z = \phi(x, 0)$.

Hence M_0 is an *equilibrium manifold* of (4.8) at $\varepsilon = 0$. If Assumption 3.2 is satisfied for all $x, z \in M_0$ then M_0 is a stable (attractive) manifold.

Remark 4.1 For the remaining discussion in this section, it is sufficient that M_0 be "conditionally stable"; that is, we can relax the requirement $\leq -c$ in (3.9) to $\neq 0$. This relaxed form of Assumption 3.2 will be used in Chapter 6 when dealing with optimal trajectories having boundary layers at both ends.

A consequence of Theorem 3.1 is that the existence of a conditionally stable equilibrium manifold M_0 of (4.8) for $\varepsilon = 0$ implies the existence of an invariant manifold M_ε of (4.1), (4.2) satisfying the manifold condition (4.6) for all $\varepsilon \in [0, \varepsilon^*]$ and converging to M_0 as $\varepsilon \to 0$; that is,

$$\phi(x, \varepsilon) \to \phi(x, 0), \quad M_\varepsilon \to M_0 \quad \text{as } \varepsilon \to 0. \tag{4.9}$$

Further insight is gained if the deviation of z from M_ε is represented by a new variable

$$\eta = z - \phi(x, \varepsilon). \tag{4.10}$$

In terms of x and η the system (4.1), (4.2) becomes

$$\dot{x} = f(x, \phi(x, \varepsilon) + \eta), \tag{4.11}$$

$$\varepsilon\dot{\eta} = g(x, \phi(x, \varepsilon) + \eta) - \varepsilon \frac{\partial \phi}{\partial x} f(x, \phi(x, \varepsilon) + \eta). \tag{4.12}$$

The invariant manifold M_ε is now characterized by the fact that $\eta = 0$ implies $\dot{\eta} = 0$ for all x for which (4.6) is true. Thus, if $\eta(t_0) = 0$, all we need to solve is

$$\dot{x} = f(x, \phi(x, \varepsilon)), \quad x(t_0) = x^0, \tag{4.13}$$

which is *the exact slow model* of (4.1), (4.2) valid for $x, z \in M_\varepsilon$. The reduced model (2.5) is its approximation in which $\phi(x, \varepsilon)$ is replaced by $\phi(x, 0)$. We call M_ε a *slow manifold* of (4.1), (4.2), stressing that to each slow manifold there corresponds a slow model (4.13). To introduce the notion of a fast manifold we examine the behavior of η in the fast time scale τ. Substituting (4.10) into (4.8) and letting $\varepsilon \to 0$, we see that η at $\varepsilon = 0$ is defined by the boundary layer system

$$\frac{d\eta}{d\tau} = g(x^0, \phi(x^0, 0) + \eta), \quad \eta^0 = z^0 - \phi(x^0, 0), \tag{4.14}$$

which we have already encountered in (3.6). This system describes the trajectories x, η, which, for every given x^0, lie in a *fast manifold* F_x defined by $x = x^0 = \text{const}$, and rapidly descend to the equilibrium manifold M_0. For ε larger than zero but small, the fast manifolds are "foliations" of solutions rapidly approaching the slow manifold M_ε. Let us illustrate this picture by two examples.

Example 4.1

For the system

$$\dot{x} = xz^3, \quad \varepsilon\dot{z} = -z - x^{4/3} + \tfrac{4}{3}\varepsilon x^{16/3}, \tag{4.15}$$

the manifold condition (4.6) is

$$\varepsilon \frac{\partial \phi}{\partial x} x\phi^3 = -\phi - x^{4/3} + \tfrac{4}{3}\varepsilon x^{16/3}, \tag{4.16}$$

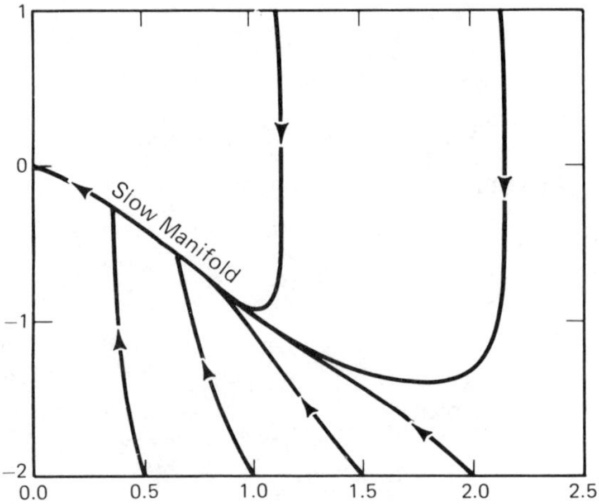

Fig. 1.6. Trajectories of the system (4.15).

and it is easily checked that it is satisfied by

$$z = \phi(x, \varepsilon) = -x^{4/3}. \tag{4.17}$$

In this very special case M_ε not only has a simple analytical expression, but it even coincides with M_0, because (4.17) does not depend on ε. Since $\partial g/\partial z = -1$, the slow manifold is stable, see the plots in Fig. 1.6. The slow model (4.13) for $z(t_0) = -[x(t_0)]^{4/3}$ and all $x(t_0)$ is $\dot{x} = -x^5$.

Example 4.2

To illustrate the possibility of several slow manifolds we consider the system

$$\dot{x} = -xz, \quad \varepsilon\dot{z} = -(z - \sin^2 x)(z - e^{\alpha x})(z - 2e^{2\alpha x}), \quad \alpha > 0, \tag{4.18}$$

which is a variation on the theme of Example 3.2. The manifold condition (4.6) is

$$\varepsilon \frac{\partial \phi}{\partial x}(-x\phi) = -(\phi - \sin^2 x)(\phi - e^{\alpha x})(\phi - 2e^{2\alpha x}) \tag{4.19}$$

and is not easy to solve. However, at $\varepsilon = 0$ we have three equilibrium manifolds defined by

$$M_0^1 : \phi_1(x, 0) = \sin^2 x, \quad M_0^2 : \phi_2(x, 0) = e^{\alpha x}, \quad M_0^3 : \phi_3(x, 0) = 2e^{2\alpha x}, \tag{4.20}$$

Since they intersect for some negative values of x, let us assume that only $x \geq 0$ is of interest. From

$$\frac{\partial g}{\partial z} = -(z - e^{\alpha x})(z - 2e^{2\alpha x}) - (z - \sin^2 x)(z - 2e^{2\alpha x}) - (z - \sin^2 x)(z - e^{\alpha x}), \tag{4.21}$$

it is observed that M_0^1 and M_0^3 are stable, while M_0^2 is unstable.

We close this section with a brief look at the linear system

$$\dot{x} = A_{11}x + A_{12}z \tag{4.22}$$

$$\varepsilon\dot{z} = A_{21}x + A_{22}z \tag{4.23}$$

which will be studied in detail in Chapter 2. The manifold condition (4.6), in this case

$$\varepsilon \frac{\partial \phi}{\partial x}(A_{11}x + A_{12}\phi) = A_{21}x + A_{22}\phi, \tag{4.24}$$

can be satisfied by a ϕ linear in x,

$$z = \phi(x, \varepsilon) = -L(\varepsilon)x, \tag{4.25}$$

which, when substituted in (4.24), results in

$$-\varepsilon L(A_{11} - A_{12}L) = A_{21} - A_{22}L. \tag{4.26}$$

It is shown in Section 2.2 that a solution $L(\varepsilon)$ of this equation exists and is close to $L(0) = A_{22}^{-1}A_{21}$. The substitution

$$\eta = z + L(\varepsilon)x \tag{4.27}$$

into the system (4.22), (4.23) reduces it to the block-triangular form

$$\dot{x} = (A_{11} - A_{12}L)x + A_{12}\eta \tag{4.28}$$

$$\varepsilon\dot{\eta} = (A_{22} + \varepsilon LA_{12})\eta \tag{4.29}$$

and separates (4.29) as the exact fast subsystem. The slow manifold M_ε defined by $z = L(\varepsilon)x$ is the eigenspace corresponding to the slow eigenvalues of (4.22), (4.23), that is, the eigenvalues that remain finite as $\varepsilon \to 0$.

1.5 Construction of Approximate Models

It is in the spirit of engineering, control engineering in particular, to simplify the models used for design. Apart from obvious practical reasons, there are important conceptual advantages in this seemingly pragmatic approach. Phenomena that may be hidden in an exact but complex model sometimes appear in transparent clarity in a simplified model. Simplified two-time-scale models used throughout this book are based on explicit approximations of the manifold condition (4.6). To each approximation of $\phi(x, \varepsilon)$ there corresponds an exact transformed model analogous to (4.11), (4.12). Approximate slow and fast models, analogous to (4.13) and (4.14), are then constructed by neglecting various ε-terms in the exact transformed model. In this section the main steps of such a model building procedure are outlined. The neglect of ε-terms, which at this stage is *ad hoc*, will be subsequently validated in Chapter 2 for linear time-invariant systems, in Chapter 5 for linear time-varying systems, and in Chapter 7 for nonlinear systems.

The approximation procedure starts by substituting into the manifold condition (4.6) a power series for $\phi(x, \varepsilon)$,

$$\phi(x, \varepsilon) = \varphi_0(x) + \varepsilon\varphi_1(x) + \varepsilon^2\varphi_2(x) + \dots \tag{5.1}$$

and calculating $\varphi_0(x)$, $\varphi_1(x)$ etc. by equating terms in like powers of ε. This requires that the functions f and g also be expanded as power series of ε. The more important of the two is the series for g:

$$g(x, \varphi_0(x) + \varepsilon\varphi_1(x) + \dots) = g(x, \varphi_0(x)) + \frac{\partial g}{\partial z}\varepsilon\varphi_1(x) + \dots \tag{5.2}$$

where $\partial g/\partial z$ and all higher derivatives are evaluated at $x, z = \varphi_0(x)$. An immediate consequence of (4.6), (4.7), (5.1) and (5.2) is

$$g(x, \varphi_0(x)) = 0, \quad \phi(x, 0) = \varphi_0(x). \tag{5.3}$$

With $\varphi_0(x)$ known, we proceed to equate the ε-terms of power one, ε^1, and obtain

$$\varphi_1(x) = \left(\frac{\partial g}{\partial z}\right)^{-1}_{z=\varphi_0(x)} \frac{\partial \varphi_0(x)}{\partial x} f(x, \varphi_0(x)). \tag{5.4}$$

Although with $\varphi_0(x)$ and $\varphi_1(x)$ known, one can equate ε-terms of power two, ε^2, and obtain $\varphi_2(x)$, etc., the approximation $\phi(x, \varepsilon) \cong \varphi_0(x) + \varepsilon\varphi_1(x)$ will suffice for most applications. The substitution of

$$\eta = z - \varphi_0(x) - \varepsilon\varphi_1(x) \tag{5.5}$$

into (4.1), (4.2) results in an exact transformed model analogous to (4.11), (4.12), namely,

$$\dot{x} = f(x, \varphi_0(x) + \varepsilon\varphi_1(x) + \eta), \tag{5.6}$$

$$\varepsilon\dot{\eta} = g(x, \varphi_0(x) + \varepsilon\varphi_1(x) + \eta) - \varepsilon\left(\frac{\partial \varphi_0}{\partial x} + \varepsilon\frac{\partial \varphi_1}{\partial x}\right)$$

$$\times f(x, \varphi_0(x) + \varepsilon\varphi_1(x) + \eta). \tag{5.7}$$

The essential difference between (4.11), (4.12) and (5.6), (5.7) is that $\eta = 0$ is not an invariant manifold of (5.6), (5.7). However, from (5.5) and a detailed examination of (5.7), we see that $\eta = 0$ describes a manifold $O(\varepsilon^2)$ close to the invariant manifold of (5.6), (5.7). Furthermore, it will be shown later that when η is $O(1)$ its effect on x in (5.6) is only $O(\varepsilon)$. Thus neglecting η in the slow model (5.6) will cause an $O(\varepsilon)$ error in x and, if η itself is $O(\varepsilon)$, then this error will be $O(\varepsilon^2)$. This suggests that (5.6) without η can be used as an *approximate slow model*. We note that (5.6) improves the reduced model (2.5) by adding the correction term $\varepsilon\varphi_1(x)$, which is conceptually important, because it allows us to analyze the dependence of the properties of the slow model on ε. This point will later be illustrated by an adaptive control example. As for the construction of an *approximate fast model*, (5.7) offers several possibilities, one of which is to neglect the ε-terms of power two, ε^2, and higher, to obtain

$$\varepsilon\dot{\eta} = g(x, \varphi_0(x) + \varepsilon\varphi_1(x) + \eta) - \varepsilon\frac{\partial \varphi_0}{\partial x} f(x, \varphi_0(x) + \eta). \tag{5.8}$$

Specializing (5.3) and (5.4) for the linear system (4.22), (4.23), we obtain

$$\varphi_0(x) = -A_{22}^{-1}A_{21}x, \quad \varphi_1(x) = -A_{22}^{-2}A_{21}A_0x, \quad A_0 = A_{11} - A_{12}A_{22}^{-1}A_{21}, \tag{5.9}$$

and hence the exact transformed model (5.6), (5.7) is

$$\dot{x} = (I - \varepsilon A_{12}A_{22}^{-2}A_{21})A_0 x + A_{12}\eta, \tag{5.10}$$

$$\varepsilon\dot{\eta} = (A_{22} + \varepsilon A_{22}^{-1}A_{21}A_{12})\eta + \varepsilon^2\alpha(x, \eta, \varepsilon), \tag{5.11}$$

where

$$\alpha(x, \eta, \varepsilon) \triangleq -A_{22}^{-1}A_{21}A_{12}A_{22}^{-2}A_{21}A_0 x + A_{22}^{-2}A_{21}A_0 \times [A_{12}\eta + (I - \varepsilon A_{12}A_{22}^{-2}A_{21})A_0 x].$$

Neglecting $\varepsilon^2\alpha$ in (5.11), this model is written compactly as

$$\dot{x} = A_s x + A_{12}\eta, \quad x(t_0) = x^0, \quad z(t_0) = z^0, \tag{5.12}$$

$$\varepsilon\dot{\eta} = A_f\eta, \quad \eta(t_0) = \eta^0 = z^0 - (A_{22}^{-1}A_{21} + \varepsilon A_{22}^{-2}A_{21}A_0)x^0, \tag{5.13}$$

where A_s and A_f denote the corresponding matrices in (5.10), (5.11), and η^0 is expressed in terms of the original initial conditions x^0 and z^0, according to (5.5). Neglecting η in (5.12), one can use $\dot{x} = A_s x$ as a slow approximate model, and $\varepsilon\dot{\eta} = A_f\eta$ as a fast approximate model. However, this is a convenient place to show, first, that the effect of η on x is $O(\varepsilon)$ and, second, that it can be further reduced. Expressing η in terms of $\varepsilon\dot{\eta}$, using (5.13), and substituting it into (5.12), one obtains

$$\dot{x} - \varepsilon A_{12}A_f^{-1}\dot{\eta} = A_s x \tag{5.14}$$

which suggests that a new slow variable should be

$$\xi = x - \varepsilon A_{12}A_f^{-1}\eta. \tag{5.15}$$

Using this variable, (5.14) becomes

$$\dot{\xi} = A_s\xi + \varepsilon A_s A_{12}A_f^{-1}\eta \tag{5.16}$$

and the η-term is now $O(\varepsilon)$. This process can be continued and the presence of η in the slow model can be progressively reduced to any desired power of ε. Further details are presented in Chapter 2.

With almost the same ease we can develop slow and fast approximate models for nonlinear systems that are linear in z, that is, for

$$\dot{x} = a_{11}(x) + A_{12}z, \tag{5.17}$$

$$\varepsilon\dot{z} = a_{21}(x) + A_{22}z, \tag{5.18}$$

where A_{12} and A_{22} can also be functions of x, but, to simplify the derivations, we treat them as constant matrices. In this case

$$\left.\begin{aligned} \varphi_0(x) &= -A_{22}^{-1}a_{21}(x), \quad \varphi_1(x) = -A_{22}^{-2}A_{21}(x)a_0(x), \\ A_{21}(x) &= \frac{\partial a_{21}}{\partial x}, \quad a_0(x) = a_{11}(x) - A_{12}A_{22}^{-1}a_{21}(x), \end{aligned}\right\} \tag{5.19}$$

and the substitution of (5.5) and (5.19) into (5.17), (5.18) gives

$$\dot{x} = [I - \varepsilon A_{12}A_{22}^{-2}A_{21}(x)]a_0(x) + A_{12}\eta, \tag{5.20}$$

$$\varepsilon\dot{\eta} = [A_{22} + \varepsilon A_{22}^{-1}A_{21}(x)A_{12}]\eta + \varepsilon^2\beta(x, \eta, \varepsilon), \tag{5.21}$$

where

$$\begin{aligned} \beta(x, \eta, \varepsilon) = -&A_{22}^{-1}A_{21}(x)A_{12}A_{22}^{-2}A_{21}(x)a_0(x) \\ &+ A_{22}^{-2}\frac{\partial}{\partial x}[A_{21}(x)a_0(x)] \\ &\times [A_{12}\eta + (I - \varepsilon A_{12}A_{22}^{-2}A_{21}(x))a_0(x)]. \end{aligned} \tag{5.22}$$

The remarkable analogy of (5.20), (5.21) and the linear system (5.10), (5.11) shows how, in systems with *slow nonlinearities*, the linearity of the fast phenomena is preserved. Adaptive systems with fast linear parasitics belong to this class, and are now illustrated by an example.

Example 5.1

The equations of the adaptive system in Fig. 1.7 are

$$\dot{y} = ay + z, \tag{5.23}$$

$$\dot{k} = y^2, \tag{5.24}$$

$$\varepsilon\dot{z} = -z + u = -z - ky, \tag{5.25}$$

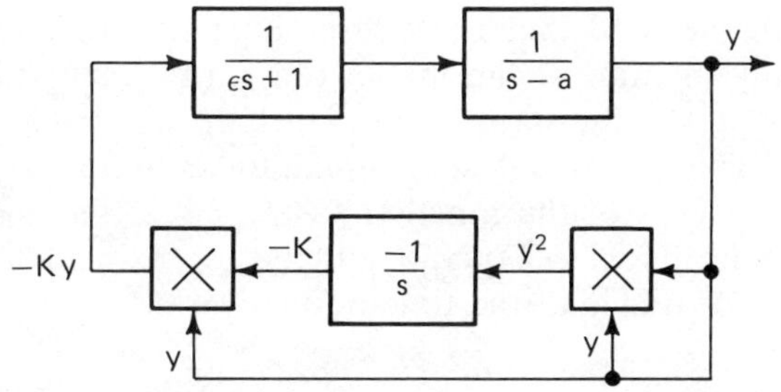

Fig. 1.7. Adaptive system with parasitics.

where the plant parameter a and the small parasitic time constant $\varepsilon > 0$ are unknown. A result of adaptive control theory (Narendra, Lin and Valavani, 1980) is that the update law (5.24) for the adjustable gain k stabilizes the system for all a = constant in the parasitic-free case, that is, when $\varepsilon = 0$. A task of the so-called robustness analysis is to determine how the presence of the parasitic time constant ε affects the stability of the system (5.23)–(5.25). To express (5.23)–(5.25) in the form (5.17), (5.18), we let

$$x = \begin{bmatrix} y \\ k \end{bmatrix}, \quad a_{11}(x) = \begin{bmatrix} ay \\ y^2 \end{bmatrix}, \quad A_{12} = \begin{bmatrix} 1 \\ 0 \end{bmatrix}, \quad a_{21}(x) = -ky, \quad A_{22} = -1, \tag{5.26}$$

and evaluate the quantities needed for the model (5.20), (5.21), that is,

$$A_{21}(x) = \frac{\partial a_{21}(x)}{\partial x} = [-k \quad -y], \quad a_0(x) = \begin{bmatrix} (a-k)y \\ y^2 \end{bmatrix}. \tag{5.27}$$

Finally, we obtain the model (5.20), (5.21) as

$$\dot{y} = [(1 + \varepsilon k)(a - k) + \varepsilon y^2]y + \eta, \tag{5.28}$$

$$\dot{k} = y^2, \tag{5.29}$$

$$\varepsilon\dot{\eta} = (-1 + \varepsilon k)\eta + \varepsilon^2\beta(y, k, \eta, \varepsilon). \tag{5.30}$$

For $\varepsilon = 0$ and $\eta = 0$, (5.28), (5.29) represent a globally stable adaptive regulation. The presence of ε-terms in (5.28) and (5.30) shows that the parasitic time constant may lead to a loss of stability, which illustrates how a transformed model directs our attention to possible causes of non-robustness. The robustness analysis of adaptive control problems will be pursued in Chapters 5 and 7.

In this and the preceding section, the discussion has been restricted to the case of time-invariant models without inputs; see (4.1), (4.2). Let us now show how to remove these restrictions by introducing appropriate modifications in the modeling procedure. First, if the functions f and g depend explicitly on time t, the model (4.1), (4.2) can still be used by extending the x-part of the state space to include $x_{n+1} = t$ as an additional state variable. The additional state equation is $\dot{x}_{n+1} = 1$ and the corresponding component of the function f is $f_{n+1} = 1$. The most important new property of the model (5.6), (5.7), and the approximate models derived from it, is the appearance of the time derivatives

$$\frac{\partial \varphi_0}{\partial x_{n+1}} = \frac{\partial \varphi_0}{\partial t}, \quad \frac{\partial \varphi_1}{\partial x_{n+1}} = \frac{\partial \varphi_1}{\partial t}, \tag{5.31}$$

which are required to be $O(1)$ or less, so that the term $\varepsilon\varphi_1(x)$ in the slow model (5.6) and the term $\varepsilon\,\partial\varphi_0/\partial x$ in the fast model (5.8) remain $O(\varepsilon)$. In other words, for a "quasi-steady-state" to be "slow", it should not contain frequencies higher than $O(1)$ and it should certainly not contain terms such as $\sin(t/\varepsilon)$.

Our next two remarks concern the effects of an inclusion of time-dependent inputs in the model (4.1), (4.2). The effects of slow inputs on slow models and fast inputs on fast models are treated without further simplification. However, the effects of slow inputs on fast models and fast inputs on slow models can be approximated using the development in this and the preceding section. Let us consider the case of linear time-invariant systems for which the procedure is straightforward.

The presence of a slow input $u_s(t)$ into a fast model, namely,

$$\varepsilon\dot{z} = A_f z + B_f u_s, \quad z(t_0) = z^0, \tag{5.32}$$

implies that the "quasi-steady-state" $\bar{z}(t)$ of z is $\bar{z}(t) = -A_f^{-1}B_f u_s(t)$. The discrepancy between $\bar{z}$ and z is

$$\eta_1 = z + A_f^{-1}B_f u_s, \tag{5.33}$$

which, when substituted into (5.32), yields

$$\varepsilon\dot{\eta}_1 = A_f\eta_1 + \varepsilon A_f^{-1}B_f\dot{u}_s. \tag{5.34}$$

Requiring that $\dot{u}_s(t)$ be $O(1)$ or less, that is, bounded by a constant independent of ε, we assure that the "quasi-steady-state" of η_1 is only $O(\varepsilon)$ or less. If this accuracy is satisfactory, we can treat η_1 as our fast variable and (5.34) as the fast model. For higher accuracy, the step (5.33) can be repeated by introducing η_2 and transforming (5.34) as follows:

$$\eta_2 = \eta_1 + \varepsilon A_f^{-2}B_f\dot{u}_s, \quad \varepsilon\dot{\eta}_2 = A_f\eta_2 + \varepsilon^2 A_f^{-2}B_f\ddot{u}_s, \tag{5.35}$$

with a further requirement that $\ddot{u}_s$ be $O(1)$ or less. This procedure can be continued, each step generating a term in a series expansion for z. To relate the model (5.35) to (5.11), we can assume that u_s is generated by a slow model $\dot{u}_s = A_s u_s$. Then the substitution of $\ddot{u}_s = A_s^2 u_s$ into (5.35) will make (5.35) appear in the form (5.11).

The effect of a fast input u_f on a slow model, that is,

$$\dot{x} = A_s x + B_s u_f, \quad x(t_0) = x^0, \tag{5.36}$$

can be reduced to our previous discussion of (5.14)–(5.16) by considering that u_f is the output of a stable fast model;

$$\varepsilon\dot{\eta} = A_f\eta, \quad u_f = C_f\eta. \tag{5.37}$$

Letting $B_s C_f = A_{12}$, and using (5.15), we obtain

$$\xi_1 = x - \varepsilon B_s C_f A_f^{-1} \eta, \quad \dot{\xi}_1 = A_s \xi_1 + \varepsilon A_s B_s C_f A_f^{-1} \eta. \tag{5.38}$$

This procedure can be continued by defining ξ_2, which would lead to an approximation of x by the solution of $\dot{\xi}_k = A_s \xi_k$ and a power series in ε beginning with the term $\varepsilon B_s C_f A_f^{-1} \eta$ which is $O(\varepsilon)$. This proves that if we simply disregard u_f in (5.36) and obtain x from $\dot{x} = A_s x$, the error will not be larger than $O(\varepsilon)$. It is important to note that, by construction (5.37), the signal u_f does not contain frequencies lower than $O(1/\varepsilon)$.

A more general stable fast model (5.37) can have purely oscillatory modes. Then u_f remains bounded, but, in addition to the exponentially decaying terms, it also contains the terms of the form $\sin(t/\varepsilon)$. For such inputs the system (5.36) acts as a low-pass filter, and hence their effect on x is only $O(\varepsilon)$. This conclusion will be helpful in Chapter 4, where we deal with stochastic problems. If instead of (5.36) we consider a nonlinear slow model, then A_f in (5.37) should be Hurwitz, that is, high-frequency periodic inputs such as $\sin(t/\varepsilon)$ should be excluded. This is an important caveat, because a persistent high-frequency signal may produce an effect $O(1)$ due to nonlinear averaging in the slow model. Problems of this sort are handled by averaging techniques such as those of Krylov–Bogolyubov (Bogolyubov and Mitropolsky, 1961), further extended by Hale (1980). They are not considered in this book, and the appearance of high-frequency oscillatory modes is excluded by Assumption 3.2.

1.6 From Nonstandard to Standard Forms

The two-time-scale properties of the standard model (4.1), (4.2) have been characterized in terms of the geometric properties of the same model (4.8) considered in the fast time scale τ at $\varepsilon = 0$. The latter model was shown to possess at least one equilibrium manifold M_0, defined by $z = \phi(x, 0) = \varphi_0(x)$, and a family of fast manifolds F_x on which x as a function of τ is constant, while z converges to M_0. A further characteristic of the standard form is that the fast manifolds F_x are "parallel to the z-axis"; that is, they are translates of the subspace $x = 0$. If we now perform a nonsingular change of coordinates, the transformed model may no longer be in the standard form, but will still possess the same two-time-scale properties. What may be lost is the property that some variables are predominantly slow (the former variables x), while the others exhibit a boundary layer behavior (the former variables z). Instead, all the new variables may possess boundary layers and converge to quasi-steady-states. However, this more general situation is still characterized by the existence of the equilibrium manifolds M_0 and the fast manifolds F_x, although their positions and shapes in the new coordinate system are different. In particular, F_x need no longer be parallel to any of the coordinate axes.

What has just been described is that a change of coordinates may "hide" the time-scale properties already discovered from a model in standard form. Of course, we are interested in a reverse step that would convert a nonstandard model into the standard form and thus reveal the hidden time scales.

To begin with, let us consider a fairly general linear model written, for convenience, in both slow and fast time scales, namely,

$$\varepsilon \dot{v} = F(\varepsilon)v, \quad \frac{\mathrm{d}v}{\mathrm{d}\tau} = F(\varepsilon)v, \quad v \in R^{n+m}. \tag{6.1}$$

Clearly, if $\det F(0) \neq 0$, the crucial Assumption 2.1 would be satisfied and all the components of v would be fast. Thus, for two or more time scales to appear we need $\det F(0) = 0$.

To simplify notation, let $F(\varepsilon) = F_0 + \varepsilon F_1(\varepsilon)$ and assume that the dimension of the null-space $\mathcal{N}$ of F_0 is n. Our next assumption is crucial. It states that the range-space $\mathcal{R}$ and the null-space $\mathcal{N}$ of F_0 span R^{n+m}, that is,

$$\mathcal{R}(F_0) \oplus \mathcal{N}(F_0) = R^{n+m}. \tag{6.2}$$

This means that F_0 has n zero eigenvalues with the eigenspace $\mathcal{N}(F_0)$ and m nonzero eigenvalues with the eigenspace $\mathcal{R}(F_0)$. Therefore $\mathcal{N}(F_0)$ is the equilibrium manifold M_0.

To select a fast variable η we choose m linearly independent vectors in R^{n+m} orthogonal to $\mathcal{N}(F_0)$ and arrange them as the rows of the $m \times (n+m)$ matrix Q such that

$$Qv = 0 \Leftrightarrow F_0 v = 0 \Leftrightarrow v \in \mathcal{N}(F_0). \tag{6.3}$$

In particular, Q can be formed of any m independent rows of F_0. The requirement that η be zero on M_0 is satisfied by the m-vector Qv. Thus we take Qv as a candidate for the fast variable η.

To select a slow variable x we recall that, as a function of τ, the variable x was constant at $\varepsilon = 0$. Hence we look for a constant quantity in (6.1). It is found with the help of an $n \times (n+m)$ matrix P whose rows span the left null-space of F_0, that is, $PF_0 = 0$. Then at $\varepsilon = 0$ we have

$$P\frac{\mathrm{d}v}{\mathrm{d}t} = PF_0 v = 0 \quad \forall v \in R^{n+m}. \tag{6.4}$$

Therefore the n-vector Pv, which remains constant for all $\tau \geq 0$, is our candidate for the slow variable x.

In the proposition below we show that this choice of x and η indeed transforms (6.1) into the standard form (2.1), (2.2).

Proposition 6.1

If $F(0) = F_0$ satisfies (6.2) then the change of coordinates

$$x = Pv, \quad \eta = Qv; \quad T = \begin{bmatrix} P \\ Q \end{bmatrix} \tag{6.5}$$

transforms the system (6.1) into

$$\dot{x} = A_{11}(\varepsilon)x + A_{12}(\varepsilon)\eta, \tag{6.6}$$

$$\varepsilon\dot{\eta} = \varepsilon A_{21}(\varepsilon)x + A_{22}(\varepsilon)\eta, \tag{6.7}$$

which is in standard form since $A_{22}(0)$ is nonsingular.

Proof The inverse of T is $T^{-1} = [V \quad W]$, where the columns of V and W are bases for $\mathcal{N}(F_0)$ and $\mathcal{R}(A_0)$ respectively. Since $PF_0 = 0$ and $F_0V = 0$ we have

$$T\left(\frac{1}{\varepsilon}F_0 + F_1\right)T^{-1} = \begin{bmatrix} PF_1(\varepsilon)V & PF_1(\varepsilon)W \\ QF_1(\varepsilon)V & \frac{1}{\varepsilon}QF_0W + QF_1(\varepsilon)W \end{bmatrix} \tag{6.8}$$

and hence $A_{11}(\varepsilon) = PF_1(\varepsilon)V$, $A_{12}(\varepsilon) = PF_1(\varepsilon)W$, $A_{21}(\varepsilon) = QF_1(\varepsilon)V$ and $A_{22}(\varepsilon) = QF_0W + \varepsilon QF_1(\varepsilon)W$. The $m \times m$ matrix $A_{22}(0) = QF_0W$ is nonsingular by (6.2). It is the only nonzero block of TF_0T^{-1}, and as such it contains the m nonzero eigenvalues of F_0. □

It should be noted that in (6.7) the variable x appears multiplied by ε. For this reason we have chosen the notation η, rather than z. Although $\eta = 0$ is not the slow manifold M_ε of (6.6), (6.7), it is $O(\varepsilon^2)$ close to it, because the quasi-steady-state of η is $O(\varepsilon)$.

Example 6.1

Returning to Example 2.4, a more systematic solution proceeds as follows. The F-matrix in (2.19), (2.20) is

$$F = \begin{bmatrix} -\dfrac{1}{C_1} & \dfrac{1}{C_1} \\ \dfrac{1}{C_2} & -\dfrac{1}{C_2} \end{bmatrix} + \varepsilon \begin{bmatrix} 0 & 0 \\ 0 & \dfrac{1}{R_2} \end{bmatrix} \tag{6.9}$$

and hence $Q = [-1 \quad 1]$. The matrix $P = [p_1 \quad p_2]$ is determined by $p_1/C_1 = p_2/C_2$ and we can take either $p_1 = C_1$, $p_2 = C_2$, or

$$p_1 = \frac{C_1}{C_1 + C_2}, \quad p_2 = \frac{C_2}{C_1 + C_2}, \tag{6.10}$$

as we did in (2.23). The choice (6.10) is preferable if we want the variable x to represent a voltage, while the choice of $p_1 = C_1$, $p_2 = C_2$ would make x a sum of the charges in C_1 and C_2.

A nonlinear analog of (6.1) is the system

$$\varepsilon\dot{\nu} = h(\nu, \varepsilon), \quad \frac{d\nu}{d\tau} = h(\nu, \varepsilon), \quad \nu \in R^{n+m}, \tag{6.11}$$

and its equilibrium manifold M_0 in the τ-scale is the set of all ν such that $h(\nu, 0) = 0$. Assuming that M_0 is n-dimensional, we can reduce the $n + m$ equations $h(\nu, 0) = 0$ to m independent equations $\theta(\nu) = 0$ such that

$$\nu \in M_0 \Leftrightarrow \theta(\nu) = 0 \Leftrightarrow h(\nu, 0) = 0, \quad \text{rank}\,\frac{\partial \theta}{\partial \nu} = m. \tag{6.12}$$

Again the quantity $\theta(\nu)$, which is zero when ν is in the equilibrium manifold, will be used as the fast variable. To find a candidate for our slow variable we look for a quantity that, in the fast time scale τ, is constant when $\varepsilon = 0$. Suppose that the n-vector $\sigma(\nu)$ is such a quantity, that is,

$$\frac{d\sigma(\nu)}{d\tau} = \frac{\partial \sigma}{\partial \nu} h(\nu, 0) = 0 \quad \forall \nu \in S. \tag{6.13}$$

where S is a compact set in R^{n+m} relevant to the application at hand. A crucial assumption, analogous to (6.2), is

$$\text{rank}\begin{bmatrix} \dfrac{\partial \sigma}{\partial \nu} \\[2ex] \dfrac{\partial \theta}{\partial \nu} \end{bmatrix} = n + m \quad \forall \nu \in S. \tag{6.14}$$

Under this assumption, the change of coordinates

$$x = \sigma(\nu), \quad \eta = \theta(\nu) \tag{6.15}$$

converts the system (6.11) to the standard form (4.1), (4.2). This is illustrated by two control problems of practical significance.

Example 6.2

Consider a nonlinear system with linear high-gain feedback

$$\dot{\nu} = \gamma(\nu) + Bu, \quad \nu \in R^{n+m}, \tag{6.16}$$

$$u = \frac{1}{\varepsilon} C\nu, \quad u \in R^m. \tag{6.17}$$

In this case

$$h(\nu, \varepsilon) = \varepsilon\gamma(\nu) + BC\nu, \tag{6.18}$$

and hence $C\nu = 0$ implies $h(\nu, 0) = 0$. So we will take $\eta = \theta(\nu) = C\nu$. Next we observe from

$$\frac{d\nu}{d\tau} = h(\nu, 0) = BC\nu \tag{6.19}$$

that if the $n \times (n + m)$ matrix P satisfies $PB = 0$ then the constant quantity $\sigma(\nu)$ is $P\nu$, which we take as x. In this way, our proposed "standardizing" transformation is

$$\begin{bmatrix} x \\ \eta \end{bmatrix} = \begin{bmatrix} P \\ C \end{bmatrix} \nu, \quad PB = 0, \tag{6.20}$$

and the condition (6.14) requires that it be nonsingular. This will be the case if

$$\mathcal{R}(B) \oplus \mathcal{N}(C) = R^{n+m} \tag{6.21}$$

which also ensures that $(CB)^{-1}$ exists. Our standard form of (6.16), (6.17) is therefore

$$\dot{x} = P\gamma(\nu), \tag{6.22}$$

$$\varepsilon\dot{\eta} = CB\eta + \varepsilon C\gamma(\nu), \tag{6.23}$$

where ν is to be expressed in terms of x and η using the inverse transformation of (6.20). The discussion of this problem will be continued in Chapter 3, where it will be shown that for the stability of the feedback loop we need that $\operatorname{Re}\lambda(CB)$ be negative, as also required by Assumption 3.2.

Example 6.3

Energy modeling of aircraft flight has been behind many applications of singular perturbation techniques to guidance and control (see e.g. Kelley,

1973; Calise, 1981; Ardema 1979). The point-mass model of symmetric flight in the fast time-scale τ is

$$\frac{\mathrm{d}H}{\mathrm{d}\tau} = V \sin\gamma, \tag{6.24}$$

$$\frac{\mathrm{d}V}{\mathrm{d}\tau} = g(\varepsilon\Delta - \sin\gamma), \tag{6.25}$$

$$\frac{\mathrm{d}\gamma}{\mathrm{d}\tau} = \frac{g}{V}(l - \cos\gamma), \tag{6.26}$$

where $H =$ geometric altitude, $V =$ velocity and $\gamma =$ flight path angle are conventional state variables, while g is the acceleration due to gravity, $l = L/w$ is the lift per unit weight w and $\varepsilon\Delta$ is the difference of thrust minus drag per unit weight; both l and Δ are functions of H and V. More elaborate models include an equation for the change of weight, but for our purposes the assumption that the weight is constant is acceptable.

The system (6.24)–(6.26) is in the nonstandard form (6.11), and the condition $h(\nu, 0) = 0$ defines the equilibrium manifold M_0 as $\sin\gamma = 0$, $l = 1$, which is a level flight with lift balancing the weight. A constant quantity at $\varepsilon = 0$ in τ-scale is provided by the fact that without thrust and drag the energy is conserved. Thus, multiplying (6.24) by g and (6.25) by V and adding them together, we get at $\varepsilon = 0$, for all $\tau \geq 0$ and all H, V and γ,

$$g\frac{\mathrm{d}H}{\mathrm{d}\tau} + V\frac{\mathrm{d}V}{\mathrm{d}\tau} = 0, \quad gH + \tfrac{1}{2}V^2 = \text{const.} \tag{6.27}$$

Hence as our slow variable we take $E = \sigma(\nu) = gH + \frac{1}{2}V^2$, and obtain, in t-scale

$$\dot{E} = g\Delta[2(E - gH)]^{1/2}, \tag{6.28}$$

$$\varepsilon\dot{H} = [2(E - gH)]^{1/2} \sin\gamma, \tag{6.29}$$

$$\varepsilon\dot{\gamma} = \frac{g}{[2(E - gH)]^{1/2}}(l - \cos\gamma). \tag{6.30}$$

Although we have not defined the new fast variables yet, this model is in standard form (4.1), (4.2). Our development suggests that $\eta_1 = \sin\gamma$ and $\eta_2 = l - 1$ be the fast variables. For $\eta_1 = \gamma$, noting that $\eta_1 = \sin\gamma \cong \gamma$ for small γ, an interesting change, according to (6.15), would be to eliminate

H using $\eta_2 = l - \cos\gamma$. However, we can retain the classical model (6.28)–(6.30), which in the slow time scale shows, for example, how to choose Δ (i.e. thrust) and altitude H to minimize fuel. The boundary layer part of the solution corresponds to climb, dive or landing.

Our presentation thus far has concentrated on two-time-scale phenomena. More complex models may involve three or more time scales. For example, a standard three-time-scale model is

$$\dot{x} = f(x, z, v), \tag{6.31}$$

$$\varepsilon\dot{z} = g(x, z, v), \tag{6.32}$$

$$\varepsilon^2\dot{v} = p(x, z, v), \tag{6.33}$$

where z is faster than x, and v is faster than z. The corresponding time scales are

$$\tau = \frac{t}{\varepsilon}, \quad \theta = \frac{\tau}{\varepsilon} = \frac{t}{\varepsilon^2}. \tag{6.34}$$

In the fastest time-scale θ the system (6.31)–(6.33) is

$$\frac{\mathrm{d}x}{\mathrm{d}\theta} = \varepsilon^2 f(x, z, v), \tag{6.35}$$

$$\frac{\mathrm{d}z}{\mathrm{d}\theta} = \varepsilon g(x, z, v), \tag{6.36}$$

$$\frac{\mathrm{d}v}{\mathrm{d}\theta} = p(x, z, v), \tag{6.37}$$

and, as $\varepsilon \to 0$, both x and z are constant, while $v(\theta)$ is determined from the sublayer system (6.37). In the intermediate scale τ the same system is

$$\frac{\mathrm{d}x}{\mathrm{d}\tau} = \varepsilon f(x, z, v), \tag{6.38}$$

$$\frac{\mathrm{d}z}{\mathrm{d}\tau} = g(x, z, v), \tag{6.39}$$

$$\varepsilon\frac{\mathrm{d}v}{\mathrm{d}\tau} = p(x, z, v), \tag{6.40}$$

and, as $\varepsilon \to 0$, only $x(\tau)$ is constant, while $z(\tau)$ and $v(\tau)$ are determined from the standard singularly perturbed system (6.39), (6.40). This system is the boundary layer subsystem of the original system (6.31)–(6.33), and has (6.40) as its own boundary layer subsystem.

1.7 Case Studies in Scaling

It has been assumed thus far that the time-scale behavior is dictated by the smallness of a scalar parameter $\varepsilon \geq 0$. Although in the limit as $\varepsilon \to 0$ fast solutions exhibit boundary layer "jumps", that is, their limits are not uniform, a uniform approximation (3.7) is achieved by an additive composition of the solutions of the fast and slow models. The fact that, say, $\bar{x}(t)$ is a uniform approximation of $x(t, \varepsilon)$ over an interval of t is expressed by the statement that over this interval the difference $x(t, \varepsilon) - \bar{x}(t)$ is $O(\varepsilon)$. The "big O" symbol is defined as follows.

Definition of $O(\varepsilon)$

A vector function $f(t, \varepsilon) \in R^n$ is said to be $O(\varepsilon)$ over an interval $[t_1, t_2]$ if there exist positive constants k and ε^* such that

$$\|f(t, \varepsilon)\| \leq k\varepsilon \quad \forall \varepsilon \in [0, \varepsilon^*], \quad \forall t \in [t_1, t_2], \tag{7.1}$$

where $\|\cdot\|$ is the Euclidean norm.

In the asymptotic analysis the "small o" symbol is also common and means that $f(t, \varepsilon)$ is $o(\varepsilon)$ if

$$\lim_{\varepsilon \downarrow 0} \frac{\|f(t, \varepsilon)\|}{\varepsilon} = 0. \tag{7.2}$$

For simplicity, we will use only the "big O" symbol. In some cases we will be able to give estimates of constants k and ε^* in (7.1) and thus quantify the corresponding $O(\varepsilon)$ approximations. Otherwise, we will be satisfied by $O(\varepsilon)$ being an "order of magnitude relation", valid for "ε sufficiently small." As an example, let us compare 2ε and $8\varepsilon^2$, considering 2ε as $O(\varepsilon)$ and $8\varepsilon^2$ as $O(\varepsilon^2)$. However, $8\varepsilon^2 \leq 2\varepsilon$ only if $\varepsilon \in [0, \frac{1}{4}]$, that is, ε is "sufficiently small" only if it is less than $\frac{1}{4}$. Thus, applying the results of an asymptotic analysis to a problem with a given value of ε, requires *a priori* knowledge that this value is indeed "sufficiently small".

But what is the meaning of ε and when is it small? This raises the perennial issue of *scaling*, which we now address in three short case studies.

Case Study 7.1 Dimensionless ε in the DC-Motor Model

In some models the time and state variables can be scaled to exhibit ε as a dimensionless ratio of two physical parameters such as time constants. The simplest illustration is again the armature controlled DC-motor model (2.7), (2.8), rewritten here as

$$\frac{J\Omega}{kI}\frac{d\omega_r}{dt} = i_r, \tag{7.3}$$

$$\frac{LI}{k\Omega}\frac{di_r}{dt} = -\omega_r - \frac{RI}{k\Omega} i_r + u_r, \tag{7.4}$$

where

$$\omega_r = \frac{\omega}{\Omega}, \quad i_r = \frac{i}{I}, \quad u_r = \frac{u}{k\Omega}. \tag{7.5}$$

The relative unit for current is chosen such that in the steady state when $u_r = 1$ and $\omega_r = 0$, then from (7.4) $i_r = 1$. This gives $I = k\Omega/R$, so that (7.3), (7.4) become

$$T_m \frac{d\omega_r}{dt} = i_r, \tag{7.6}$$

$$T_e \frac{di_r}{dt} = -\omega_r - i_r + u_r, \tag{7.7}$$

where

$$T_e = \frac{L}{R}, \quad T_m = \frac{JR}{k^2}, \tag{7.8}$$

are the well known electrical and mechanical time constants respectively. Since $T_m \gg T_e$, we let T_m be the time unit, that is, $t_r = t/T_m$, and rewrite (7.6), (7.7) as

$$\frac{d\omega_r}{dt_r} = i_r, \tag{7.9}$$

$$\frac{T_e}{T_m}\frac{di_r}{dt_r} = -\omega_r - i_r + u_r. \tag{7.10}$$

This scaling has brought us to a physically meaningful dimensionless parameter

$$\varepsilon = \frac{T_e}{T_m} = \frac{Lk^2}{JR^2}. \tag{7.11}$$

Model (7.9), (7.10) will be used for illustrative purposes throughout the remaining text.

Case Study 7.2 Parameter Scaling in an Airplane Model

A slow "phugoid mode" and a fast "short-period mode" are well known time-scale characteristics of the longitudinal motion of an airplane. We now show how these characteristics are exhibited by first scaling and then applying the modeling methodology of Section 1.6.

Assuming, as in Fig. 1.8, that the longitudinal motion consists of pitching and displacement in the plane of symmetry (Etkin 1972), we apply Newton's law and obtain the following nonlinear model:

$$m\frac{\mathrm{d}v}{\mathrm{d}\tau} = -D - mg\sin\gamma + T\cos\alpha \triangleq mX, \tag{7.12}$$

$$-mv\frac{\mathrm{d}\gamma}{\mathrm{d}\tau} = -L + mg\cos\gamma = T\sin\alpha \triangleq -mvZ, \tag{7.13}$$

$$I\frac{\mathrm{d}^2\theta}{\mathrm{d}\tau^2} \triangleq \bar{M} = IM, \tag{7.14}$$

where T, L, D, mg and $\bar{M}$ are the thrust, lift, drag, weight and moment, while α, γ and θ are the angle of attack, flight-path angle and pitch angle, respectively; I is the moment of inertia, τ is the time corresponding to the

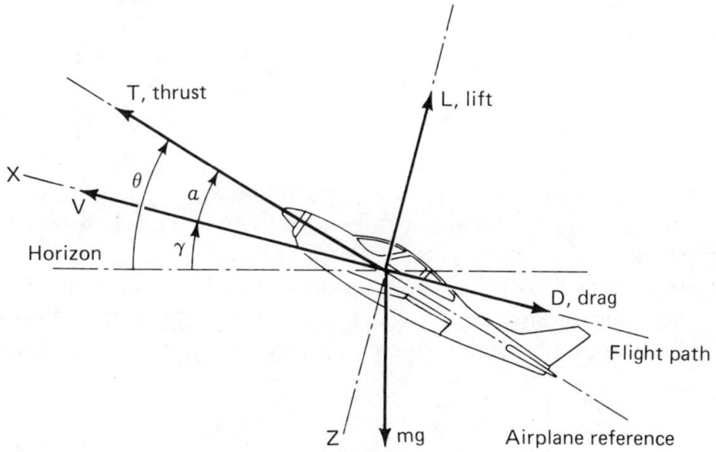

Fig. 1.8. Parameter scaling in an airplane model.

fast motion and $\theta = \alpha + \gamma$. The linearization of this model for the straight steady flight with velocity V_0 yields, after neglecting some small terms,

$$\frac{\mathrm{d}}{\mathrm{d}\tau}\begin{bmatrix} v \\ \theta \\ \alpha \\ q \end{bmatrix} = \begin{bmatrix} X_v & -\dfrac{g}{V_0} & \dfrac{X_\alpha}{V_0} & 0 \\ 0 & 0 & 0 & 1 \\ Z_v V_0 & 0 & Z_\alpha & 1 \\ M_v V_0 & 0 & M_\alpha & M_q \end{bmatrix} \begin{bmatrix} v \\ \theta \\ \alpha \\ q \end{bmatrix} + \begin{bmatrix} \dfrac{X_\delta}{V_0} \\ 0 \\ Z_\delta \\ M_\delta \end{bmatrix} \delta \qquad (7.15)$$

where θ, α, q and δ are respectively incremental pitch angle, angle of attack, pitch rate and elevator position, while $v = (V - V_0)/V_0$ is the normalized incremental velocity and the subscripts denote the partial derivatives.

There is no obvious candidate for ε in (7.15). However, for typical numerical values (Elliott, 1977) the small quantities are

$$X_v, \quad \frac{g}{V_0}, \quad \frac{X_\alpha}{V_0}, \quad Z_v V_0, \quad M_v V_0 \ll 1 \qquad (7.16)$$

while the quantities

$$Z_\alpha, M_\alpha, M_q \text{ are } O(1). \qquad (7.17)$$

This suggests that ε be introduced as the ratio of the largest of the small quantities (7.16) versus the smallest of the large quantities (7.17). To achieve this we rewrite the 4×4 system matrix of (7.15) in the form $F(\varepsilon) = F_0 + \varepsilon F_1$ and partition it into 2×2 matrices as folows:

$$F(\varepsilon) = \begin{bmatrix} \varepsilon F_{111} & F_{120} \\ \varepsilon F_{211} & F_{220} \end{bmatrix} = \begin{bmatrix} 0 & F_{120} \\ 0 & F_{220} \end{bmatrix} + \varepsilon \begin{bmatrix} F_{111} & 0 \\ F_{211} & 0 \end{bmatrix}. \qquad (7.18)$$

In this way the nonzero entries of F_{111}, namely X_v/ε, $-g/\varepsilon V_0$, and those of F_{211}, namely $Z_v V_0/\varepsilon$, $M_v V_0/\varepsilon$ are all scaled to be $O(1)$. It is also important to note that the entry X_α/V_0 of F_{120} is $O(\varepsilon)$.

With the scaling (7.18) the linearized model (7.15) has become a special case of the nonstandard model (6.1). Thus we apply Proposition 6.1 to (7.15) scaled by (7.18), which, for generality, we first rewrite as follows:

$$\begin{bmatrix} \varepsilon \dot{\nu}_1 \\ \varepsilon \dot{\nu}_2 \end{bmatrix} = \begin{bmatrix} \varepsilon F_{111} & F_{120} \\ \varepsilon F_{211} & F_{220} \end{bmatrix} \begin{bmatrix} \nu_1 \\ \nu_2 \end{bmatrix}, \quad \nu_1 \in R^n, \quad \nu_2 \in R^m, \qquad (7.19)$$

where F_{220}^{-1} exists. The matrix $P = [P_1 \quad P_2]$ of Proposition 6.1 must satisfy

$$P_1 F_{120} + P_2 F_{220} = 0. \tag{7.20}$$

If we chose $P_1 = I_{n\times n}$ then $P_2 = -F_{120}F_{220}^{-1}$. Thus our slow variable is

$$x = \nu_1 - F_{120}F_{220}^{-1}\nu_2. \tag{7.21}$$

Since F_{220} is nonsingular the expression $Q\nu = 0$ reduces to $\nu_2 = 0$, so we can take $Q = [0 \quad I_{m\times m}]$, and hence our fast variable $\eta = \nu_2$. The transformed model is (6.6), (6.7), where

$$\left.\begin{aligned} A_{11} &= F_{111} - F_{120}F_{220}^{-1}F_{211}, \quad A_{12} = A_{11}F_{120}F_{220}^{-1}, \\ A_{21} &= F_{211}, \quad A_{22} = F_{220} + \varepsilon F_{211}F_{120}F_{220}^{-1}. \end{aligned}\right\} \tag{7.22}$$

Specializing the steps (7.19)–(7.22) for the airplane model (7.15) scaled as in (7.18), we have $n = 2$, $m = 2$ and

$$\nu_1 = \begin{bmatrix} v \\ \theta \end{bmatrix}, \quad \nu_2 = \begin{bmatrix} \alpha \\ q \end{bmatrix}. \tag{7.23}$$

Thus $\eta = \nu_2$ implies that α and q are our fast variables, while the slow variables can be evaluated from (7.21).

In the traditional analysis (Etkin 1972) of the longitudinal model (7.15), the slow "phugoid mode" and the fast "short-period mode" are approximated to reveal the factors influencing natural frequencies ω_{ni} and damping ratios ζ_i, which provide important stability information. It is of interest to compare one of these approximations with the approximations offered by our model matrices (7.22). From the standard model (6.6), (6.7) we know that the eigenvalues of the matrix

$$\varepsilon A_{11} = \begin{bmatrix} X_v - \dfrac{Z_v M_q - M_v}{Z_\alpha M_q - M_\alpha} X_\alpha & -\dfrac{g}{V_0} \\ \dfrac{M_\alpha Z_v - M_v Z_\alpha}{Z_\alpha M_q - M_\alpha} V_0 & 0 \end{bmatrix} \tag{7.24}$$

approximate the slow mode. (Note that A_{11} is multiplied by ε in order to return to the fast time scale τ in which the original model (7.15) was defined.) Matrix (7.24) shows that the slow mode's ω_{n}^2 is

$$\omega_{\mathrm{n}}^2 = \frac{M_\alpha Z_v - M_v Z_\alpha}{Z_\alpha M_q - M_\alpha} g, \tag{7.25}$$

and its damping ratio ζ can be evaluated from

$$2\zeta\omega_{\mathrm{n}} = -X_v + \frac{Z_v M_q - M_v}{Z_\alpha M_q - M_\alpha} X_\alpha. \tag{7.26}$$

This approximation is more accurate than a simpler approximation

$$\omega_{\mathrm{n}}^2 = -Z_v g, \quad 2\zeta\omega_{\mathrm{n}} = -X_v, \tag{7.27}$$

found in standard texts such as Etkin (1972). It is of conceptual importance, however, that, while Etkin's approximation (7.27) is obtained via several heuristic steps, the approximation (7.25), (7.26) follows from the analytic procedure summarized in Proposition 6.1. Moreover, in the common situation when

$$|(Z_v M_q - M_v)X_\alpha| \ll |(Z_\alpha M_q - M_\alpha)X_v|, \tag{7.28}$$

$$Z_\alpha M_q - M_\alpha \sim -M_\alpha, \quad M_\alpha Z_v - M_v Z_\alpha \sim M_\alpha Z_v, \tag{7.29}$$

the approximation (7.27) follows from the singular perturbation approximation (7.25), (7.26). For the NASA F-8 numerical values given by Elliot (1977), the error of (7.25), (7.26) is 0.8% and that of (7.27) is 1.9%.

The fast-mode matrix A_{22} defined by (7.15), (7.18) and (7.22) is

$$A_{22} = \begin{bmatrix} Z_\alpha + \dfrac{Z_v X_\alpha M_q}{Z_\alpha M_q - M_\alpha} & 1 - \dfrac{Z_v X_\alpha}{Z_\alpha M_q - M_\alpha} \\ M_\alpha + \dfrac{M_v X_\alpha M_q}{Z_\alpha M_q - M_A} & M_q - \dfrac{M_v X_\alpha}{Z_\alpha M_q - M_\alpha} \end{bmatrix}$$

which, using (7.16) and (7.17), may be approximated according to

$$A_{22} = \begin{bmatrix} Z_\alpha & 1 \\ M_\alpha & M_q \end{bmatrix} + O(\varepsilon) \tag{7.30}$$

and then the fast mode approximation

$$\omega_{\mathrm{n}}^2 = Z_\alpha M_q - M_\alpha, \quad 2\zeta\omega_{\mathrm{n}} = -(Z_\alpha + M_q) \tag{7.31}$$

is the same as in Etkin (1972).

Case Study 7.3 State Scaling in a Voltage Regulator

While in Case Study 7.2 only system parameters are scaled by ε, we now consider a model in which a state variable is scaled by ε. This scaling is fundamentally different from the scaling by relative units used in Case

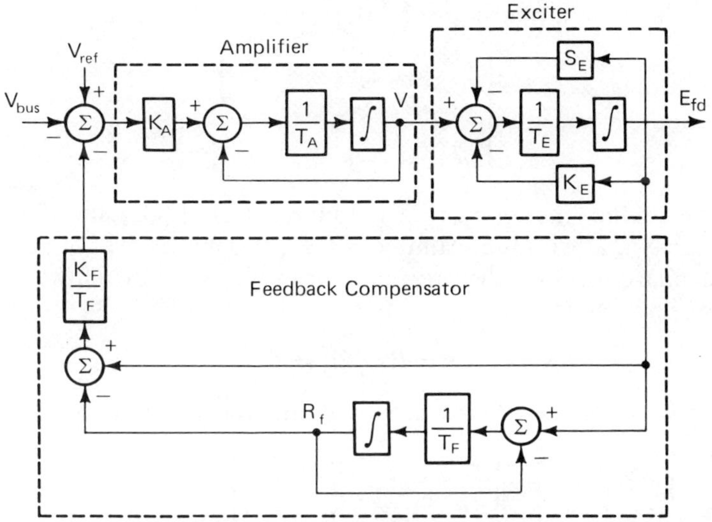

Fig. 1.9. State scaling in a voltage regulator.

Study 7.1. The IEEE Type 2 voltage regulator is shown in Fig. 1.9, and, neglecting saturation S_{E}, is described by

$$\left.\begin{aligned}\frac{\mathrm{d}R_{\mathrm{f}}}{\mathrm{d}t} &= -\frac{1}{T_{\mathrm{F}}}(R_{\mathrm{f}} - E_{\mathrm{fd}}),\\ \frac{\mathrm{d}E_{\mathrm{fd}}}{\mathrm{d}t} &= -\frac{1}{T_{\mathrm{E}}}(K_{\mathrm{E}}E_{\mathrm{fd}} - V),\\ \frac{\mathrm{d}V}{\mathrm{d}t} &= \frac{1}{T_{\mathrm{A}}}\left(\frac{K_{\mathrm{A}}K_{\mathrm{F}}}{T_{\mathrm{F}}}R_{\mathrm{f}} - \frac{K_{\mathrm{A}}K_{\mathrm{F}}}{T_{\mathrm{F}}}E_{\mathrm{fd}} - V + K_{\mathrm{A}}\Delta\right),\end{aligned}\right\} \tag{7.32}$$

where $\Delta = V_{\mathrm{ref}} - V_{\mathrm{bus}}$. To represent this model in a singular perturbation form we examine typical numerical values of its parameters:

$$\left.\begin{aligned}T_{\mathrm{A}} &= 0.06\ \mathrm{s}, & K_{\mathrm{A}} &= 25,\\ T_{\mathrm{E}} &= 0.5\ \mathrm{s}, & K_{\mathrm{E}} &= -0.0445,\\ T_{\mathrm{F}} &= 1.0\ \mathrm{s}, & K_{\mathrm{F}} &= 0.16.\end{aligned}\right\} \tag{7.33}$$

First attempt: We assume that the fast phenomena are caused, as in Example 2.2, solely by high gain K_{A}, and exhibit this assumption by rewriting the system matrix of (7.32) as

$$\begin{bmatrix} a_1 & a_2 & 0 \\ 0 & a_3 & a_4 \\ K_A a_5 & -K_A a_5 & a_6 \end{bmatrix} \tag{7.34}$$

where the coefficients $a_1, a_2, \ldots,$ are known from (7.32) and (7.33). It is easy to verify that with the scaling $\varepsilon = 1/K_A$ this matrix cannot be brought into a standard form via the procedure of Proposition 6.1. However, if we define ε and the states as

$$\varepsilon = K_A^{-1/2}, \quad x = R_f, \quad z_1 = E_{fd}, \quad z_2 = \varepsilon V, \tag{7.35}$$

that is, if *we scale one of the states by* ε then the system (7.32) becomes

$$\begin{bmatrix} \dot{x} \\ \varepsilon \dot{z}_1 \\ \varepsilon \dot{z}_2 \end{bmatrix} = \begin{bmatrix} a_1 & a_2 & 0 \\ 0 & \varepsilon a_3 & a_4 \\ a_5 & -a_5 & \varepsilon a_6 \end{bmatrix} \begin{bmatrix} x \\ z_1 \\ z_2 \end{bmatrix} + \begin{bmatrix} 0 \\ 0 \\ a_7 \end{bmatrix} \Delta \tag{7.36}$$

Although it would seem that we have reached a standard form, this is not so because our crucial Assumption 3.2 is not met since

$$\operatorname{Re} \lambda \begin{bmatrix} 0 & a_4 \\ -a_5 & 0 \end{bmatrix} = 0 \tag{7.37}$$

violates the requirement (3.9).

Second attempt: Recognizing that, in addition to high gain K_A, the small time constant T_A also contributes to the appearance of the fast phenomena, we scale T_A by ε and define

$$\varepsilon = K_A^{-1}, \quad T_A = 1.5\varepsilon, \quad x = R_F, \quad z_1 = E_{fd}, \quad z_2 = \varepsilon V. \tag{7.38}$$

Although $z_2 = \varepsilon V$ as in (7.35), the scaling (7.38) is different, because ε is different. With the parameter and state scaling (7.38), the system (7.32) becomes

$$\begin{bmatrix} \dot{x} \\ \varepsilon \dot{z}_1 \\ \varepsilon \dot{z}_2 \end{bmatrix} = \begin{bmatrix} a_1 & a_2 & 0 \\ 0 & \varepsilon a_3 & a_4 \\ a_8 & -a_8 & a_9 \end{bmatrix} \begin{bmatrix} x \\ z_1 \\ z_2 \end{bmatrix} + \begin{bmatrix} 0 \\ 0 \\ a_{10} \end{bmatrix} \Delta, \tag{7.39}$$

which is in the standard form satisfying Assumption 3.2, because using the parameter values of (7.33)

$$\operatorname{Re}\lambda\begin{bmatrix} 0 & a_4 \\ -a_8 & a_9 \end{bmatrix} < 0. \tag{7.40}$$

The two models, (7.36) and (7.39), behave differently insofar as their fast modes (7.37) and (7.40) are concerned. As $\varepsilon \to 0$ the fast eigenvalues of (7.36) become purely oscillatory, while those of (7.39) remain well damped.

An important consequence of the state scaling $z_2 = \varepsilon V$ is that it allows the original variable $V = z_2/\varepsilon$ to be $O(1/\varepsilon)$ large, as is to be expected of the output of the high-gain amplifier in Fig. 1.9.

The last case study illustrates a state scaling by ε compatible with the range of the actual physical variable. A state scaling could have been used in the airplane model (7.19) by treating ν_1 as x and ν_2 as εz. However, the initial condition for z would then be $z^0 = (1/\varepsilon)\nu_2^0$, and, in order to have the properties of a standard form, ν_2^0 would not be allowed to be larger than $O(\varepsilon)$. The methodology of Proposition 6.1 has avoided this restriction.

Although we have given guidelines for modeling two-time-scale systems in a standard singular perturbation form, it is appropriate to close this section and the whole chapter with the conclusion that, in general, this modeling task requires considerable *a priori* knowledge about the physical nature of the system.

1.8 Exercises

Exercise 1.1

Generalize the derivations (6.18)–(6.23) to include the case of ν-dependent B-matrices, $B = B(\nu)$, and apply them to the system (2.11), (2.12) of Example 2.2 with $\nu_1 = x - r$, $\nu_2 = e^z - 1$, $y = \nu_1 + \nu_2$ and $u = y/\varepsilon$.

Exercise 1.2

A scalar PI-controller

$$u(s) = \left(k_1 + \varepsilon k_2 \frac{1}{s}\right)e(s),$$

of the type analyzed in Case Study 3.1, is now applied to a nonlinear plant

$$\frac{dz}{d\tau} = a(z(\tau), u(\tau)), \quad z \in R^m, \quad \tau = \frac{t}{\varepsilon}.$$

The error $e = y - r$ between its scalar output $y = h(z)$ and a constant set-point r is to be regulated to zero. Introducing an additional state variable x via the equation

$$\frac{dx}{d\tau} = \varepsilon k_2 e(\tau).$$

represent this nonlinear feedback system in a singular perturbation form. When will this form satisfy Assumptions 2.1, 3.1 and 3.2?

Exercise 1.3

Extend the invariant manifold methodology of Sections 1.4 and 1.5 to allow a dependence of f and g in (4.1), (4.2) on ε. Using this methodology, obtain an improved slow model and an improved fast model for the system (2.11), (2.12) of Example 2.2.

Exercise 1.4

In the coupled circuit in Fig. 1.10 the ratio of leakage inductances l_1, l_2 and the self-inductances L_1, L_2 is assumed to be the same small parameter

$$\frac{l_1}{L_1} = \frac{l_2}{L_2} = \mu.$$

Using the flux linkages

$$\lambda_1 = (1 + \mu)L_1 i_1 - \frac{N_1}{N_2} L_2 i_2,$$

$$\lambda_2 = -\frac{N_2}{N_1} L_1 i_1 + (1 + \mu)L_2 i_2$$

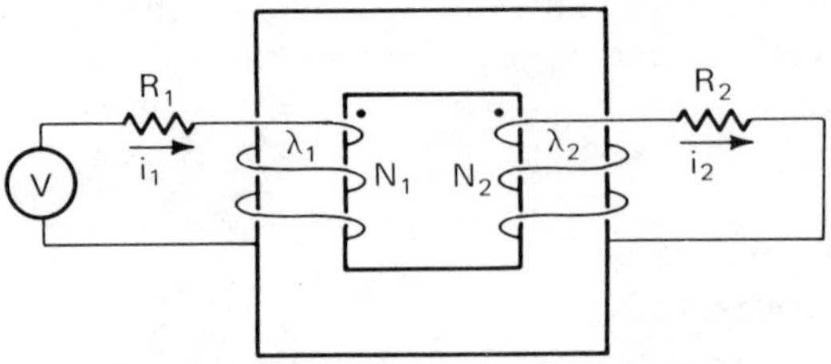

Fig. 1.10. Coupled circuit of Exercise 1.4.

as the state variables and eliminating i_1, i_2 from

$$\frac{d\lambda_1}{dt} = -R_1 i_1 + v, \quad \frac{d\lambda_2}{dt} = -R_2 i_2,$$

show that the following non-standard model is obtained:

$$\varepsilon \frac{d\lambda_1}{dt} = -\frac{(1+\varepsilon)^{1/2}}{T_1}\lambda_1 - \frac{h}{T_2}\lambda_2 + \varepsilon v,$$

$$\varepsilon \frac{d\lambda_2}{dt} = -\frac{1}{hT_1}\lambda_1 - \frac{(1+\varepsilon)^{1/2}}{T_2}\lambda_2,$$

where

$$\varepsilon = 2\mu + \mu^2, \quad T_1 = \frac{L_1}{R_1}, \quad T_2 = \frac{L_2}{R_2}, \quad h = \frac{T_2 N_1}{T_1 N_2}.$$

Apply the methodology of Section 1.6 to transform this model into a standard form and interpret the meaning of its slow and fast systems.

Exercise 1.5

Paralleling Case Study 7.1, perform a scaling to obtain a dimensionless form of the *RC*-circuit model (2.15), (2.16) of Example 2.3. Show that ε can be introduced as the dimensionless ratio of two time constants.

1.9 Notes and References

Basic references for an in-depth study of the method of asymptotic expansions applied to the standard singular perturbation model are the articles by Vasil'eva (1963) and Hoppensteadt (1971), and the books by Wasow (1965) and O'Malley (1974c), in English, and by Vasil'eva and Butuzov (1973), in Russian. Our presentation is self-contained in that it does not assume familiarity with the above sources, which are recommended as additional reading, especially for higher-order expansions and proofs of their asymptotic validity. A broader view on applications to systems and control problems can be obtained from surveys by Kokotović, O'Malley and Sannuti (1976), Saksena, O'Reilly and Kokotović (1984) and Kokotović (1984). These surveys contain references to singular perturbation models of electromechanical networks and power systems, chemical kinetics, nuclear reactors, heat exchangers and to some of the modeling issues dealt with in this chapter. A selection from these references is reprinted in a volume edited by Kokotović and Khalil (1986).

A class of models, not explicitly covered in this book, are Markov chains in which low transition probabilities lead to slow phenomena and high transition probabilities to fast phenomena. Gaitsgori and Pervozvanskii (1975), Delebecque and Quadrat (1981) and Phillips and Kokotović (1981) develop two-time-scale models of Markov chains and apply singular perturbation techniques to queueing networks. As shown by Kokotović (1981), such problems are analogous to problems of aggregate modeling of large electromechanical networks, treated by the methodology of Section 1.6. This methodology was developed and applied to power systems in the monograph edited by Chow (1982). It was further extended to a hierarchy of time scales by Delebecque (1983), Coderch *et al.* (1983) and Khalil (1984b). Its application to a nonlinear model in economics is found in Peponides and Kokotović (1983), and to networks with dense and sparse connections in Chow and Kokotović (1985).

In our presentation, the modeling methodology of Section 1.6 follows a discussion of slow manifolds. This notion is introduced in Section 1.4 and used as a tool for construction of appropriate models in Section 1.5, a generalization of the iterative modeling procedure proposed in Kokotović *et al.* (1980). Although the slow manifold concept is implicit in every separation of slow and fast models, its explicit use as a geometric framework for a singular perturbation analysis is recent; see Fenichel (1979) and Sobolev (1984). Invariant or integral manifolds more commonly appear in the averaging method of Bogolyubov and Mitropolsky (1961) and Hale (1980) and in the reduction method of Pliss (1966a, b), which extends the classical conditional stability results of Lyapunov and Perron. Monographs by Mitropolsky and Lykova (1973), in Russian, and Carr (1981), in English, dedicate special sections to integral and center manifolds of singularly perturbed systems.

Only single-parameter perturbations are considered in this book. A class of multiparameter perturbations of different orders of magnitude can be treated as nested single-parameter perturbations giving rise to multiple time scales. They were analyzed by Tikhonov (1952), Hoppensteadt (1971) and O'Malley (1974c), and applied to a rolling mill problem by Jamshidi (1974). Multiparameter perturbations of the same order of magnitude represent a new class of problems. As shown by Khalil and Kokotović (1978, 1979a, b), they are of interest in decentralized control when each controller uses a different simplified model of the same large-scale system.

2 LINEAR TIME-INVARIANT SYSTEMS

2.1 Introduction

Linear time-invariant models are of interest in "local" or "small-signal" approximations of more realistic nonlinear models of dynamic systems. Let x_e, z_e be the equilibrium state of the nonlinear singularly perturbed system

$$\dot{x} = f(x, z, u), \quad x(t_0) = x^0, \quad x \in R^n, \quad u \in R^r, \tag{1.1}$$

$$\varepsilon \dot{z} = g(x, z, u), \quad z(t_0) = z^0, \quad z \in R^m, \tag{1.2}$$

for a constant control u_e, that is

$$f(x_e, z_e, u_e) = 0, \quad g(x_e, z_e, u_e) = 0, \tag{1.3}$$

and consider the small deviations

$$\delta x = x - x_e, \quad \delta z = z - z_e, \quad \delta u = u - u_e \tag{1.4}$$

about the equilibrium (x_e, z_e, u_e). These small deviations are approximately described by the linear time-invariant model

$$\delta \dot{x} = \frac{\partial f}{\partial x}\,\delta x + \frac{\partial f}{\partial z}\,\delta z + \frac{\partial f}{\partial u}\,\delta u, \quad \delta x(t_0) = x^0 - x_e, \tag{1.5}$$

$$\varepsilon \delta \dot{z} = \frac{\partial g}{\partial x}\,\delta x + \frac{\partial g}{\partial z}\,\delta z + \frac{\partial g}{\partial u}\,\delta u, \quad \delta z(t_0) = z^0 - z_e, \tag{1.6}$$

where the partial derivatives $\partial f/\partial x$, $\partial f/\partial z$, etc., evaluated at $x = x_e$, $z = z_e$, $u = u_e$, are constant matrices. Most of the local properties of the nonlinear system (1.1), (1.2) can be deduced from the linear system (1.5), (1.6). For example, if the linear system (1.5), (1.6) is asymptotically stable with $\delta u = 0$, then the equilibrium x_e, z_e of the nonlinear system (1.1), (1.2) is asymptotically stable. If this is not so for $\delta u = 0$, but a feedback control $\delta u = F_1 \delta x + F_2 \delta z$ can be found to make the linear system (1.5), (1.6) asymptotically stable, then the same feedback control will make the equilibrium x_e, z_e of the nonlinear system (1.1), (1.2) asymptotically stable. These and similar properties explain why this chapter and the next two chapters of this book are devoted to a separate study of the system

$$\dot{x} = A_{11}x + A_{12}z + B_1 u, \quad x \in R^n, \quad u \in R^r, \tag{1.7}$$

$$\varepsilon \dot{z} = A_{21}x + A_{22}z + B_2 u, \quad z \in R^m, \tag{1.8}$$

which differs from (1.5), (1.6) only in the obvious change of notation. In addition to (1.7), (1.8), our study will also include the system output y defined by

$$y = C_1 x + C_2 z. \tag{1.9}$$

All the matrices of the model (1.7)–(1.9) are assumed to be constant and independent of ε. The latter assumption is for convenience only and leads to a minor loss of generality, which we avoid as follows. If some of the matrices depend on ε, say $A_{22} = A_{22}(\varepsilon)$, then we assume that they are sufficiently many times continuously differentiable with respect to ε at $\varepsilon = 0$, and that for $\varepsilon = 0$ they possess the properties required from the matrices whose dependence on ε is suppressed. For example, when we assume that A_{22} is nonsingular, then this can be extended to mean that $A_{22}(\varepsilon)$ is nonsingular at $\varepsilon = 0$ and for $\varepsilon = 0$ its derivative $dA_{22}/d\varepsilon$ exists and is bounded. We will also use the Laplace transform of (1.7)–(1.9), namely,

$$sx(s) - x^0 = A_{11}x(s) + A_{12}z(s) + B_1 u(s), \tag{1.10}$$

$$\varepsilon s z(s) - \varepsilon z^0 = A_{21}x(s) + A_{22}z(s) + B_2 u(s), \tag{1.11}$$

$$y(s) = C_1 x(s) + C_2 z(s), \tag{1.12}$$

and represent the model by the block diagram in Fig. 2.1. The deterministic and stochastic designs of feedback controls for (1.7)–(1.9) are presented in Chapters 3 and 4 respectively. In this chapter, we concentrate on the analysis of two-time-scale properties of the model (1.7)–(1.9) in the time domain, and the two-frequency-scale properties of the model (1.10)–(1.12) in the frequency domain. In Sections 2.2–2.5, we disregard the input u and the output y and study the properties of the free system (1.7), (1.8) with $u = 0$. The so-called actuator and sensor forms derived in Section 2.2 make some of the time-scale properties of (1.7), (1.8) not only easy to analyze,

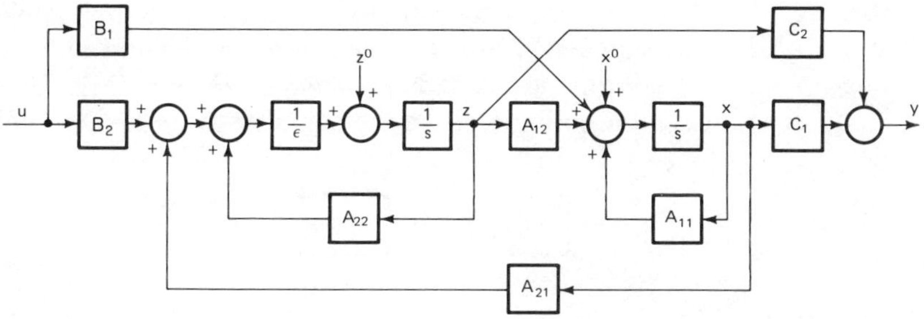

Fig. 2.1. Block diagram of standard open-loop singularly perturbed model.

but also more clearly related to applications where fast transients are due to physical actuators and sensors. Among these properties are the eigenvalue properties analyzed in Section 2.3. A second step in the transformation of (1.7), (1.8) brings it to a block-diagonal form in which all the fast phenomena are described by one block and all the slow phenomena by another block. Since each block corresponds to an invariant subspace of (1.7), (1.8), this step gives a geometric interpretation of the time-scale properties alluded to in the discussion of the slow manifold in Chapter 1. In Section 2.5 we show that the results of Sections 2.2, 2.3 and 2.4 offer a great flexibility in constructing and validating approximate models with known asymptotic properties. In Section 2.6 we return to the model (1.7)–(1.9) with the control input u and the output y and study its controllability and observability properties. Also, this model provides the basis for a characterization of approximate models in the frequency domain in Section 2.7.

2.2 The Block-Triangular Forms

In order to study two-time-scale properties of the system

$$\dot{x} = A_{11}x + A_{12}z, \quad x(t_0) = x^0, \tag{2.1}$$

$$\varepsilon\dot{z} = A_{21}x + A_{22}z, \quad z(t_0) = z^0, \tag{2.2}$$

or, equivalently, to study two-frequency-scale properties of its Laplace transform

$$sx(s) - x^0 = A_{11}x(s) + A_{12}z(s), \tag{2.3}$$

$$\varepsilon sz(s) - \varepsilon z^0 = A_{21}x(s) + A_{22}z(s), \tag{2.4}$$

we introduce two different sets of coordinates in which the system appears in two distinct block-triangular forms. For ease of reference and with some intuitive appeal, these forms are called the *actuator form* and the *sensor form*, and are represented in Fig. 2.2. In the actuator form the fast block,

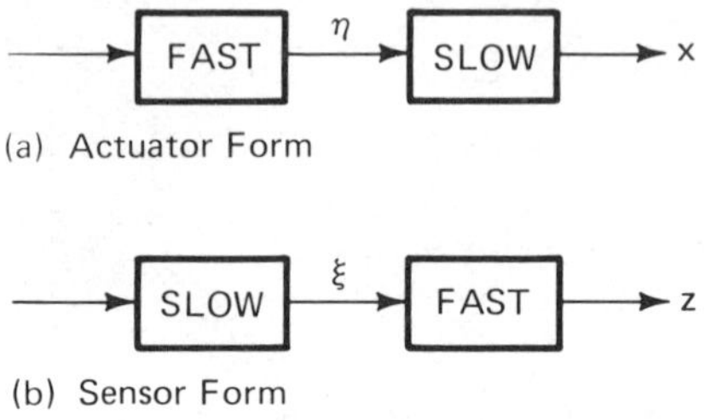

Fig. 2.2. The two block-triangular forms of (2.1), (2.2).

whose state is the new variable η, "drives" the slow block, whose state is the original variable x. In the sensor form the roles are reversed: the slow block, whose state is the new variable ξ, is followed by the fast block, whose state is the original variable z, which can be thought of as the sensor information about the slow variable ξ. To avoid repetition of similar derivations, we deal mostly with the actuator form and relegate the details about the sensor form for problems and exercises.

The actuator form is the result of the change of variables

$$\eta(t) = z(t) + L(\varepsilon)x(t), \tag{2.5}$$

introduced in Section 1.4, that is, the similarity transformation

$$\begin{bmatrix} x(t) \\ \eta(t) \end{bmatrix} = \begin{bmatrix} I_n & 0 \\ L(\varepsilon) & I_m \end{bmatrix} \begin{bmatrix} x(t) \\ z(t) \end{bmatrix} \tag{2.6}$$

which transforms (2.1), (2.2) into

$$\begin{bmatrix} \dot{x}(t) \\ \varepsilon\dot{\eta}(t) \end{bmatrix} = \begin{bmatrix} A_{11} - A_{12}L & A_{12} \\ R(L, \varepsilon) & A_{22} + \varepsilon L A_{12} \end{bmatrix} \begin{bmatrix} x(t) \\ \eta(t) \end{bmatrix}. \tag{2.7}$$

The requirement that the matrix $R(L, \varepsilon)$ be zero, that is, the requirement that the $m \times n$ matrix $L(\varepsilon)$ satisfy the algebraic equation

$$R(L, \varepsilon) = A_{21} - A_{22}L + \varepsilon L A_{11} - \varepsilon L A_{12} L = 0 \tag{2.8}$$

results in the system (2.7) assuming the upper block-triangular form

$$\begin{bmatrix} \dot{x}(t) \\ \varepsilon\dot{\eta}(t) \end{bmatrix} = \begin{bmatrix} A_{11} - A_{12}L & A_{12} \\ 0 & A_{22} + \varepsilon L A_{12} \end{bmatrix} \begin{bmatrix} x(t) \\ \eta(t) \end{bmatrix} \tag{2.9}$$

which is the actuator form of Fig. 2.2(a). What has been achieved is that the system (2.1), (2.2) has been partially decoupled in (2.9) to provide a separate fast subsystem

$$\varepsilon\dot{\eta}(t) = (A_{22} + \varepsilon L A_{12})\eta(t), \tag{2.10}$$

rewritten in the fast time scale τ as

$$\frac{d\eta(\tau)}{d\tau} = (A_{22} + \varepsilon L A_{12})\eta(\tau), \quad \tau = \frac{t - t_0}{\varepsilon}, \tag{2.11}$$

with, by (2.5), the initial condition

$$\eta(0) = z^0 + Lx^0. \tag{2.12}$$

The block-diagram of the actuator form (2.9), shown in Fig. 2.3, is a detailed representation of the diagram in Fig. 2.2(a). Note that the feedback loop of the fast subsystem is a high-gain loop, that is, its gain tends to infinity as $\varepsilon \to 0$. This is a characteristic of the fast part of every singularly perturbed system in the standard form, that is, when A_{22} is nonsingular. This would not be so if A_{22} were singular, because then in some component loops of the vector loop the gain $1/\varepsilon$ would be multiplied by ε coming from the term $\varepsilon L A_{12}$. Before proceeding with a discussion of the properties of the actuator form, we first need to establish that an $L(\varepsilon)$ satisfying (2.8) exists.

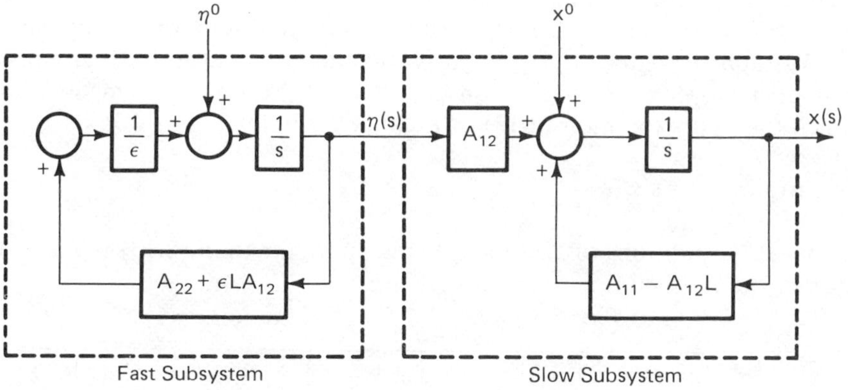

Fig. 2.3. Actuator form.

Lemma 2.1

If A_{22} is nonsingular, there is an $\varepsilon^* \geq 0$ such that for all $\varepsilon \in [0, \varepsilon^*]$, there exists a solution $L(\varepsilon)$ to the matrix quadratic equation (2.8) which is approximated according to

$$L(\varepsilon) = A_{22}^{-1}A_{21} + \varepsilon A_{22}^{-2}A_{21}A_0 + O(\varepsilon^2), \tag{2.13}$$

where

$$A_0 = A_{11} - A_{12}A_{22}^{-1}A_{21}. \tag{2.14}$$

Proof The unique solution of (2.8) for $\varepsilon = 0$ is

$$L(0) = A_{22}^{-1}A_{21}. \tag{2.15}$$

Differentiating (2.8) with respect to ε, we obtain

$$[A_{22} + \varepsilon L(\varepsilon)A_{12}]\frac{\mathrm{d}L}{\mathrm{d}\varepsilon} - \varepsilon\frac{\mathrm{d}L}{\mathrm{d}\varepsilon}[A_{11} - A_{12}L(\varepsilon)] = L(\varepsilon)A_{11} - L(\varepsilon)A_{12}L(\varepsilon) \tag{2.16}$$

and see that the unique solution of this equation at $\varepsilon = 0$ is

$$\left.\frac{\mathrm{d}L}{\mathrm{d}\varepsilon}\right|_{\varepsilon=0} = A_{22}^{-1}L(0)[A_{11} - A_{12}L(0)] = A_{22}^{-2}A_{21}A_0, \tag{2.17}$$

which proves that (2.13) represents the first two terms of the MacLaurin series for $L(\varepsilon)$. □

It should be pointed out that $L(\varepsilon)$ defined by (2.13) is unique in the sense that, although equation (2.8) may have several real solutions, only one of them is approximated by (2.13). It is sometimes useful to have an upper bound, however conservative, on the value of the small parameter ε for which the approximation (2.13) of Lemma 2.1 holds.

Lemma 2.2

Under the conditions of Lemma 2.1, the approximation (2.13) is valid for all ε in the interval $0 \leqq \varepsilon < \varepsilon_1$, where

$$\varepsilon_1 = \frac{1}{\|A_{22}^{-1}\|(\|A_0\| + \|A_{12}\|\|A_{22}^{-1}A_{21}\| + 2(\|A_0\|\|A_{12}\|\|A_{22}^{-1}A_{21}\|)^{1/2})}. \tag{2.18}$$

Proof Denoting $L(0) = A_{22}^{-1}A_{21} = L_0$ and introducing $D = L - L_0$, we write (2.8) as

$$D = \varepsilon A_{22}^{-1}(L_0A_0 + DA_0 - L_0A_{12}D - DA_{12}D) \triangleq f(D) \qquad (2.19)$$

and show that, for sufficiently small $c = \|\varepsilon A_{22}^{-1}\|$, $f(D)$ is a contraction mapping. Let

$$\mathcal{S} = \{D : \|D\| \leqq (al/b)^{1/2}\}$$

where $a = \|A_0\|$, $b = \|A_{12}\|$ and $l = \|L_0\|$. For any $D \in \mathcal{S}$ we have

$$\begin{aligned}\|f(D)\| &\leqq c\left[al + (a + bl)\left(\frac{al}{b}\right)^{1/2} + b\frac{al}{b}\right] \\ &= c[a + bl + 2(abl)^{1/2}]\left(\frac{al}{b}\right)^{1/2} \\ &\leqq \left(\frac{al}{b}\right)^{1/2}\end{aligned}$$

provided that

$$c \leqq \frac{1}{a + bl + 2(abl)^{1/2}}$$

For all $D, \tilde{D} \in \mathcal{S}$ we obtain

$$\begin{aligned}\|f(D) - f(\tilde{D})\| &= \|\varepsilon A_{22}^{-1}[(D - \tilde{D})A_0 - L_0A_{12}(D - \tilde{D}) \\ &\qquad - (D - \tilde{D})A_{12}D - \tilde{D}A_{12}(D - \tilde{D})]\| \\ &\leqq c\left[a + bl + 2b\left(\frac{al}{b}\right)^{1/2}\right]\|D - \tilde{D}\| \\ &< 1\end{aligned}$$

provided that

$$c < \frac{1}{a + bl + 2(abl)^{1/2}} \qquad (2.20)$$

Thus, by the contraction mapping principle, (2.19) has a unique solution in $\mathcal{S}$ which can be reached by successive approximation starting with any D_0 satisfying $\|D_0\| \leqq (al/b)^{1/2}$. The bound (2.18) follows from (2.20). □

A by-product of Lemma 2.2 is an iterative scheme for solving (2.8), namely,

$$L_{k+1} = A_{22}^{-1}A_{21} + \varepsilon A_{22}^{-1}L_k(A_{11} - A_{12}L_k), \quad L_0 = A_{22}^{-1}A_{21}. \qquad (2.21)$$

After k iterations the exact solution $L(\varepsilon)$ is approximated to within $O(\varepsilon^k)$ error. For the purposes of the model development and design, the first iteration giving (2.13) is usually sufficient.

Examination of Lemma 2.2 and its proof reveals that the small parameter ε is associated with A_{22}^{-1} only: it does not appear in L_0, A_0 and A_{12}. Were we to subsume ε in A_{22}, the proof of Lemma 2.2 would go through as before with the bound (2.18) replaced by

$$\|A_{22}^{-1}\| < \frac{1}{\|A_0\| + \|A_{12}\|\|A_{22}^{-1}A_{21}\| + 2(\|A_0\|\|A_{12}\|\|A_{22}^{-1}A_{21}\|)^{1/2}}. \tag{2.22}$$

This bound guarantees that the system

$$\left.\begin{aligned} \dot{x} &= A_{11}x + A_{12}z, \quad x(t_0) = x^0, \\ \dot{z} &= A_{21}x + A_{22}z, \quad z(t_0) = z^0 \end{aligned}\right\} \tag{2.23}$$

has a two-time-scale property, analogous to that of a singularly perturbed system. In other words, it can be transformed to the upper block-triangular form

$$\begin{bmatrix} \dot{x}(t) \\ \dot{\eta}(t) \end{bmatrix} = \begin{bmatrix} A_{11} - A_{12}L & A_{12} \\ 0 & A_{22} + LA_{12} \end{bmatrix} \begin{bmatrix} x(t) \\ \eta(t) \end{bmatrix} \tag{2.24}$$

where the smallest eigenvalue of $A_{22} + LA_{12}$ is larger than the largest eigenvalue of $A_{11} - A_{12}L$, that is,

$$|\lambda_{\min}(A_{22} + LA_{12})| > |\lambda_{\max}(A_{11} - A_{12}L)|, \tag{2.25}$$

since (2.22) can be shown to imply

$$\|(A_{22} + LA_{12})^{-1}\|^{-1} > \|(A_{11} - A_{12}L)\|$$

As we shall see later in this section, the parameter ε represents the order of magnitude of the ratio of the large and small eigenvalues (2.25). As a time-scale factor, ε conveniently allows the designer to determine reduced-order models by solving L iteratively to within "order of $O(\varepsilon^{k+1})$ error. In terms of the condition (2.22), where explicit use of ε is suppressed, the error diminishes as $\|A_{22}^{-1}\|$ and $\|A_0\|$ decrease; that is, as the ill-conditioning of the system (2.23) increases.

Finally, we note that Lemma 2.2 furnishes us with a bound on the range of ε, albeit a conservative one, for which the validity of the theorems in succeeding sections is guaranteed.

Example 2.1

The matrix in the DC-motor model of Case Study 7.1 in Chapter 1, and the corresponding L-equation (2.8) are

$$\begin{bmatrix} 0 & 1 \\ -1 & -1 \end{bmatrix} = \begin{bmatrix} A_{11} & A_{12} \\ A_{21} & A_{22} \end{bmatrix}, \qquad -1 + L - \varepsilon L^2 = 0, \tag{2.26}$$

and hence $L_0 = 1$, $A_0 = -1$, and the approximate solution (2.13) is

$$L(\varepsilon) = 1 + \varepsilon + O(\varepsilon^2), \tag{2.27}$$

which can be readily verified by evaluating the exact solution. Note in this simple example that the L-equation (2.26) also has an $O(1/\varepsilon)$ root. The actuator form for the DC motor is as in Fig. 2.4, where Δ_1 is $O(\varepsilon^2)$ and Δ_2 is $O(\varepsilon^3)$.

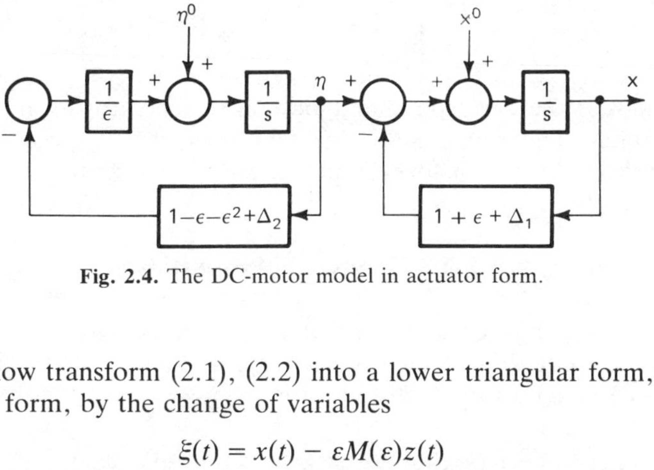

Fig. 2.4. The DC-motor model in actuator form.

Let us now transform (2.1), (2.2) into a lower triangular form, that is, the sensor form, by the change of variables

$$\xi(t) = x(t) - \varepsilon M(\varepsilon) z(t) \tag{2.28}$$

and the requirement that the matrix $M(\varepsilon)$ satisfy the algebraic equation

$$A_{12} - MA_{22} + \varepsilon A_{11} M - \varepsilon M A_{21} M = 0. \tag{2.29}$$

Then the system (2.1), (2.2) becomes

$$\begin{bmatrix} \dot{\xi} \\ \varepsilon \dot{z} \end{bmatrix} = \begin{bmatrix} A_{11} - MA_{21} & 0 \\ A_{21} & A_{22} + \varepsilon A_{21} M \end{bmatrix} \begin{bmatrix} \xi \\ z \end{bmatrix} \tag{2.30}$$

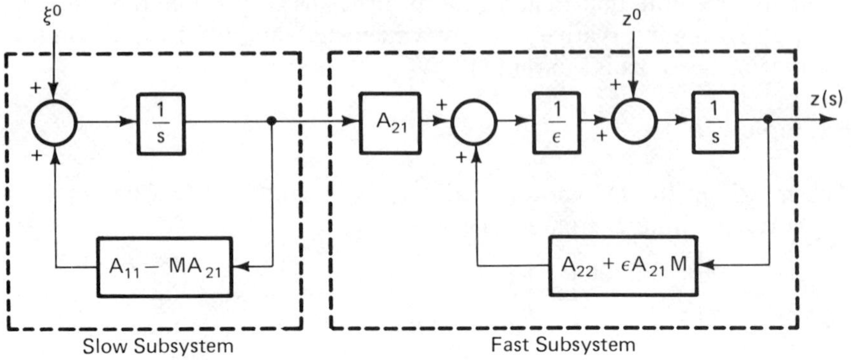

Fig. 2.5. Sensor form.

and is represented by the block diagram of Fig. 2.5. Since (2.28) is equivalent to the similarity transformation

$$\begin{bmatrix} \xi \\ z \end{bmatrix} = \begin{bmatrix} I_n & -\varepsilon M(\varepsilon) \\ 0 & I_m \end{bmatrix} \begin{bmatrix} x \\ z \end{bmatrix} \tag{2.31}$$

the slow and the fast block in the sensor form (2.30) have the same eigenvalues as the corresponding blocks in the actuator form (2.9). Their state variables and matrices are different, however. Exercise 2.1 asks for an analysis of the sensor form similar to that of the actuator form.

2.3 Eigenvalue Properties

In view of the block-triangular nature of the actuator form (2.9), its characteristic equation, which is the same as that of the original system (2.1), (2.2), assumes the factored form

$$\psi(s, \varepsilon) = \frac{1}{\varepsilon^m} \psi_s(s, \varepsilon) \psi_f(p, \varepsilon) = 0, \tag{3.1}$$

where

$$\psi_s(s, \varepsilon) \triangleq \det[sI_n - (A_{11} - A_{12}L)] \tag{3.2}$$

is the characteristic polynomial of the slow subsystem, and

$$\psi_f(p, \varepsilon) \triangleq \det[pI_m - (A_{22} + \varepsilon LA_{12})] \tag{3.3}$$

is the characteristic polynomial of the fast subsystem (2.11) exhibited in the high-frequency scale $p = \varepsilon s$. Thus, of the $n + m$ eigenvalues of the original system (2.1), (2.2), the first n eigenvalues $\{\lambda_1, \ldots, \lambda_n\}$ are the roots of the slow characteristic equation

$$\psi_s(s, \varepsilon) = 0 \tag{3.4}$$

and the remaining m eigenvalues are

$$\lambda_i = \frac{\pi_j}{\varepsilon}, \quad i = n + j, \quad j = 1, \ldots, m, \tag{3.5}$$

where π_j are the roots of the fast characteristic equation

$$\psi_f(p, \varepsilon) = 0. \tag{3.6}$$

The roots of the slow characteristic equation at $\varepsilon = 0$, namely,

$$\psi_s(s, 0) = \det[sI_n - (A_{11} - A_{12}L(0))] = 0, \tag{3.7}$$

are the eigenvalues of the matrix A_0 defined by (2.14), and the roots of fast characteristic equation at $\varepsilon = 0$,

$$\psi_f(p, 0) = \det[pI_m - A_{22}] = 0, \tag{3.8}$$

in the high-frequency scale p are the eigenvalues of the matrix A_{22}. This leads to the issue as to how closely the slow eigenvalues and the fast eigenvalues, evaluated at $\varepsilon = 0$, approximate the eigenvalues of the actual system (2.1), (2.2). The next theorem provides an answer to this question.

Theorem 3.1

If A_{22}^{-1} exists, then as $\varepsilon \to 0$ the first n eigenvalues of the system (2.1), (2.2) tend to fixed positions in the complex plane defined by the eigenvalues of A_0, namely,

$$\lambda_i(A_0), \quad i = 1, 2, \ldots, n; \tag{3.9}$$

while the remaining m eigenvalues of the system (2.1), (2.2) tend to infinity, with the rate $1/\varepsilon$, along asymptotes defined by the eigenvalues of A_{22}, namely,

$$\frac{1}{\varepsilon}\lambda_j(A_{22}), \quad i = n + j, \quad j = 1, \ldots, m. \tag{3.10}$$

Furthermore, if the n eigenvalues $\lambda_i(A_0)$ are distinct and the m eigenvalues $\lambda_j(A_{22})$ are distinct, where $\lambda_i(A_0) = \lambda_j(A_{22})$ is allowed, then the eigenvalues of the original system (2.1), (2.2) are approximated as

$$\lambda_i = \lambda_i(A_0) + O(\varepsilon), \qquad i = 1, \ldots, n; \tag{3.11}$$

$$\lambda_i = [\lambda_j(A_{22}) + O(\varepsilon)]/\varepsilon, \quad i = n + j, \quad j = 1, \ldots, m. \tag{3.12}$$

Proof The limiting behavior of (3.9) and (3.10) follows from the continuity of the coefficients of the polynomials (3.2) and (3.3) with respect to ε. The approximation (3.11) follows from the total derivative of the exact characteristic equation (3.4) with respect to ε:

$$\frac{\partial \psi_s}{\partial s}\frac{\mathrm{d}s}{\mathrm{d}\varepsilon} + \frac{\partial \psi_s}{\partial \varepsilon} = 0. \tag{3.13}$$

Since at $\varepsilon = 0$ and $s = \lambda_i(A_0)$ the fact that λ_i, $i = 1, \ldots, n$, are distinct guarantees that $\partial \psi_s/\partial s \neq 0$, the derivative $\mathrm{d}\lambda_i/\mathrm{d}\varepsilon$ exists at $\varepsilon = 0$, which proves (3.11). Analogously, taking the total derivative of the exact characteristic equation (3.6) and using $\partial \psi_f/\partial p \neq 0$ at $\varepsilon = 0$ and $p = \lambda_j(A_{22})$ proves (3.12). □

The presence of two disjoint sets of eigenvalues of different orders of magnitude, $O(1)$ and $O(1/\varepsilon)$, is an inherent property of singularly perturbed linear time-invariant systems. The smaller the value of ε, the wider the "eigenvalue gap" and the greater the separation of time scales in the original system (2.1), (2.2). When, as in the model (2.23), the parameter ε is not explicit, but the presence of eigenvalues of different orders of magnitude is known, this should alert us to the two-time-scale or singularly perturbed nature of that system.

The eigenvalue approximations of Theorem 3.1 make no assumption as to the stability of the system (2.1), (2.2). On the other hand, stability is a crucial system requirement in control system design and other fields. In this respect, Theorem 3.1 affords us the following useful two-time-scale stability result.

Corollary 3.1

If A_{22}^{-1} exists, and if A_0 and A_{22} are Hurwitz matrices, then there exists an $\varepsilon^* > 0$ such that for all $\varepsilon \in (0, \varepsilon^*]$ the system (2.1), (2.2) is asymptotically stable.

The fact that the stability of the actual system (2.1), (2.2) can be inferred from the analysis of lower-order systems in separate time scales is of great practical significance for analysis and design.

It is sometimes useful to achieve closer eigenvalue approximations through consideration of *corrected models* in separate time scales. This is particularly true when ε is not very small or when the eigenvalues of A_0 or A_{22} are close to the imaginary axis of the complex plane. For an $O(\varepsilon^2)$ approximation, we substitute (2.13) into the characteristic equation (3.2), and obtain a corrected slow characteristic equation

$$\psi_{sc}(s, \varepsilon) = \det[sI_n - A_{0c}] = 0, \tag{3.14}$$

where

$$A_{0c} \triangleq A_0 - \varepsilon A_{12}A_{22}^{-2}A_{21}A_0. \tag{3.15}$$

Also, substituting (2.14) into (3.3), we obtain a corrected fast characteristic equation

$$\psi_{fc}(p, \varepsilon) = \det[pI_m - A_{22c}] = 0, \tag{3.16}$$

where

$$A_{22c} \triangleq A_{22} + \varepsilon A_{22}^{-1}A_{21}A_{12}. \tag{3.17}$$

Then, in analogy with Theorem 3.1, an $O(\varepsilon^2)$ approximation to the eigenvalues of the original system (2.1), (2.2) is provided by

$$\lambda_i = \lambda_i(A_{0c}) + O(\varepsilon^2), \qquad i = 1, \ldots, n; \tag{3.18}$$

$$\lambda_i = [\lambda_j(A_{22c}) + O(\varepsilon^2)]/\varepsilon, \quad i = n + j, \quad j = 1, \ldots, m. \tag{3.19}$$

Notice that (3.18) and (3.19) can be used in combination to provide different orders of approximation to the system eigenvalues. For example, an $O(\varepsilon)$ accuracy of the "fast" eigenvalues may be sufficient in a particular application, in which an $O(\varepsilon^2)$ accuracy of the slow eigenvalues is desirable. Higher-order approximations of the eigenvalues can be obtained by way of evaluating more terms of the MacLaurin series beyond those of (2.13), and substituting the higher-order approximation for $L(\varepsilon)$ into the characteristic polynomials (3.2) and (3.3).

Example 3.1

Continuing with the DC-motor Example 2.1, where $A_0 = -1$ and $A_{22} = -1$, we see that its eigenvalues are approximated by

$$\lambda_1 = -1 + O(\varepsilon), \quad \lambda_2 = -\frac{1}{\varepsilon}(1 + O(\varepsilon)) \tag{3.20}$$

where λ_1 is slow and λ_2 is fast. Should greater accuracy be required, the corrected eigenvalue approximations (3.18) and (3.19) yield

$$\lambda_1 = -(1 + \varepsilon) + O(\varepsilon^2), \quad \lambda_2 = -\frac{1}{\varepsilon}(1 - \varepsilon - O(\varepsilon^2)). \tag{3.21}$$

As a check, the roots of the exact characteristic equation $\varepsilon\lambda^2 + \lambda + 1 = 0$ are

$$\lambda_1 = \frac{-1 + (1 - 4\varepsilon)^{1/2}}{2\varepsilon}, \quad \lambda_2 = \frac{-1 - (1 - 4\varepsilon)^{1/2}}{2\varepsilon}, \tag{3.22}$$

which, in view of $(1 - 4\varepsilon)^{1/2} = 1 - 2\varepsilon - 2\varepsilon^2 + O(\varepsilon^3)$, coincides with (3.21). The bound from Lemma 2.2 is in this case $\varepsilon = 1/4$. For $\varepsilon = 1/6$ the approximate and the correct expressions (3.21) and (3.22) yield respectively $\lambda_1 = -1.17$ instead of $\lambda_1 = -1.27$ and $\lambda_2 = -5$ instead of $\lambda_2 = -4.73$. As it happens, typical values of motor constants in Example 2.1 result in a singular perturbation parameter of $\varepsilon = 0.1$ for which the corrected eigenvalues $\lambda_1 = -1.10$, $\lambda_2 = -9$ compare favorably with the actual eigenvalues of -1.13 and -8.87.

2.4 The Block-Diagonal Form: Eigenspace Properties

From either the actuator or the sensor block-triangular form, one more change of variables leads to a complete separation of the fast and slow states of the system (2.1), (2.2). Starting with the actuator form (2.9), the change of variables

$$\xi(t) = x(t) - \varepsilon H\eta(t) \tag{4.1}$$

results in

$$\begin{bmatrix} \dot{\xi}(t) \\ \varepsilon\dot{\eta}(t) \end{bmatrix} = \begin{bmatrix} A_{11} - A_{12}L & S(H, \varepsilon) \\ 0 & A_{22} + \varepsilon LA_{12} \end{bmatrix} \begin{bmatrix} \xi(t) \\ \eta(t) \end{bmatrix} \tag{4.2}$$

and the requirement that the $n \times m$ matrix H satisfy the linear algebraic equation

$$S(H, \varepsilon) = \varepsilon(A_{11} - A_{12}L)H - H(A_{22} + \varepsilon LA_{12}) + A_{12} = 0 \tag{4.3}$$

allows the system (4.2) to finally assume the block-diagonal or decoupled form

$$\begin{bmatrix} \dot{\xi}(t) \\ \varepsilon\dot{\eta}(t) \end{bmatrix} = \begin{bmatrix} A_{11} - A_{12}L & 0 \\ 0 & A_{22} + \varepsilon LA_{12} \end{bmatrix} \begin{bmatrix} \xi(t) \\ \eta(t) \end{bmatrix}. \tag{4.4}$$

As in the actuator form, (2.11) defines the exact fast subsystem, while the new block

$$\dot{\xi}(t) = (A_{11} - A_{12}L)\xi(t) \tag{4.5}$$

defines the slow subsystem with, by (4.1) and (2.12), the initial condition

$$\xi(t_0) = (I_n - \varepsilon HL)x^0 - \varepsilon Hz^0. \tag{4.6}$$

That the unique solution $H(\varepsilon)$ of (4.3) exists and is approximated by

$$H(\varepsilon) = A_{12}A_{22}^{-1} + O(\varepsilon) \tag{4.7}$$

for all ε for which (2.13) holds, is readily established by the linearity of (4.3).

Taken together, the change of variables (2.5) leading to the actuator form and the present change of variables (4.1) constitute a similarity transformation T:

$$\begin{bmatrix} x(t) \\ z(t) \end{bmatrix} = \begin{bmatrix} I_n & \varepsilon H \\ -L & I_m - \varepsilon LH \end{bmatrix} \begin{bmatrix} \xi(t) \\ \eta(t) \end{bmatrix} = T \begin{bmatrix} \xi(t) \\ \eta(t) \end{bmatrix}. \tag{4.8}$$

Among many conveniences that this transformation offers is that *for all L and H*, and not only those satisfying (2.8) and (4.3), the inverse transformation is explicitly given by

$$\begin{bmatrix} \xi(t) \\ \eta(t) \end{bmatrix} = \begin{bmatrix} I_n - \varepsilon HL & -\varepsilon H \\ L & I_m \end{bmatrix} \begin{bmatrix} x(t) \\ z(t) \end{bmatrix} = T^{-1} \begin{bmatrix} x(t) \\ z(t) \end{bmatrix}. \tag{4.9}$$

This is extremely helpful when, instead of the exact solution $L(\varepsilon)$ of (2.8) and $H(\varepsilon)$ of (4.3), we use their approximations. Although with approximate solutions of (2.8) and (4.3) the transformed system will not be in the exact block-diagonal form (4.4), it will be close to it because its off-diagonal terms will be $O(\varepsilon)$ or smaller. To illustrate this point let us use the approximations

$$L_0 = A_{22}^{-1}A_{21}, \quad H_0 = A_{12}A_{22}^{-1} \tag{4.10}$$

to form the transformation

$$T_0 = \begin{bmatrix} I_n & \varepsilon H_0 \\ -L_0 & I_m - \varepsilon L_0 H_0 \end{bmatrix}, \quad T_0^{-1} = \begin{bmatrix} I_n - \varepsilon H_0 L_0 & -\varepsilon H_0 \\ L_0 & I_m \end{bmatrix}. \tag{4.11}$$

When T_0 is applied instead of T in (4.8), the system (2.1), (2.2) is transformed to

$$\begin{bmatrix} \dot{\xi}_0 \\ \varepsilon\dot{\eta}_0 \end{bmatrix} = \begin{bmatrix} (I - \varepsilon H_0 L_0)A_0 & \varepsilon(A_0 H_0 - H_0 L_0 A_{12}) - \varepsilon^2 H_0 L_0 A_0 H_0 \\ \varepsilon L_0 A_0 & A_{22} + \varepsilon L_0 A_{12} + \varepsilon^2 L_0 A_0 H_0 \end{bmatrix} \begin{bmatrix} \xi_0 \\ \eta_0 \end{bmatrix} \tag{4.12}$$

where the subscript "$_0$" distinguishes the state variables in (4.12) from the exact slow and fast variables ξ and η in (4.4), although the system (4.12) is still an exact similarity transformation of (2.1), (2.2). If the $O(\varepsilon)$ off-diagonal terms and the $O(\varepsilon^2)$ diagonal term $\varepsilon^2 L_0 A_0 H_0$ are neglected, and L_0 and H_0 are as expressed in (4.10), the approximate slow and fast models obtained from (4.12) are

$$\dot{\xi}_{0s} = (I - \varepsilon A_{12} A_{22}^{-2} A_{21}) A_0 \xi_{0s}, \tag{4.13}$$

$$\varepsilon \dot{\eta}_{0f} = (A_{22} + \varepsilon A_{22}^{-1} A_{21} A_{12}) \eta_{0s}. \tag{4.14}$$

We have already encountered these slow and fast corrected models in Chapter 1 (page 24), and we will have more to say about them in the next section. In the discussion above, the emphasis is on the use of different L and H in the transformation (4.8) as additional tools for the construction of approximate models. In Chapter 5, a time-varying generalization of the model (4.12) is used to investigate the stability and robustness properties of an adaptive system. A simplified time-invariant version of the same system is briefly analyzed in the following example.

Example 4.1

Equations of an adaptive system are

$$\text{model:} \quad \dot{y}_m = -y_m + g_m r, \tag{4.15}$$

$$\text{plant:} \quad \dot{y} = -y + z, \tag{4.16}$$

$$\varepsilon \dot{z} = -z + kr, \tag{4.17}$$

$$\text{adaptation:} \quad \dot{k} = -\gamma r(y - y_m), \tag{4.18}$$

where k is an adjustable parameter and the goal of adaptation is to achieve $y(t) - y_m(t) \to 0$ for arbitrary bounded reference inputs $r(t)$. This goal can be achieved if the order of the model and the plant are the same, which is not the case here. The model–plant mismatch is due to a parasitic time constant ε, which can destroy the stability properties of the ideal scheme. While a general time-varying analysis is postponed until Chapter 5, we investigate the stability of (4.16)–(4.18) when $r = \text{const}$. The homogeneous part of (4.16)–(4.18) is

$$\begin{bmatrix} \dot{y} \\ \dot{k} \end{bmatrix} = \begin{bmatrix} -1 & 0 \\ -\gamma r & 0 \end{bmatrix} \begin{bmatrix} y \\ k \end{bmatrix} + \begin{bmatrix} 1 \\ 0 \end{bmatrix} z, \tag{4.19}$$

$$\varepsilon \dot{z} = [0 \quad r] \begin{bmatrix} y \\ k \end{bmatrix} + [-1] z. \tag{4.20}$$

To develop the approximate model (4.13), (4.14) we use

$$L_0 = [0 \quad -r], \quad H_0 = \begin{bmatrix} -1 \\ 0 \end{bmatrix} \tag{4.21}$$

and the change of variables $\eta = z - rk$, $\xi_1 = y + \varepsilon\eta$, $\xi_2 = k$. To simplify notation we let $\xi_1 = \xi$ and keep k instead of ξ_2. The exact transformed model (4.12) is

$$\begin{bmatrix} \dot{\xi} \\ \dot{k} \end{bmatrix} = \begin{bmatrix} -1 + \varepsilon\gamma r^2 & r \\ -\gamma r & 0 \end{bmatrix} \begin{bmatrix} \xi \\ k \end{bmatrix} + \varepsilon \begin{bmatrix} 1 \\ \gamma r \end{bmatrix} \eta, \tag{4.22}$$

$$\varepsilon\dot{\eta} = \varepsilon\gamma r^2 \xi - (1 + \varepsilon^2 \gamma r^2)\, \eta, \tag{4.23}$$

and the approximate model (4.13), (4.14) is obtained by neglecting the $O(\varepsilon)$ off-diagonal term and the $O(\varepsilon^2)$ term in (4.22), (4.23), that is,

$$\begin{bmatrix} \dot{\xi}_0 \\ \dot{k}_0 \end{bmatrix} = \begin{bmatrix} -1 + \varepsilon\gamma r^2 & r \\ -\gamma r & 0 \end{bmatrix} \begin{bmatrix} \xi_0 \\ k_0 \end{bmatrix}, \quad \varepsilon\dot{\eta}_0 = -\eta_0. \tag{4.24}$$

From the slow model's characteristic equation

$$\lambda^2 + (1 - \varepsilon\gamma r^2)\lambda + \gamma r^2 = 0 \tag{4.25}$$

we see that the slow model will be unstable if $\varepsilon\gamma r^2 > 1$. Thus the parasitic parameters ε sets a limit on the amplitude of the input r and the adaptation gain γ. Although extremely simple, this example illustrates the use of approximate models in discovering destabilizing effects of parasitics.

Whenever a similarity transformation T brings a system to a block-diagonal form, the two sets of columns of T corresponding to the diagonal blocks constitute bases of the eigenspaces represented by the blocks. In the case of the transformation (4.8), this means that the first n columns of T are a basis for the slow eigenspace of (2.1), (2.2), and the remaining m columns are a basis for the fast eigenspace of (2.1), (2.2). Let us make this precise in the folowing lemma.

Lemma 4.1

Let $v_\xi \in R^n$ be an eigenvector of the matrix A_s of the slow system (4.5) and $v_\eta \in R^m$ be an eigenvector of the matrix A_f of the fast system (2.11), where, as in (4.4),

$$A_s = A_{11} - A_{12}L, \quad A_f = A_{22} + \varepsilon LA_{12}. \tag{4.26}$$

Then the corresponding slow eigenvector $v_s \in R^{n+m}$ and the fast eigenvector $v_f \in R^{n+m}$ of the original system (2.1), (2.2) are

$$v_s = \begin{bmatrix} I_n \\ -L \end{bmatrix} v_\xi, \quad v_f = \begin{bmatrix} \varepsilon H \\ I_m - \varepsilon LH \end{bmatrix} v_\eta. \tag{4.27}$$

Proof The application of the transformation T in (4.8), with L and H satisfying (2.8) and (4.3) respectively, results in

$$\begin{bmatrix} A_{11} & A_{12} \\ \dfrac{A_{21}}{\varepsilon} & \dfrac{A_{22}}{\varepsilon} \end{bmatrix} T = T \begin{bmatrix} A_s & 0 \\ 0 & \dfrac{1}{\varepsilon} A_f \end{bmatrix}, \tag{4.28}$$

or, in the expanded form,

$$\begin{bmatrix} A_{11} & A_{12} \\ \dfrac{A_{21}}{\varepsilon} & \dfrac{A_{22}}{\varepsilon} \end{bmatrix} \begin{bmatrix} I_n \\ -L \end{bmatrix} = \begin{bmatrix} I_n \\ -L \end{bmatrix} A_s, \tag{4.29}$$

$$\begin{bmatrix} A_{11} & A_{12} \\ \dfrac{A_{21}}{\varepsilon} & \dfrac{A_{12}}{\varepsilon} \end{bmatrix} \begin{bmatrix} \varepsilon H \\ I - \varepsilon LH \end{bmatrix} = \begin{bmatrix} \varepsilon H \\ I - \varepsilon LH \end{bmatrix} \frac{1}{\varepsilon} A_f. \tag{4.30}$$

Noting that $A_s v_\xi = \lambda_s v_\xi$ and $A_f v_\eta = \lambda_f v_\eta$, where λ_s and λ_f are the eigenvalues of A_s and A_f corresponding to the eigenvectors v_ξ and v_η respectively, we postmultiply (4.29) by v_ξ and (4.30) by v_η to prove (4.27). □

An immediate consequence of this lemma and the eigenvalue approximation theorem is the following useful eigenvector approximation.

Corollary 4.1

Under the conditions of distinct eigenvalues in Theorem 3.1, the slow and the fast eigenvectors of the system (2.1), (2.2) are approximated according to

$$v_s = \begin{bmatrix} I_n \\ -A_{22}^{-1} A_{21} \end{bmatrix} \bar{v}_\xi + O(\varepsilon), \quad v_f = \begin{bmatrix} 0 \\ I_m \end{bmatrix} \bar{v}_\eta + O(\varepsilon), \tag{4.31}$$

where $\bar{v}_\xi$ is an eigenvector of $A_0 = A_{11} - A_{12} A_{22}^{-1} A_{21}$ and $\bar{v}_\eta$ is an eigenvector of A_{22}.

When defining approximations (4.31), we should take into account the fact that for an eigenvector only the angle is specified, while its length is free. Therefore the $O(\varepsilon)$ refers to an angular displacement of an eigenvector whose length is normalized to one. Comparing v_s from (4.27) and (4.31) with the expressions (4.3) and (4.7) of Chapter 1 for the slow manifold M_ε and the equilibrium manifold M_0, we see that $v_s \in M_\varepsilon$, that is, the slow eigenspace is the slow manifold M_ε as expected, while the approximation (4.31) of v_s lies in M_0. As for the fast eigenspace spanned by the eigenvectors v_f, the exact expression (4.27) shows that its angle with respect to the subspace of the original z-variables is $O(\varepsilon)$, and the approximation (4.31) of v_f is in the z-subspace, that is, orthogonal to all the x coordinates. Using $H_0 = A_{12}A_{22}^{-1}$, a closer $O(\varepsilon^2)$ approximation of the angle that v_f in (4.27) makes with the z-subspace is achieved by

$$\bar{v}_f = \begin{bmatrix} \varepsilon A_{12}A_{22}^{-1} \\ I_m \end{bmatrix} \bar{v}_\eta . \tag{4.32}$$

Since we have not corrected the term I_m in (4.32) by a corresponding ε-correction from (4.27), it is not obvious that the approximation (4.32) is $O(\varepsilon^2)$ as claimed. This is left as Exercise 2.6.

In summary, the exact slow eigenspace M_ε and the exact fast eigenspace F_0 of (2.1), (2.2) and their approximations are sketched in Fig. 2.6.

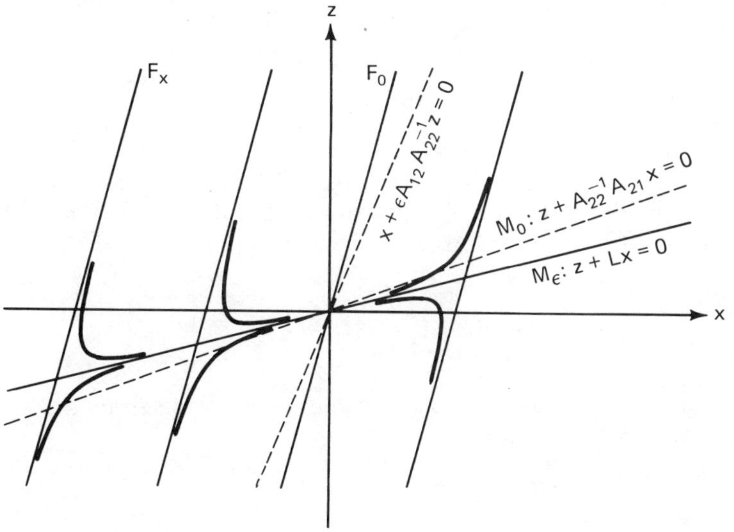

Fig. 2.6. Exact and approximate eigenspaces.

Recalling the discussion of fast manifolds F_x from Section 1.4 of Chapter 1, we see that, in the case of linear time-invariant systems, they are the translates of the fast eigenspace F_0 to the value of x prescribed by the initial condition $x^0 = x(t_0)$. Fast manifolds are represented in Fig. 2.6 by lines parallel to F_0 and hence inclined with respect to the z-axis by an $O(\varepsilon)$ angle. It is observed that the fast state $\eta(t)$ moves rapidly along F_x towards or away from the equilibrium manifold M_ε depending on whether M_ε is stable or unstable. Fast manifolds F_x appear as "foliations" off the unique slow manifold M_ε.

Example 4.2

For the DC motor of Example 2.1 described by

$$\begin{bmatrix} \dot{x}(t) \\ \varepsilon\dot{z}(t) \end{bmatrix} = \begin{bmatrix} 0 & 1 \\ -1 & -1 \end{bmatrix} \begin{bmatrix} x(t) \\ z(t) \end{bmatrix} \tag{4.33}$$

the slow and the fast eigenspaces, in this case the slow and the fast eigenvectors v_s and v_f, are approximated by

$$v_s \sim \begin{bmatrix} I_n \\ -A_{22}^{-1}A_{21} \end{bmatrix} = \begin{bmatrix} 1 \\ -1 \end{bmatrix}, \quad v_f \sim \begin{bmatrix} \varepsilon A_{12}A_{22}^{-1} \\ I_m \end{bmatrix} = \begin{bmatrix} -\varepsilon \\ 1 \end{bmatrix}. \tag{4.34}$$

As a check, the exact expressions are

$$v_s = \begin{bmatrix} 1 \\ \dfrac{-1 + (1-4\varepsilon)^{1/2}}{2\varepsilon} \end{bmatrix}, \quad v_f = \begin{bmatrix} \dfrac{-\varepsilon}{(1-4\varepsilon)^{1/2}} \\ \dfrac{1}{2}\left(1 + \dfrac{1}{(1-4\varepsilon)^{1/2}}\right) \end{bmatrix}, \tag{4.35}$$

and coincide with (4.34) in view of the expression $(1-4\varepsilon)^{1/2} = 1 - 2\varepsilon + O(\varepsilon^2)$. Let $\hat{\alpha}$ and $\bar{\alpha}$ be the angles that

$$\hat{v}_f = \begin{bmatrix} -\varepsilon \\ 1 + c\varepsilon \end{bmatrix} \quad \text{and} \quad \bar{v}_f = \begin{bmatrix} -\varepsilon \\ 1 \end{bmatrix}$$

respectively make with the z-axis, where 1 is corrected by the ε-correction $c\varepsilon$ in $\hat{v}_f$:

$$\hat{\alpha} = \arctan\left(\frac{-\varepsilon}{1 + c\varepsilon}\right) = \arctan\left(-\varepsilon + c\varepsilon^2 + O(\varepsilon^3)\right)$$

$$\sim -\varepsilon + c\varepsilon^2 + O(\varepsilon^3) \tag{4.36}$$

It is clear that $\hat{\alpha}$ is an $O(\varepsilon^2)$ approximation of the exact angle of v_{f} in (4.27) and both the exact angle and $\hat{\alpha}$ are $O(\varepsilon)$ with the z-axis. An $O(\varepsilon^2)$ approximation is also achieved by $\bar{\alpha}$ since

$$\hat{\alpha} - \bar{\alpha} \sim -\varepsilon + c\varepsilon^2 - \arctan(-\varepsilon) = O(\varepsilon^2), \tag{4.37}$$

whereas the angle of the eigenvector $[0 \quad 1]^{\mathrm{T}}$ of (4.31) is zero and approximates the angle of v_{f} by $O(\varepsilon)$. In this example $c = 1$.

2.5 Validation of Approximate Models

The results of preceding sections, as well as offering a repertoire of approximate models, also justify some common *ad hoc* model simplification procedures. One of those is the reduction of the $(n + m)$th-order model (2.1), (2.2) to the nth order model

$$\dot{x}_{\mathrm{s}} = A_{11}x_{\mathrm{s}} + A_{12}z_{\mathrm{s}}, \quad x_{\mathrm{s}}(t_0) = x^0 \tag{5.1}$$

$$0 = A_{21}x_{\mathrm{s}} + A_{22}z_{\mathrm{s}} \tag{5.2}$$

by formally setting $\varepsilon = 0$ in (2.1), (2.2) and disregarding the initial condition for z. When A_{22}^{-1} exists

$$z_{\mathrm{s}} = -A_{22}^{-1}A_{21}x_{\mathrm{s}} \tag{5.3}$$

is the unique solution of (5.2), whose substitution into (5.1) results in the unique nth order uncorrected slow model

$$\dot{x}_{\mathrm{s}} = A_0 x_{\mathrm{s}}, \quad x_{\mathrm{s}}(t_0) = x^0. \tag{5.4}$$

An *ad hoc* derivation of the "fast" subsystem assumes that the slow variables are constant during fast transients, that is, $\dot{z}_{\mathrm{s}} = 0$ and $x = x_{\mathrm{s}} = \text{const.}$, and that the only fast variations are the deviations of z from its quasi-steady-state z_{s}. Denoting these fast variations by

$$z_{\mathrm{f}} \triangleq z - z_{\mathrm{s}}, \tag{5.5}$$

one obtains from (2.2), (5.3), (5.5) and $\varepsilon\dot{z}_{\mathrm{s}} = 0$ the mth-order uncorrected fast model

$$\varepsilon\dot{z}_{\mathrm{f}}(t) = A_{22}z_{\mathrm{f}}(t), \quad z_{\mathrm{f}}(t_0) = z^0 + A_{22}^{-1}A_{21}x^0, \tag{5.6}$$

rewritten in the fast time scale τ as

$$\frac{\mathrm{d}z_{\mathrm{f}}}{\mathrm{d}\tau}(\tau) = A_{22}z_{\mathrm{f}}(\tau), \quad z_{\mathrm{f}}(0) = z^0 + A_{22}^{-1}A_{21}x^0. \tag{5.7}$$

This *ad hoc* procedure was justified in Chapter 1 by Theorem 3.1, establishing that if $\operatorname{Re}\lambda(A_{22}) < 0$ then the solution x, z of the original system

(2.1), (2.2) is approximated, for ε sufficiently small, by

$$x(t) \sim x_s(t), \tag{5.8}$$

$$z(t) \sim -A_{22}^{-1}A_{21}x_s(t) + z_f\left(\frac{t-t_0}{\varepsilon}\right), \tag{5.9}$$

where $z_f(\tau)$ is expressed in the t time scale. Here we establish the validity of these uncorrected state approximations by way of a direct comparison of the uncorrected candidate models (5.4) and (5.7) with the *exact* decomposed two-time-scale model (4.4).

Theorem 5.1

If $\operatorname{Re}\lambda(A_{22}) < 0$, there exists an $\varepsilon^* > 0$ such that, for all $\varepsilon \in (0, \varepsilon^*]$, the states of the original system (2.1), (2.2) starting from any bounded initial conditions x^0 and z^0, $\|x^0\| \le c_1$, $\|z^0\| \le c_2$, where c_1 and c_2 are constants independent of ε, are approximated for all finite $t \ge t_0$ by

$$x(t) = x_s(t) + O(\varepsilon), \tag{5.10}$$

$$z(t) = -A_{22}^{-1}A_{21}x_s(t) + z_f(\tau) + O(\varepsilon), \tag{5.11}$$

where $x_s(t)$ and $z_f(\tau)$ are the respective states of the slow model (5.4) and the fast model (5.7). If also $\operatorname{Re}\lambda(A_0) < 0$ then (5.10) and (5.11) hold for all $t \in [t_0, \infty)$.

Moreover, the "boundary layer" correction $z_f(\tau)$ is significant only during the initial short interval $[t_0, t_1]$, $t_1 - t_0 = O(\varepsilon \ln \varepsilon)$, after which

$$z(t) = -A_{22}^{-1}A_{21}x_s(t) + O(\varepsilon). \tag{5.12}$$

Proof From (4.8) and (4.4), the exact solution of the system (2.1)–(2.2) is

$$x(t) = \exp[(A_{11} - A_{12}L)(t-t_0)]\,\xi^0 + \varepsilon H \exp\left[(A_{22} + \varepsilon L A_{12})\left(\frac{t-t_0}{\varepsilon}\right)\right]\eta^0, \tag{5.13}$$

$$z(t) = -Lx(t) + \exp\left[(A_{22} + \varepsilon L A_{12})\left(\frac{t-t_0}{\varepsilon}\right)\right]\eta^0, \tag{5.14}$$

where, by (4.9)

$$\xi^0 = x^0 - \varepsilon H z^0 - \varepsilon H L x^0, \tag{5.15}$$

$$\eta^0 = z^0 + Lx^0. \tag{5.16}$$

First, note that the initial conditions x^0 and z^0, and hence ξ^0 and η^0, are no larger than $O(1)$. Accordingly, it follows from the continuous dependence of the matrix exponential function e^M on the coefficients of the matrix M that (5.13) to (5.16) and (2.13) imply

$$x(t) = e^{A_0(t-t_0)}x^0 + O(\varepsilon), \tag{5.17}$$

$$z(t) = -A_{22}^{-1}A_{21}e^{A_0(t-t_0)}x^0 + e^{A_{22}(t-t_0)/\varepsilon}(z^0 + A_{22}^{-1}A_{21}x^0) + O(\varepsilon), \tag{5.18}$$

where $\operatorname{Re}\lambda(A_{22}) < 0$. This proves (5.10) and (5.11). If in addition $\operatorname{Re}\lambda(A_0) < 0$, (5.17) and (5.18) hold for all $t \in [t_0, \infty)$. To show that the "thickness" $t_1 - t_0$ of the boundary layer is $O(\varepsilon \ln \varepsilon)$, we note that

$$\|e^{A_{22}(t-t_0)/\varepsilon}\| \le ke^{-\alpha(t-t_0)/\varepsilon}, \tag{5.19}$$

where k and α are some positive constants independent of ε, and that $e^{-(t-t_0)/\varepsilon} < \varepsilon$ for $t_1 - t_0 \gtrless \varepsilon \ln \varepsilon$. □

It is observed that for the eigenvalue approximations of Theorem 3.1 to hold well, all that was assumed was that $\det A_{22} \neq 0$. For the time-domain state approximations of Theorem 5.1 it is assumed that the fast system (5.7) also be asymptotically stable. This condition, in addition to being a sufficient one, is *almost* a necessary one since if any $\operatorname{Re}\lambda(A_{22}) > 0$ then Theorem 5.1 no longer holds in that $e^{A_{22}(t-t_0)/\varepsilon}$ grows without bound, invalidating the $O(\varepsilon)$ approximation in (5.17) and (5.18). As before, the slow subsystem may be either stable or unstable.

Returning to the model (4.12), obtained using the transformation (4.11), Theorem 5.1 can be used to justify neglecting ε in the right-hand side of (4.12). In doing so, we retain ε multiplying the derivative of η_0, so that as before, $\varepsilon\dot{\eta}_0 = d\eta_0/d\tau$. This is a characteristic step when, in the transformation (4.8), the matrices L and H do not exactly satisfy (2.8) and (4.3) respectively, and the exact transformed models are not in the block-diagonal form (4.4). On the other hand, note that the proof of Theorem 5.1 uses the block-diagonal form (4.4) and the fact that its solution ξ in the time scale t, and η in the time scale τ, are continuous and differentiable functions of ε. Observe that the same is not true about η as a function of t, because as $\varepsilon \to 0$ in t-scale η experiences a discontinuity ("boundary layer jump") at $t = t_0$. This jump distinguishes singular perturbations from regular perturbations. The transfer from t to τ has made the fast subsystem of (4.4) regularly perturbed by ε. A convenience provided by the block-diagonal form (4.4) is that the transfer from t to τ can be performed in the fast subsystem while letting the slow subsystem remain in the t-scale. For models where the fast and slow parts interact, such as nonlinear models, the proof of more general versions of Theorem 3.1 of Chapter 1 is more complicated. However, the spirit of the proof is the same in that the discontinuity in the fast variable is removed by analyzing it in the fast time scale.

An algebraic counterpart of the removal of the time-domain discontinuity is the removal of the ill-conditioning in the system matrix of (2.1), (2.2). In other words, the ill-conditioning due to the presence of the singular perturbation parameter ε,

$$\begin{bmatrix} A_{11} & A_{12} \\ \dfrac{A_{21}}{\varepsilon} & \dfrac{A_{22}}{\varepsilon} \end{bmatrix}, \tag{5.20}$$

is alleviated in the lower-order calculations associated with A_0 and A_{22} for the slow and fast models (5.4) and (5.7) respectively.

Example 5.1

The dynamic behavior of the PID control system in Case Study 3.1 of Chapter 1 with $d = 0$ is described in the time scale of the integral action by the state-space model

$$\begin{bmatrix} \dot{x}(t) \\ \varepsilon\dot{z}_1(t) \\ \varepsilon\dot{z}_2(t) \end{bmatrix} = \left[\begin{array}{c|cc} 0 & 1 & 0 \\ \hline 0 & 0 & 1 \\ -k_3 & -k_1 & -k_2 \end{array}\right] \begin{bmatrix} x(t) \\ z_1(t) \\ z_2(t) \end{bmatrix}, \quad \begin{bmatrix} x(t_0) \\ z_1(t_0) \\ z_2(t_0) \end{bmatrix} = \begin{bmatrix} 1 \\ 1 \\ 0 \end{bmatrix}, \tag{5.21}$$

where $x(t)$ is the integral action state variable, $z_1(t)$ and $z_2(t)$ are the system state variables, and the parameter ε scales the integral action. As in (3.34) of Chapter 1, the uncorrected slow model (5.4), obtained by setting $\varepsilon = 0$ in (5.21), is

$$\dot{x}_s(t) = \frac{-k_3}{k_1} x_s(t), \quad x_s(0) = 1 \tag{5.22}$$

with eigenvalue

$$\lambda_1 = -k_3/k_1. \tag{5.23}$$

The uncorrected fast model (5.7) is

$$\begin{bmatrix} \varepsilon\dot{z}_{1f}(t) \\ \varepsilon\dot{z}_{2f}(t) \end{bmatrix} = \begin{bmatrix} 0 & 1 \\ -k_1 & -k_2 \end{bmatrix} \begin{bmatrix} z_{1f}(t) \\ z_{2f}(t) \end{bmatrix}, \quad \begin{bmatrix} z_{1f}(t_0) \\ z_{2f}(t_0) \end{bmatrix} = \begin{bmatrix} 1 + \dfrac{k_3}{k_1} \\ 0 \end{bmatrix}. \tag{5.24}$$

In the fast frequency scale $p = \varepsilon s$, the characteristic equation of (5.24) is

$$\psi_f(p, 0) = \det[pI_2 - A_{22}] = p^2 + k_2 p + k_1 = 0. \tag{5.25}$$

PD control design is carried out on the uncorrected fast system model (5.24) by choosing k_1 and k_2 so as that (5.25) is equal to the desired characteristic equation

$$p^2 + 2\zeta\omega_n p + \omega_n^2 = 0 \tag{5.26}$$

for some specified damping ratio ζ and natural frequency ω_n. Equating coefficients in (5.25) and (5.26),

$$k_1 = \omega_n^2, \quad k_2 = 2\zeta\omega_n. \tag{5.27}$$

Also, in $s = p/\varepsilon$ scale, the desired eigenvalues from solving (5.26) are

$$\lambda_2, \lambda_3 = [-\zeta\omega_n \pm j\omega_n(1 - \zeta^2)^{1/2}]/\varepsilon. \tag{5.28}$$

Integral action control design is completed on the uncorrected slow model (5.22). Since, in the fast design (5.27) $k_1 > 0$, it is necessary to choose $k_3 > 0$ so as to obtain a stable slow system (5.22).

In a numerical design, ζ, ω_n and ε were chosen to be

$$\zeta = 0.707, \quad \omega_n = 1\,\text{rad/s}, \quad \varepsilon = 0.2. \tag{5.29}$$

The gains k_1 and k_2 of (5.27) and the desired fast system eigenvalues are respectively

$$k_1 = 1, \quad k_2 = 1.414 \tag{5.30}$$

and

$$\lambda_2, \lambda_3 = -3.535 \pm j3.535. \tag{5.31}$$

Given k_1 in (5.30), the choice in (5.22) of

$$k_3 = 1 \tag{5.32}$$

results in a stable uncorrected slow system with the eigenvalue

$$\lambda_1 = -1. \tag{5.33}$$

But what of the actual performance of the full closed-loop system (5.21) using the uncorrected design values of k_1, k_2 and k_3 in (5.30) and (5.32)? The eigenvalues of the full closed-loop system (5.21) are

$$\{-1.503, -2.784 \pm j2.980\}. \tag{5.34}$$

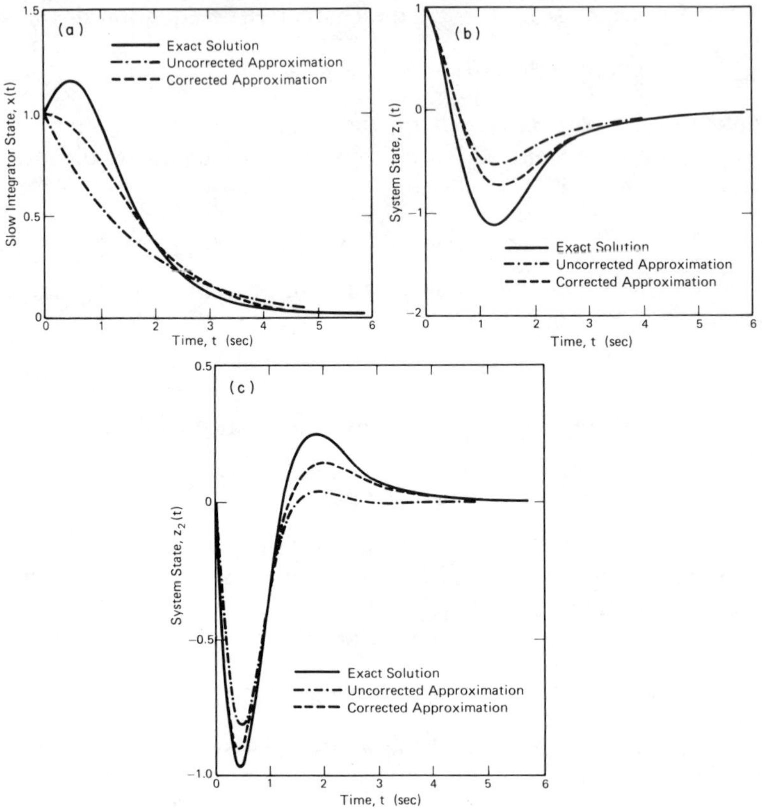

Fig. 2.7. Uncorrected and corrected state approximations for a PID control system.

The state trajectories and their $O(\varepsilon)$ approximations of Theorem 5.1, based on the uncorrected slow and fast models of Theorem 5.1, are given in Fig. 2.7. Evaluating (2.18) using the induced norm

$$\|A\|_\infty \triangleq \max_i \sum_j |a_{ij}|,$$

we have an upper bound on ε given by

$$\varepsilon_1 = 0.104. \tag{5.35}$$

Clearly, this bound is a conservative one, as the uncorrected slow and fast eigenvalues and state trajectories are still within $O(\varepsilon)$ approximation of the time eigenvalues and state trajectories for the design value of $\varepsilon = 0.2$.

In general, the reasons for seeking corrected models are threefold. First, the parameter ε may not be very small. Second, the presence of eigenvalues close to the imaginary axis may mean that more reliable stability information about the model is necessary. Last, more accurate analysis of steady-state system behavior requires a more accurate model.

Again, the instrument at hand is the exact block-diagonal form (4.4). Typically, the most important, and often sufficient, approximation is the correction which includes the ε-terms and neglects $O(\varepsilon^2)$ and higher-order terms. This has been done in the model (4.13), (4.14), which is rewritten here as

$$\left.\begin{aligned} \dot{x}_{sc}(t) &= A_{0c}x_{sc}(t), \quad A_{0c} \triangleq A_0 - \varepsilon A_{12}A_{22}^{-2}A_{21}A_0, \\ x_{sc}(t_0) &= x^0 - \varepsilon A_{12}A_{22}^{-1}(z^0 + A_{22}^{-1}A_{21}x^0), \end{aligned}\right\} \tag{5.36}$$

$$\left.\begin{aligned} \frac{dz_{fc}}{d\tau}(\tau) &= A_{22c}z_{fc}(\tau), \quad A_{22c} \triangleq A_{22} + \varepsilon A_{22}^{-1}A_{21}A_{12}, \\ z_{fc}(0) &= z^0 + (A_{22}^{-1}A_{21} + \varepsilon A_{22}^{-2}A_{21}A_0)x^0. \end{aligned}\right\} \tag{5.37}$$

It should be recalled that the roots of the corrected "slow" characteristic equation (3.14) and the roots of the corrected "fast" characteristic equation (3.16) are the eigenvalues of A_{0c} in (5.36) and A_{22c} in (5.37) respectively.

In analogy with the $O(\varepsilon)$ approximations of Theorem 5.1, an $O(\varepsilon^2)$ approximation to the actual system states is obtained through the use of the corrected slow and fast models (5.36) and (5.37) in separate time scales.

Theorem 5.2

Under the conditions of Theorem 5.1, the approximations

$$x(t) = x_{sc}(t) + \varepsilon A_{12}A_{22}^{-1}z_{fc}(\tau) + O(\varepsilon^2), \tag{5.38}$$

$$\begin{aligned} z(t) = &-(A_{22}^{-1}A_{21} + \varepsilon A_{22}^{-2}A_{21}A_0)x_{sc}(t) \\ &+ (I_m - \varepsilon A_{22}^{-1}A_{21}A_{12}A_{22}^{-1})z_{fc}(\tau) + O(\varepsilon^2) \end{aligned} \tag{5.39}$$

hold for all finite time $t \geq t_0$ where $x_{sc}(t)$ and $z_{fc}(\tau)$ are the respective solutions of the corrected slow model (5.36) and fast model (5.37). If A_0 is also Hurwitz, then (5.38) and (5.39) hold for all $t \in [t_0, \infty)$.

Proof Similarly to the proof of Theorem 5.1, (5.13)–(5.16) and (2.13) imply

$$\begin{aligned} x(t) = &\, e^{A_{0c}(t-t_0)}[(I_n - \varepsilon H_0L_0)x^0 - \varepsilon H_0z^0] \\ &+ \varepsilon H_0 e^{A_{22c}(t-t_0)/\varepsilon}[z^0 + L_1x^0] + O(\varepsilon^2), \end{aligned} \tag{5.40}$$

$$z(t) = -L_1x(t) + e^{A_{22c}(t-t_0)/\varepsilon}[z^0 + L_1x^0] + O(\varepsilon^2), \tag{5.41}$$

where

$$L_0 \triangleq A_{22}^{-1} A_{21}, \quad H_0 \triangleq A_{12} A_{22}^{-1} \tag{5.42}$$

$$L_1 \triangleq A_{22}^{-1} A_{21} + \varepsilon A_{22}^{-2} A_{21} A_0, \tag{5.43}$$

etc. □

Comparing Theorem 5.2 with Theorem 5.1, it is seen that the approximation to the actual state response of the system (2.1), (2.2) is improved for two reasons. First, corrective fast terms in $z_{fc}(t - t_0)/\varepsilon$ are introduced on the right-hand sides of (5.38) and (5.39) which are, however, only active during the initial short interval $[t_0, t_1]$. Second, and more important from the point of view of application, the "slow" state $x_{sc}(t)$ is itself the state of a model (5.36) which corrects the reduced model (5.4), through the addition of the term $\varepsilon A_{12} A_{22}^{-2} A_{21} A_0$.

Example 5.2

Pursuing the PID control design of Example 5.1, from (5.36) the corrected slow system model is

$$\dot{x}_{sc}(t) = \frac{-k_3}{k_1}\left(1 + \frac{\varepsilon k_3 k_3}{k_1^2}\right) x_{sc}(t), \tag{5.44}$$

$$x_{sc}(0) = 1 + \frac{\varepsilon k_2}{k_1}\left(1 + \frac{k_3}{k_1}\right), \tag{5.45}$$

whereas, from (5.37), the corrected fast (normal) system model is

$$\begin{bmatrix} \varepsilon \dot{z}_{1fc}(t) \\ \varepsilon \dot{z}_{2fc}(t) \end{bmatrix} = \begin{bmatrix} \varepsilon \dfrac{k_3}{k_1} & 1 \\ -k_1 & -k_2 \end{bmatrix} \begin{bmatrix} z_{1fc}(t) \\ z_{2fc}(t) \end{bmatrix}, \tag{5.46}$$

$$\begin{bmatrix} z_{1fc}(0) \\ z_{2fc}(0) \end{bmatrix} = \begin{bmatrix} 1 + \dfrac{k_3}{k_1} \\ 0 \end{bmatrix} - \frac{\varepsilon k_3^2}{k_1^2} \begin{bmatrix} \dfrac{k_2 k_3}{k_1} \\ 1 \end{bmatrix}. \tag{5.47}$$

The eigenvalue

$$\lambda_{1c} = \frac{-k_3}{k_1}\left(1 + \frac{\varepsilon k_2 k_3}{k_1^2}\right) \tag{5.48}$$

of the corrected slow system model (5.44) indicates that for $k_3 > 0$, by uncorrected design, the stability of corrected slow system enjoys an $O(\varepsilon)$ improvement over that of the uncorrected model (5.22). Consider now the characteristic equation of the corrected fast system model (5.46) in p-scale, where $s = p/\varepsilon$ as in (3.16):

$$\psi_{\text{fc}}(p, \varepsilon) = \det[pI_2 - A_{22c}]$$
$$= p^2 + \left[k_2 - \frac{\varepsilon k_3}{k_1}\right]p + \left[k_1 - \varepsilon\frac{k_2 k_3}{k_1}\right] = 0, \tag{5.49}$$

which, using (5.27), is rewritten as

$$\psi_{\text{fc}}(p, \varepsilon) = p^2 + \left[2\zeta\omega_{\text{n}} - \frac{\varepsilon k_3}{k_1}\right]p + \left[\omega_{\text{n}}^2 - 2\varepsilon\zeta\omega_{\text{n}}\left(\frac{k_3}{k_1}\right)\right] = 0. \tag{5.50}$$

The roots of (5.50), in $s = p/\varepsilon$ scale, are the corrected fast eigenvalues

$$\lambda_{2c}, \lambda_{3c} = \left[\!\left[-2\zeta\omega_{\text{n}} + \frac{\varepsilon k_3}{k_1} \pm \mathrm{j}\left\{4\left[\omega_{\text{n}}^2 - 2\varepsilon\zeta\omega_{\text{n}}\left(\frac{k_3}{k_1}\right)\right] - \left[2\zeta\omega_{\text{n}} - \frac{\varepsilon k_3}{k_1}\right]^2\right\}^{1/2}\right]\!\right] \Big/ 2\varepsilon \tag{5.51}$$

Since $k_1 > 0$ and $k_3 > 0$, by uncorrected design (5.27), the effect of the integral action gain εk_3 in (5.51) is seen to be *destabilizing*. Therefore, our last requirement for a successful PID design is to keep the integral action εk_3 small by choosing ε small.

For the design values (5.29) and (5.32) the corrected eigenvalues (5.48) and (5.51) are

$$\lambda_{1c} = -1.283, \quad \lambda_{2c}, \lambda_{3c} = -3.035 \pm \mathrm{j}2.952. \tag{5.52}$$

Comparisons of the state trajectories of the corrected slow and fast models with those of the uncorrected slow and fast models, and with the state trajectories of the actual full system are made in Fig. 2.7. Moreover, the corrected slow and fast models enable us to judge, without recourse to the analysis of the full model (5.21), that the original uncorrected design (5.29), (5.30) and (5.32) is satisfactory *vis-à-vis* application to the full system (5.21).

2.6 Controllability and Observability

Let us now return to the analysis of the linear system (1.7)–(1.9) with a control input $u(t) \in R^r$ and an output $y(t) \in R^p$; namely,

$$\begin{bmatrix} \dot{x} \\ \varepsilon\dot{z} \end{bmatrix} = \begin{bmatrix} A_{11} & A_{12} \\ A_{21} & A_{22} \end{bmatrix}\begin{bmatrix} x \\ z \end{bmatrix} + \begin{bmatrix} B_1 \\ B_2 \end{bmatrix}u, \quad \begin{bmatrix} x(t_0) \\ z(t_0) \end{bmatrix} = \begin{bmatrix} x^0 \\ z^0 \end{bmatrix}, \tag{6.1}$$

$$y = [C_1 \quad C_2]\begin{bmatrix} x \\ z \end{bmatrix}. \tag{6.2}$$

It should be recalled that the system (6.1) is controllable if there exists a control $u(t)$ that transfers $x(t)$, $z(t)$ from any bounded initial state $x(t_0)$, $z(t_0)$ to any bounded terminal state $x(T)$, $z(T)$ in a finite time $T - t_0$. Similarly, the system (6.1), (6.2) is observable if the initial state $x(t_0)$, $z(t_0)$ can be determined from the measurement of the input $u(t)$ and the output $y(t)$ over the period $[t_0, T]$. The presence of the two-time-scale structure offers the possibility of deducing the controllability and observability properties of the full system (6.1), (6.2) in terms of the controllability and observability properties of lower-order models in separate time scales. First, we transform the system (6.1), (6.2), using the similarity transformation (4.8), to the equivalent system

$$\begin{bmatrix} \dot{\xi}(t) \\ \dot{\eta}(t) \end{bmatrix} = A \begin{bmatrix} \xi(t) \\ \eta(t) \end{bmatrix} + Bu(t), \tag{6.3}$$

$$y(t) = C \begin{bmatrix} \xi(t) \\ \eta(t) \end{bmatrix}, \tag{6.4}$$

which represents a parallel connection in Fig. 2.8 where

$$A = \begin{bmatrix} A_s & 0 \\ 0 & \dfrac{A_f}{\varepsilon} \end{bmatrix}, \quad B = \begin{bmatrix} B_s \\ \dfrac{B_f}{\varepsilon} \end{bmatrix}, \quad C = [C_s \quad C_f] \tag{6.5}$$

and

$$\left.\begin{aligned} A_s &= A_{11} - A_{12}L, & A_f &= A_{22} + \varepsilon L A_{12}, \\ B_s &= B_1 - HB_2 - \varepsilon HLB_1, & B_f &= B_2 + \varepsilon LB_1, \\ C_s &= C_1 - C_2 L, & C_f &= C_2 + \varepsilon(C_1 - C_2 L)H. \end{aligned}\right\} \tag{6.6}$$

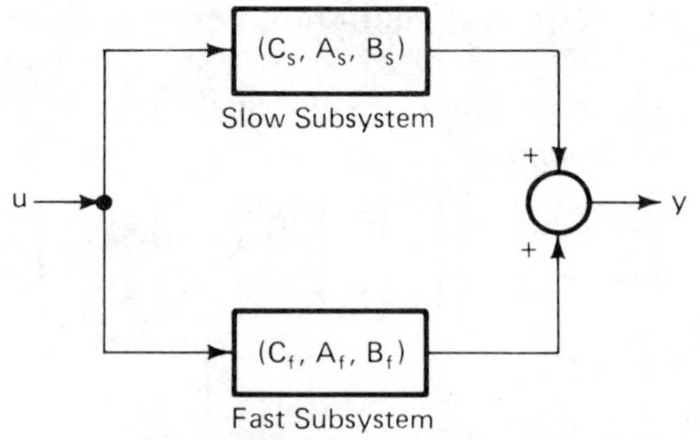

Fig. 2.8. Decomposed open-loop system.

Making use of the well-known fact that controllability and observability are invariant with respect to similarity transformation, we analyze these properties of the equivalent system (6.3)–(6.6). For linear time-invariant systems, a common characterization of controllability and observability is via the controllability and observability of the eigenvalues.

Lemma 6.1

A necessary and sufficient condition for the ith eigenvalue λ_i of the system (6.3) to be controllable is

$$\operatorname{rank}\,[\lambda_i I_{n+m} - A \ \vdots\ B] = n + m, \tag{6.7}$$

and that for the ith eigenvalue λ_i of the system (6.3), (6.4) to be observable is

$$\operatorname{rank}\begin{bmatrix} \lambda_i I_{n+m} - A \\ C \end{bmatrix} = n + m. \tag{6.8}$$

Furthermore, the system (6.3), (6.4) is completely controllable [completely observable] if and only if (6.7) [(6.8)] is satisfied for all its eigenvalues λ_i, $i = 1, \ldots, n + m$. The system (6.3), (6.4) is stabilizable [detectable] if and only if all its eigenvalues with nonnegative real parts are controllable [observable].

A crucial property of the parallel system configuration of Fig. 2.8 is that the eigenvalues of A_s are distinct from those of A_f/ε. Consequently, if an eigenvalue of the slow subsystem is controllable [observable], it will remain so in the parallel connection (6.3)–(6.6). Similarly, if an eigenvalue of the fast subsystem is controllable [observable], it will remain so in the parallel connection (6.3)–(6.6). In view of the fact that this property will not be lost for small perturbations of ε, we have the following result.

Theorem 6.1

Let λ_i^0 be the eigenvalue of $A_0 = A_{11} - A_{12}A_{22}^{-1}A_{21}$ approximating the slow eigenvalue λ_i of A in (6.3). If

$$\operatorname{rank}\,[\lambda_i^0 I_n - A_0 \ \vdots\ B_0] = n, \tag{6.9}$$

where $B_0 = B_1 - A_{12}A_{22}^{-1}B_2$, then there exists an $\varepsilon^* > 0$ such that λ_i is controllable for all $\varepsilon \in (0, \varepsilon^*]$. Similarly, let π_j^0 be the eigenvalue of A_{22} approximating the fast eigenvalue λ_i of A according to $\lambda_i = \pi_j/\varepsilon$, $i = n + j$. If

$$\operatorname{rank}\,[\pi_j^0 I_m - A_{22} \ \vdots\ B_2] = m \tag{6.10}$$

then there exists an $\varepsilon^* > 0$ such that λ_i is controllable for all $\varepsilon \in (0, \varepsilon^*]$.

Proof In view of Theorem 3.1 and the preservation of the full rank of a matrix under regular perturbation, there exists an $\varepsilon^* > 0$ such that (6.9) and (6.10) respectively imply, for all $\varepsilon \in (0, \varepsilon^*]$,

$$\operatorname{rank}[\lambda_i I_n - A_s(\varepsilon) \quad B_s(\varepsilon)] = n, \tag{6.11}$$

where λ_i is the ith eigenvalue of $A_s(\varepsilon)$, and

$$\operatorname{rank}[\pi_j I_m - A_f(\varepsilon) \quad B_f(\varepsilon)] = m, \tag{6.12}$$

where π_j is the jth eigenvalue of $A_f(\varepsilon)$. By Theorem 3.1, $A_s(\varepsilon)$ and $\dfrac{A_f(\varepsilon)}{\varepsilon}$ have no eigenvalues in common, and hence

$$\operatorname{rank}\begin{bmatrix} \lambda_i I_n - A_s(\varepsilon) & 0 & B_s(\varepsilon) \\ 0 & \lambda_i I_m - \dfrac{A_f(\varepsilon)}{\varepsilon} & \dfrac{B_f}{\varepsilon}(\varepsilon) \end{bmatrix} = n + m, \tag{6.13}$$

if λ_i is the ith eigenvalue of $A_s(\varepsilon)$ and if $\lambda_i = \pi_j/\varepsilon$, $i = n + j$, is the jth eigenvalue of $A_f(\varepsilon)/\varepsilon$. □

The observability counterpart of Theorem 6.1 is as follows.

Theorem 6.2

Let λ_i^0 be the eigenvalue of A_0 approximating the slow eigenvalue λ_i of A in (6.3). If

$$\operatorname{rank}\begin{bmatrix} \lambda_i^0 I_n - A_0 \\ C_0 \end{bmatrix} = n, \tag{6.14}$$

where $C_0 = C_1 - C_2 A_{22}^{-1} A_{21}$. Then there exists an $\varepsilon^* > 0$ such that λ_i is observable for all $\varepsilon \in (0, \varepsilon^*]$.

Similarly, let π_j^0 be the eigenvalue of A_{22} approximating the fast eigenvalue λ_i of A according to $\lambda_i = \pi_j/\varepsilon$, $i = n + j$. If

$$\operatorname{rank}\begin{bmatrix} \pi_j^0 I_m - A_{22} \\ C_2 \end{bmatrix} = m \tag{6.15}$$

then there exists an $\varepsilon^* > 0$ such that λ_i is observable for all $\varepsilon \in (0, \varepsilon^*]$.

Proof The proof is analogous to that of Theorem 6.1. □

Complete controllability [observability] and stabilizability [detectability] then follow as corollaries.

Corollary 6.1

If the pair (A_0, B_0) $[(C_0, A_0)]$ and the pair (A_{22}, B_2) $[(C_2, A_{22})]$ are each completely controllable [completely observable], then there exists an $\varepsilon^* > 0$ such that the original system (6.1), (6.2) is completely controllable [completely observable] for all $\varepsilon \in (0, \varepsilon^*]$.

Corollary 6.2

If the pair (A_0, B_0) $[(C_0, A_0)]$ and the pair (A_{22}, B_2) $[(C_2, A_{22})]$ are each stabilizable [detectable], then there exists an $\varepsilon^* > 0$ such that the original system (6.1), (6.2) is stabilizable [detectable] for all $\varepsilon \in (0, \varepsilon^*]$.

A conclusion of practical importance is that instead of testing the ε-dependent triple (C, A, B) of the full-order system (6.1), (6.2) for controllability and observability (stabilizability and detectability), one need only test the ε-independent triples (C_0, A_0, B_0) and (C_2, A_{22}, B_2) of lower-order subsystems.

Example 6.1

The model of a two-tank chemical reactor system, described in Section 3.3, is

$$\begin{bmatrix} \dot{x}(t) \\ \varepsilon \dot{z}(t) \end{bmatrix} = \left[\begin{array}{cc|cc} -\frac{1}{5} & \frac{1}{5} & 0 & 0 \\ 0 & -\frac{1}{2} & \frac{1}{2} & 0 \\ \hline 0 & 0 & 0 & 1 \\ 0 & 0 & -1 & -2 \end{array}\right] \begin{bmatrix} x(t) \\ z(t) \end{bmatrix} + \left[\begin{array}{c} 0 \\ 0 \\ \hline 0 \\ 1 \end{array}\right] u, \tag{6.16}$$

$$y = [1 \quad 0 \quad 0 \quad 0] \begin{bmatrix} x(t) \\ z(t) \end{bmatrix}, \tag{6.17}$$

where $x \in R^2$ and $z \in R^2$ respectively represent the slow reactor and actuator dynamics. The slow system model is described by the triple

$$A_0 = \begin{bmatrix} -\frac{1}{5} & \frac{1}{5} \\ 0 & -\frac{1}{2} \end{bmatrix}, \quad B_0 = \begin{bmatrix} 0 \\ \frac{1}{2} \end{bmatrix}, \quad C_0 = [1 \quad 0], \tag{6.18}$$

while the fast system model is described by the triple

$$A_{22} = \begin{bmatrix} 0 & 1 \\ -1 & -2 \end{bmatrix}, \quad B_2 = \begin{bmatrix} 0 \\ 1 \end{bmatrix}, \quad C_2 = [0 \quad 0]. \tag{6.19}$$

For controllability we check that

$$\operatorname{rank}[\lambda_i^0 I_2 - A_0 \vdots B_0] = 2, \quad i = 1, 2 \tag{6.20}$$

for $\lambda_1^0 = -\frac{1}{5}$, $\lambda_2^0 = -\frac{1}{2}$, and that

$$\operatorname{rank}[\pi_j^0 I_2 - A_{22} \vdots B_2] = 2, \quad j = 1, 2 \tag{6.21}$$

for $\pi_1^0, \pi_2^0 = -1$. Thus, by Theorem 6.1 all four eigenvalues of (6.16) are controllable for ε sufficiently small.

Testing for observability, we note that

$$\operatorname{rank}\begin{bmatrix} \lambda_i^0 I_2 - A_0 \\ C_0 \end{bmatrix} = 2, \quad i = 1, 2 \tag{6.22}$$

for $\lambda_1^0 = -\frac{1}{5}$, $\lambda_2^0 = -\frac{1}{2}$. However,

$$\operatorname{rank}\begin{bmatrix} \pi_j^0 I_2 - A_{22} \\ C_2 \end{bmatrix} = 1 < 2, \quad j = 1, 2 \tag{6.23}$$

for $\pi_1^0, \pi_2^0 = -1$. As we would expect from $C_2 = 0$ both π_1^0 and π_2^0 of the fast model are unobservable, but stable, and the fast model pair (C_2, A_{22}) is detectable.

In fact, the original system (6.16), (6.17) is completely observable for $\varepsilon > 0$. It only loses observability of the fast system modes as ε tends to zero. It illustrates the fact that Theorems 6.1 and 6.2 are only sufficient, but not necessary conditions for controllability and observability near $\varepsilon = 0$. In other words, the converse of Theorem 6.1 (and its dual Theorem 6.2) may not hold: setting $\varepsilon = 0$ in (6.11) and (6.12) does not necessarily imply (6.9) and (6.10) respectively. This motivates us to call an eigenvalue λ_i of A in (6.1) *weakly controllable* [*weakly observable*] if it is controllable [observable] for ε arbitrarily small, but the corresponding sufficiency condition (controllability at $\varepsilon = 0$) (6.9) or (6.10) [(observability at $\varepsilon = 0$) (6.14) or (6.15)] is not satisfied. Conversely, if the corresponding sufficiency condition (6.9) or (6.10) [(6.14) or (6.15)] is satisfied, the eigenvalue λ_i of A is said to be *strongly controllable* [*strongly observable*].

Example 6.2

The two systems shown in Fig. 2.9 have the same eigenvalues: one slow eigenvalue $\lambda_1 = -1$ and one fast eigenvalue $\lambda_2 = -1/\varepsilon$. They are, however,

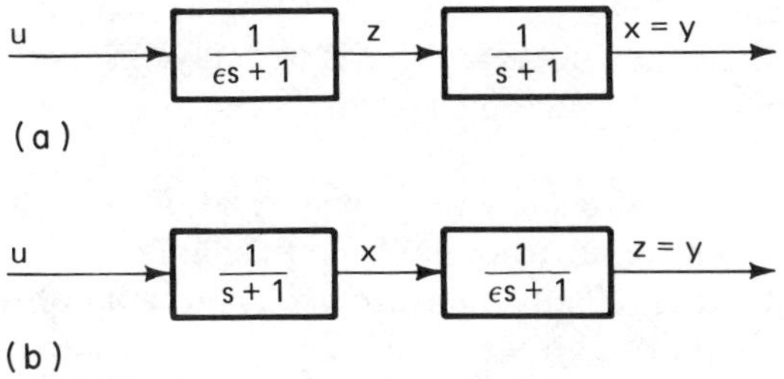

Fig. 2.9. (a) Actuator form. (b) Sensor form.

structurally different: system (a) is in actuator form while system (b) is in sensor form. The state realizations of the systems in Fig. 2.9 are

$$\text{(a)}\begin{cases} \dot{x} = -x + z, \\ \varepsilon\dot{z} = -z + u, \\ y = x, \end{cases} \quad \text{(b)}\begin{cases} \dot{x} = -x + u, \\ \varepsilon\dot{z} = x - z, \\ y = z, \end{cases} \tag{6.24}$$

while their modal forms

$$\text{(a}'\text{)}\begin{cases} \dot{\xi} = -\xi + \dfrac{1}{1-\varepsilon}u, \\ \varepsilon\dot{\eta} = -\eta + u, \\ y = \xi - \dfrac{\varepsilon}{1-\varepsilon}\eta, \end{cases} \quad \text{(b}'\text{)}\begin{cases} \dot{\xi} = -\xi + u, \\ \varepsilon\dot{\eta} = -\eta - \dfrac{\varepsilon}{1-\varepsilon}u, \\ y = \dfrac{1}{1-\varepsilon}\xi + \eta. \end{cases} \tag{6.25}$$

are parallel configurations as in Fig. 2.8. For both systems, the slow eigenvalue $\lambda_1 = -1$ and the fast eigenvalue $\lambda_2 = -1/\varepsilon$ are controllable and observable for $\varepsilon > 0$.

The realization (a′) displays the following structural characteristics of the system (a):

(a1) the slow eigenvalue $\lambda_1 = -1$ is strongly controllable and strongly observable;

(a2) the fast eigenvalue $\lambda_2 = -1/\varepsilon$ is strongly controllable, but is weakly observable through an ε-term.

Also, the realization (b′) displays the following structural characteristics of the system (b):

(b1) the slow eigenvalue $\lambda_1 = -1$ is strongly controllable and strongly observable;
(b2) the fast eigenvalue $\lambda_2 = -1/\varepsilon$ is strongly observable, but is weakly controllable through an ε-term.

It is left as Exercise 2.8 to show that, in general, for systems in sensor form the fast eigenvalues are always weakly controllable, whereas for systems in actuator form the fast eigenvalues are always weakly observable.

Weak controllability and weak observability are either a blessing or a curse when it comes to linear feedback design. They are a blessing in that stable parasitic (fast) modes, which are either weakly controllable or weakly observable, will not be affected by more than $O(\varepsilon)$ through output feedback. Accordingly, output-feedback designs, based on reduced-order slow models, are robust in the sense that stability of the fast subsystem is preserved when these designs are applied to the actual full system. On the other hand, should it be desired to shift a weakly controllable mode, $O(1/\varepsilon)$ high state-feedback gains are necessary to accomplish this task. More will be said about these problems in Chapter 3.

Two other tests for complete controllability and complete observability occasionally used, equivalent to (6.7) and (6.8), are the original rank criterion and Grammian tests due to Kalman (Kwakernaak and Sivan, 1972). Treating controllability only, since observability follows by duality, the rank criterion states that the system (6.3), (6.4) or the pair (A, B) is completely controllable if and only if

$$\operatorname{rank}[B \quad AB \quad \ldots \quad A^{n+m-1}B] = n + m. \tag{6.26}$$

Thus, again in view of the parallel structure of the system (6.3), (6.4) and the preservation of the rank of a matrix under regular perturbation, Corollary 6.1 can be restated in terms of separate lower-order rank criteria for the slow system pair (A_0, B_0) and the fast system pair (A_{22}, B_2).

Collollary 6.3

If the following two conditions

$$\operatorname{rank}[B_0, A_0B_0, \ldots A_0^{n-1}B_0] = n \tag{6.27}$$

$$\operatorname{rank}[B_2, A_{22}B_2, \ldots A_{22}^{m-1}B_2] = m \tag{6.28}$$

are satisfied, then there exists an $\varepsilon^* > 0$ such that the original system is completely controllable in the sense of (6.26) for all $\varepsilon \in (0, \varepsilon^*]$.

Even for linear time-varying systems encountered in Chapters 5 and 6, the controllability condition for the fast system is of the *algebraic* form (6.28) where $(A_{22}(t), B_2(t))$ is defined at a fixed t in τ-scale. In Chapter 7, composite control problems nonlinear in x but linear in z and u, require a similar algebraic controllability criterion only this time $(A_{22}(x), B_2(x))$ is defined for each fixed x in a certain domain.

For controllability and observability properties defined in terms of Grammian matrices, let us consider the observability Grammian

$$P = \int_0^\infty (e^{A^{T}t} C^T C\, e^{At})\, dt, \tag{6.29}$$

where the usual finite interval of integration is extended to $[0, \infty]$ on account of assuming A to be a Hurwitz matrix. Then, as is well known (Kwakernaak and Sivan 1972), P is the solution of the Lyapunov equation

$$PA + A^T P = -C^T C, \tag{6.30}$$

where

$$A = \begin{bmatrix} A_{11} & A_{12} \\ \dfrac{A_{21}}{\varepsilon} & \dfrac{A_{22}}{\varepsilon} \end{bmatrix}, \quad C = [C_1 \quad C_2]. \tag{6.31}$$

Premultiplying* (6.30) by T^T and postmultiplying by T, where T is defined by (4.8), we have

$$T^T P T(T^{-1}AT) + (T^{-1}AT)^T T^T P T = -T^T C^T C T. \tag{6.32}$$

The solution of this Lyapunov equation is

$$\tilde{P} = T^T P T = \int_0^\infty \begin{bmatrix} e^{A_s t} & 0 \\ 0 & e^{A_f t/\varepsilon} \end{bmatrix}^T \begin{bmatrix} C_s^T C_s & C_s^T C_f \\ C_f^T C_s & C_f^T C_f \end{bmatrix} \begin{bmatrix} e^{A_s t} & 0 \\ 0 & e^{A_f t/\varepsilon} \end{bmatrix} dt. \tag{6.33}$$

Equation (6.33) shows the scaling for $\tilde{P}$ that occurs in Chapters 3, 4 and 6, where we study Lyapunov equations and regular Riccati equations for singularly perturbed systems. It follows from

$$\int_0^\infty e^{A_s^T t} C_s^T C_f e^{A_f t/\varepsilon}\, dt = \varepsilon \int_0^\infty e^{A_s^T t} O(1) e^{A_f \lambda}\, d\lambda = O(\varepsilon) \tag{6.34}$$

* As a superscript T denotes matrix transposition, otherwise T is defined by the similarity transformation (4.8).

where $\lambda = t/\varepsilon$, and

$$\int_0^{\infty} e^{A_f^T t/\varepsilon} C_f^T C_f e^{A_f t/\varepsilon} \, dt = \varepsilon \int_0^{\infty} e^{A_f^T \lambda} O(1) e^{A_f \lambda} \, d\lambda = O(\varepsilon) \tag{6.35}$$

that $\tilde{P}$ in (6.33) is scaled as

$$\tilde{P} = \begin{bmatrix} \tilde{P}_1 & \varepsilon \tilde{P}_2 \\ \varepsilon \tilde{P}_2^T & \varepsilon \tilde{P}_3 \end{bmatrix}. \tag{6.36}$$

The same scaling is valid for the original $P = T^{-T}\tilde{P}T^{-1}$ of (6.29) and will be used in Chapters 3, 4 and 6. Corollary 6.1 can then be restated in terms of lower-order observability Grammians, and allied Lyapunov equations, for the slow and fast system pairs (C_0, A_0) and (C_2, A_{22}), respectively. Finally, it is left as Exercise 2.9 to show that the (dual) controllability Grammian W, satisfying the dual Lyapunov equation

$$AW + WA^T = -BB^T, \tag{6.37}$$

is scaled as

$$W = \begin{bmatrix} W_1 & W_2 \\ W_2^T & \dfrac{W_3}{\varepsilon} \end{bmatrix}. \tag{6.38}$$

2.7 Frequency-Domain Models

Singularly perturbed systems display further interesting properties when exhibited in the complex domain. The starting point in our analysis, as in the previous section, is the system (6.1), (6.2), transformed to the system (6.3), (6.4) by the similarity transformation (4.8).

The Laplace transform of (6.3), (6.4), omitting initial conditions, is

$$y(s, \varepsilon) = [C_s \quad C_f] \begin{bmatrix} sI_n - A_s & 0 \\ 0 & sI_m - \dfrac{A_f}{\varepsilon} \end{bmatrix}^{-1} \begin{bmatrix} B_s \\ \dfrac{B_f}{\varepsilon} \end{bmatrix} u(s), \tag{7.1}$$

which gives

$$y(s, \varepsilon) = [C_s \quad C_f] \begin{bmatrix} (sI_n - A_s)^{-1} & 0 \\ 0 & \varepsilon(\varepsilon sI_m - A_f)^{-1} \end{bmatrix} \begin{bmatrix} B_s \\ \dfrac{B_f}{\varepsilon} \end{bmatrix} u(s). \tag{7.2}$$

Thus the transfer-function matrix $G(s, \varepsilon)$ from the input $u(s)$ to the output $y(s, \varepsilon)$ is given by the sum of the slow and fast transfer-function matrices

$$y(s, \varepsilon) = G(s, \varepsilon)u(s) = [G_s(s, \varepsilon) + G_f(\varepsilon s, \varepsilon)]u(s). \tag{7.3}$$

The slow transfer-function matrix is a function of s:

$$G_s(s, \varepsilon) = C_s(sI_n - A_s)^{-1}B_s, \tag{7.4}$$

while the fast transfer-function matrix is a function of εs:

$$G_f(\varepsilon s, \varepsilon) = C_f(\varepsilon sI_m - A_f)^{-1}B_f. \tag{7.5}$$

Their parallel connection is illustrated in Fig. 2.10. The two frequency scales corresponding to the time scales t and t/ε are s and $p = \varepsilon s$ respectively.

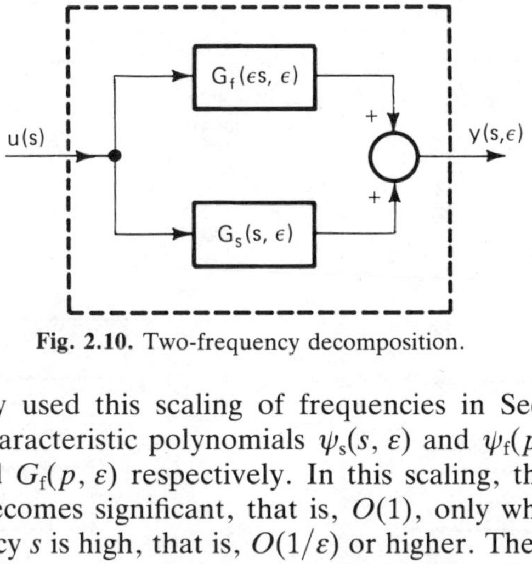

Fig. 2.10. Two-frequency decomposition.

We have already used this scaling of frequencies in Section 2.3 when discussing the characteristic polynomials $\psi_s(s, \varepsilon)$ and $\psi_f(p, s)$, associated with $G_s(s, \varepsilon)$ and $G_f(p, \varepsilon)$ respectively. In this scaling, the complex frequency $p = \varepsilon s$ becomes significant, that is, $O(1)$, only when the original complex frequency s is high, that is, $O(1/\varepsilon)$ or higher. The decomposition into the slow and fast systems now acquires the meaning of a decomposition into low-frequency and high-frequency blocks. To see this decomposition more clearly, let us restrict s to the imaginary axis, i.e., $s = j\omega$, and assume that $A_s(\varepsilon = 0)$ and $A_f(\varepsilon = 0)$ have no eigenvalues on the imaginary axis. We observe that $G_s(s, \varepsilon)$ in (7.4) is a strictly proper transfer-function matrix, and thus has the low-pass property

$$G_s(s, \varepsilon) = O(\varepsilon) \quad \text{for } s \in [j\omega_1, \infty) \tag{7.6}$$

for some positive fixed constant ω_1. On the other hand, for low frequencies $s = O(1)$, that is, $p = O(\varepsilon)$, the fast transfer-function matrix $G_f(p, \varepsilon)$ in (7.5) can be approximated by its DC-gain, that is,

$$G_f(p, \varepsilon) = G_f(0, \varepsilon) + O(\varepsilon) \quad \text{for } p \in [0, j\varepsilon\omega_2] \tag{7.7}$$

for some positive fixed constant ω_2. This suggests that a low-frequency approximation of $G(s, \varepsilon)$ in (7.3) is

$$G(s, \varepsilon) = G_s(s, \varepsilon) + G_f(0, \varepsilon) + O(\varepsilon), \quad s \in [0, j\omega_2], \tag{7.8}$$

and is depicted in Fig. 2.11(a). The corresponding high-frequency approximation is

$$G\left(\frac{p}{\varepsilon}, \varepsilon\right) = G_f(p, \varepsilon) + O(\varepsilon), \quad p \in [j\omega_1, \infty), \tag{7.9}$$

and is depicted in Fig. 2.11(b). In the state-space representation (7.2), the low-frequency model (7.8) is

$$y(s, \varepsilon) = [C_s(sI_n - A_s)^{-1}B_s - C_f A_f^{-1} B_f]u(s), \tag{7.10}$$

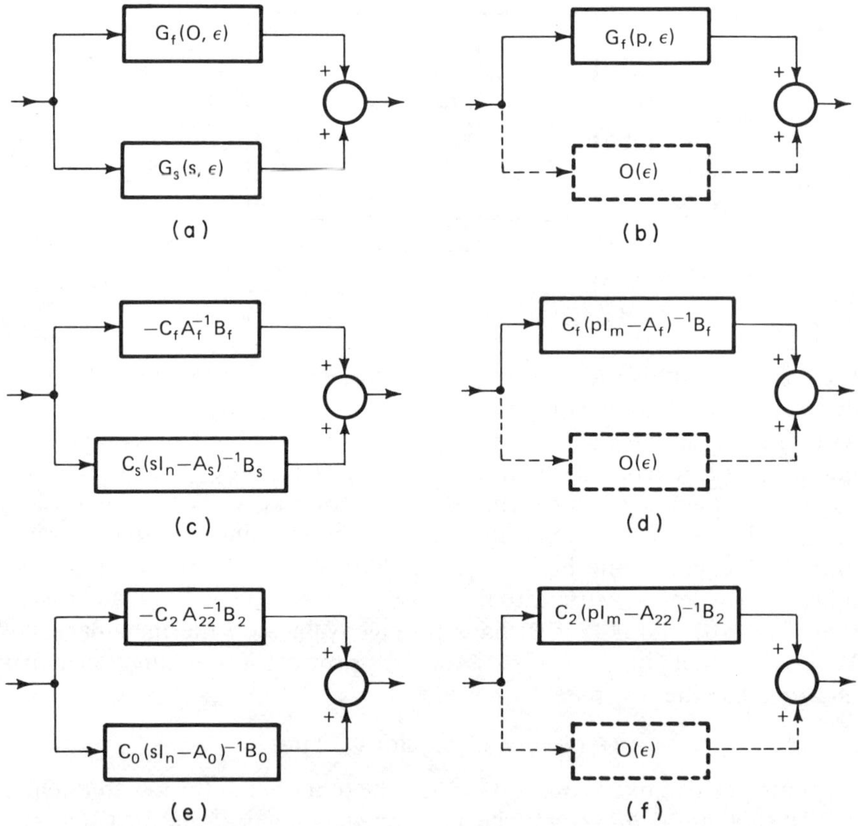

Fig. 2.11. (a) Low-frequency approximation; (b) high-frequency approximation; (c) low-frequency approximation; (d) high-frequency approximation; (e) low-frequency approximation; (f) high-frequency approximation.

as shown in Fig. 2.11(c), and the high-frequency model (7.9) is

$$y\left(\frac{p}{\varepsilon}, \varepsilon\right) = C_{\mathrm{f}}[pI_m - A_{\mathrm{f}}]^{-1}B_{\mathrm{f}}u\left(\frac{p}{\varepsilon}\right), \tag{7.11}$$

as shown in Fig. 2.11(d). It should be recalled from Section 2.1 that every matrix in (7.10) and (7.11) is a continuously differentiable function of ε, and, by (2.13) and (4.7), for ε sufficiently small can be approximated by its value at $\varepsilon = 0$. Thus a further simplification of the low-frequency model (7.10) is

$$y(s, \varepsilon) = [C_0(sI_n - A_0)^{-1}B_0 - C_2A_{22}^{-1}B_2]u(s), \tag{7.12}$$

as shown in Fig. 2.11(e), while the corresponding simplification of the high-frequency model (7.11) is

$$y\left(\frac{p}{\varepsilon}, \varepsilon\right) = C_2[pI_m - A_{22}]^{-1}B_2u\left(\frac{p}{\varepsilon}\right), \tag{7.13}$$

as shown in Fig. 2.11(f).

Example 7.1

Let us return to the system (6.24a) in actuator form of Example 6.2. Taking Laplace transforms of the model realization (a') in (6.25), the exact slow and fast transfer functions (7.4) and (7.5) are

$$G_{\mathrm{s}}(s, \varepsilon) = \frac{1}{1-\varepsilon}\frac{1}{(s+1)} \tag{7.14}$$

and

$$G_{\mathrm{f}}(\varepsilon s, \varepsilon) = \frac{-\varepsilon}{1-\varepsilon}\frac{1}{(\varepsilon s+1)} \tag{7.15}$$

in the parallel configuration of Fig. 2.10. From (7.15), the low-frequency "equivalent" of $G_{\mathrm{f}}(p, \varepsilon)$ is

$$G_{\mathrm{f}}(0, \varepsilon) = \frac{-\varepsilon}{1-\varepsilon}. \tag{7.16}$$

Taking $G_{\mathrm{f}}(0, \varepsilon)$ in (7.16) and combining it with the exact slow transfer function $G_{\mathrm{s}}(s, \varepsilon)$ in (7.14), we have

$$\frac{1}{1-\varepsilon}\frac{1}{(s+1)} - \frac{\varepsilon}{1-\varepsilon} = \frac{1 - \dfrac{\varepsilon}{1-\varepsilon}s}{s+1} \tag{7.17}$$

as the low-frequency approximation of the complete transfer function in

Fig. 2.11(a). Setting $\varepsilon = 0$ further simplifies (7.17) to

$$\frac{1}{s+1}, \tag{7.18}$$

which is the low-frequency model obtained by neglecting the actuator dynamics. It can be seen that, in this particular example, the model (7.17) yields $O(\varepsilon^2)$ approximation at low frequencies, while (7.18) is only an $O(\varepsilon)$ approximation. The presence of an unstable zero in (7.17) is significant as it shows the destabilizing effect of ε.

2.8 Exercises

Exercise 2.1

Complete the block-diagonalization of the system (2.1), (2.2) by application of the change of variables $\eta = z + N(\varepsilon)\xi$ to the sensor form (2.30) where $N(\varepsilon)$ satisfies a Lyapunov-type equation which is the adjoint of (4.3). Although the eigenvalues of the sensor form and the eigenvalues of the actuator form are the same examine in what sense the state variables are different.

Exercise 2.2

Derive the corrected slow and fast models (4.13) and (4.14) from (2.1), (2.2) using the following separation of time scales procedure. Set $\varepsilon = 0$ to obtain the quasi-steady-state $\bar{z}$; derive the fast subsystem in terms of the fast state $\eta_1 = z - \bar{z}$; remove the fast part from x to obtain the slow subsystem in $\xi_1 = x - \varepsilon A_{12}A_{22}^{-1}\eta_1$. Is this procedure equivalent to $L_0 - H_0$ transformation in (4.10), (4.11)? Suggest how it might form part of an iterative scheme to obtain corrected models to any order of accuracy (see also Kokotović *et al.*, 1980).

Exercise 2.3

For the adjoint quadratic equation (2.29), deduce the analogs of Lemma 2.1, Lemma 2.2 and equations (3.14)–(3.19).

Exercise 2.4

Consider the discrete-time system

$$x(n+1) = (I + \varepsilon A_{11})x(n) + \varepsilon A_{12}z(n),$$

$$z(n+1) = A_{21}x(n) + A_{22}z(n),$$

where $I_m - A_{22}$ is nonsingular. Use the transformation $\eta(n) = z(n) + Lx(n)$ to show that the system is asymptotically stable if A_{11} is Hurwitz in a continuous-time sense; i.e. its eigenvalues have negative real parts, and A_{22} is Hurwitz in a discrete-time sense; i.e. its eigenvalues are inside the unit circle (for more details see Litkouhi and Khalil, 1984).

Exercise 2.5

Compute the corrected slow and fast eigenvalues and eigenvectors for the system

$$\begin{bmatrix} \dot{x}(t) \\ \varepsilon\dot{x}(t) \end{bmatrix} = \begin{bmatrix} -1 & 1 \\ 0 & -1 \end{bmatrix} \begin{bmatrix} x(t) \\ z(t) \end{bmatrix}$$

and compare the exact slow and exact fast eigenvalues and eigenvectors for a value of ε less than that given by the upper bound of Lemma 2.2.

Exercise 2.6

Show that the angle of the fast eigenvector

$$\bar{v}_{\mathrm{f}} = \begin{bmatrix} \varepsilon A_{12} A_{22}^{-1} \\ I_m \end{bmatrix} \bar{v}$$

in (4.32) makes with the z-subspace is an $O(\varepsilon^2)$ approximation of the angle that v_{f} in (4.27) makes with the z-subspace by considering the effect of an ε-correction of the term I_m on its angle.

Hint: Proceed similarly to Example 4.2 by calculating the angular difference

$$\alpha - \bar{\alpha}, \text{ where } \cos(\alpha - \bar{\alpha}) = \langle v_{\mathrm{f}}, \bar{v}_{\mathrm{f}} \rangle / \|v_{\mathrm{f}}\| \|\bar{v}_{\mathrm{r}}\|.$$

Exercise 2.7

Analyse the controllability and observability properties of the system

$$\dot{x} = -x + z - u,$$

$$\varepsilon\dot{z} = -z + u,$$

$$y = x.$$

Exercise 2.8

Show that, in general, the fast eigenvalues of systems in the actuator form of Fig. 2.3 [sensor form of Fig. 2.5] are strongly controllable and weakly observable [weakly controllable and strongly observable], lending further weight to the adjectives actuator and sensor.

Exercise 2.9

Establish Corollary 6.1 in terms of lower-order controllability Grammians, while at the same time verifying that the controllability Grammian

$$W = \int_0^\infty (e^{At}BB^{\mathrm{T}}e^{A^{\mathrm{T}}t})\,dt,$$

satisfying the Lyapunov equation

$$AW + WA^{\mathrm{T}} = -BB^{\mathrm{T}},$$

where A is a Hurwitz matrix, must be scaled as

$$W = \begin{bmatrix} W_1 & W_2 \\ W_2^{\mathrm{T}} & \dfrac{W_3}{\varepsilon} \end{bmatrix}.$$

Exercise 2.10

In Example 7.1, alternatively determine the low-frequency "equivalent" of the system (6.24a) through performing a partial-fraction expansion of the model in Fig. 2.9(a).

2.9 Notes and References

The transformation of a singularly perturbed system model into block-triangular form, in its more general time-varying version, is due to Chang (1972). Following Kokotović (1975), Lemma 2.1 and Lemma 2.2 specialize Chang's result to linear time-invariant systems. In Avramović (1979), Kokotović *et al.* (1980) and O'Malley and Anderson (1982), the determination of the matrix L is iterative. When the separation of time scales is large (ε is small), the convergence rate is quite rapid, typically a couple of iterations. Even when ε is not small, as for example in the power system decomposition of Winkelman *et al.* (1980), the iterative scheme is still applicable, provided that the original choice of slow and fast states in the model is valid. The iterations for solving the matrix quadratic equation (2.8) are closely related to the eigenspace power iterations of Avramović

(1979) and the quasi-steady state iterations of Kokotović *et al.* (1980) and Winkelman *et al.* (1980). Phillips (1983) establishes the equivalences between these iterative procedures and the earlier asymptotic series methods of Vasil'eva and Butuzov (1978).

Theorem 3.1, which characterizes the essential property of singularly perturbed systems, is a time-invariant version of a theorem of Klimushchev and Krasovskii (1962) (see also Wilde and Kokotović, 1972a), and has been applied to networks with parasitics by Desoer and Shensa (1970) and to control systems by Porter (1974) and Kokotović and Haddad (1975a). A more detailed stability analysis leads to an upper bound on ε (Zien, 1973), different from that of Lemma 2.2, while another upper bound using singular values is given by Sandell (1979).

The second change of variables in Section 2.4 completes the exact decomposition of the system into slow and fast states. Determination of an H-matrix, satisfying the linear equation (4.3), like that of L is also iterative (Avramović, 1979, Kokotović *et al.*, 1980). Lemma 4.1 on the eigenspace properties of linear time-invariant systems is based on Avramović (1979) and Kokotović (1981) and can be extended to include generalized eigenvectors as well. It forms the basis of a grouping algorithm for power systems (Avramović *et al.*, 1980; Chow, 1982) and for Markov chains (Delebeque, Quadrat and Kokotović, 1984; Kokotović, 1984; Saksensa *et al.*, 1984).

The initial value theorem, Theorem 5.1, for linear time-invariant systems, is contained in Kokotović and Haddad (1975a) and Kototović *et al.* (1976). As Example 4.1 illustrates, corrected models such as (5.36) and (5.37) are useful in pinpointing the destabilizing effects of parasitics in adaptive control (Ioannou and Kokotović, 1983). Corrected models have also been used to advantage in demonstrating the stabilizing effects of the damper windings in synchronous machine models (Sauer, Ahmed-Zaid and Pai, 1984a, b) and improving the stability of large-scale power system models (Pai, Sauer and Khorasani, 1984). The $O(\varepsilon^2)$ state approximations of Theorem 5.2 for corrected models are new. Higher-order approximations can readily be obtained through the use of higher-order terms in the power-series expansions of L and H.

The controllability result of Theorem 6.1 was introduced in Kokotović and Haddad (1975a). Analogous results apply to continuous-time and discrete-time linear two-time-scale systems (Saksensa *et al.*, 1984). The observability conditions follow from duality with those of controllability. The notions of weak and strong controllability (observability), introduced in Section 2.6, are due to Chow (1977b).

In Section 2.7 the introductory discussion of singularly perturbed systems from a frequency-domain point of view is new. An early paper by Porter and Shenton (1975) uses a transfer-function approach, while Luse and

Khalil (1985) and Luse (1984), more recently, develop a two-frequency analysis of transfer-function matrices and their properties. A suggestion for two-frequency design is presented in Fossard and Magni (1980), and a frequency-domain robustness condition appears in Sandell (1979). Otherwise, considering how much is presently known about frequency-domain methods for general multivariable systems, surprisingly little is known about two-frequency-scale systems. It is expected that future analysis will, among other things, extend our understanding of system robustness in the mid-to-high frequency (parasitic) range. Steps in this direction have recently been taken by Silva-Madriz and Sastry (1984) and O'Reilly (1985).

3 LINEAR FEEDBACK CONTROL

3.1 Introduction

Our attention in this chapter is focused on linear feedback design for deterministic linear time-invariant systems containing both slow and fast dynamic phenomena. Normally, any feedback design, like the system it seeks to control, will suffer from the higher dimensionality and ill-conditioning resulting from the interaction of slow and fast dynamic modes.

In the singular perturbation or, more generally, the two-time-scale approach to feedback design, these stiffness properties are taken advantage of by decomposing the original ill-conditioned system into two subsystems in separate time scales. State-feedback design may then proceed for each lower-order subsystem, and the results combined to yield a composite state-feedback control for the original system. At the same time, the composite controller is required to achieve an asymptotic approximation to the closed-loop system performance that would have been obtained had a state-feedback controller been designed without the use of singular perturbation methods.

This is the substance of the two-time-scale design approach of Section 3.2, adopted for eigenvalue assignment, and the optimal linear regulator in Sections 3.3 and 3.4. A feature of these composite state-feedback controllers is that they do not require knowledge of the singular perturbation parameter ε. Hence, they are applicable to systems where ε represents small uncertain parameters.

Should ε be known, closer "order of ε^2" approximations are obtained using composite state-feedback control designs based upon corrected slow and fast models. A related corrected linear–quadratic design is presented in Section 3.5. In Section 3.6, it is shown how two-time-scale methods of state-feedback control provide a convenient framework within which to approach the analysis and solution of high-gain feedback and cheap control problems.

Finally, the effect of designing feedback control strategies using a model that neglects the unknown parasitic elements is considered. It is shown in Section 3.2 that state-feedback controllers are robust in the sense of retaining overall system stability, provided that the disregarded fast modes are asymptotically stable. Unlike state feedback, it is of fundamental importance to note that, in general, no such robustness property holds for output-feedback controllers. In Section 3.7, however, conditions are identified whereby closed-loop instabilities resulting from output-feedback design are avoided.

3.2 Composite State-Feedback Control

It is desired to construct a state-feedback control for the singularly perturbed linear time-invariant system model analyzed in Section 2.6:

$$\dot{x} = A_{11}x + A_{12}z + B_1 u, \quad x(t_0) = x^0, \tag{2.1}$$

$$\varepsilon\dot{z} = A_{21}x + A_{22}z + B_2 u, \quad z(t_0) = z^0. \tag{2.2}$$

Preliminary to any separation of slow and fast designs, the system (2.1), (2.2) is approximately decomposed into a slow system model with n small eigenvalues and a fast system model with m large eigenvalues, as in Section 2.6. The nth-order slow system is

$$\dot{x}_s(t) = A_0 x_s(t) + B_0 u_s(t), \quad x_s(t_0) = x^0, \tag{2.3}$$

$$z_s(t) = -A_{22}^{-1}(A_{21}x_s(t) + B_2 u_s(t)), \tag{2.4}$$

where

$$A_0 \triangleq A_{11} - A_{12}A_{22}^{-1}A_{21}, \quad B_0 \triangleq B_1 - A_{12}A_{22}^{-1}B_2 \tag{2.5}$$

and the vectors x_s, z_s and u_s are the slow parts of the corresponding variables x, z and u in the original system (2.1), (2.2). Also, the mth-order fast system is

$$\varepsilon\dot{z}_f(t) = A_{22}z_f(t) + B_2 u_f(t), \quad z_f(t_0) = z^0 - z_s(t_0), \tag{2.6}$$

where $z_f = z - z_s$ and $u_f = u - u_s$ denote the fast parts of the corresponding variables in (2.1), (2.2).

It is appropriate to consider the following decomposition of feedback controls where

$$u_s = G_0 x_s, \quad u_f = G_2 z_f \tag{2.7}$$

are separately designed for the slow and fast systems (2.3) and (2.6). A *composite* control for the full system (2.1), (2.2) might then plausibly be taken as

$$u_s + u_f = G_0 x_s + G_2 z_f. \tag{2.8}$$

However, a realizable composite control requires that the system states x_s and z_f be expressed in terms of the actual system states x and z. This can be achieved by replacing x_s by x and z_f by $z - z_s$, so that the composite control (2.8), in view of (2.4), takes the realizable feedback form

$$\begin{aligned} u &= G_0 x + G_2[z + A_{22}^{-1}(A_{21}x + B_2 G_0 x)] \\ &= G_1 x + G_2 z, \end{aligned} \tag{2.9}$$

where

$$G_1 = (I_r + G_2 A_{22}^{-1} B_2) G_0 + G_2 A_{22}^{-1} A_{21}. \tag{2.10}$$

The above design procedure is a decomposed one in that the gain matrices G_0 and G_2 are separately designed according to slow and fast mode performance specifications, resulting in the physically realizable composite control law (2.9) as shown in Fig. 3.1. It remains to establish the asymptotic validity of this composite control (2.9), as $\varepsilon \to 0$, when applied to the system (2.1), (2.2). This we now do in the following theorem.

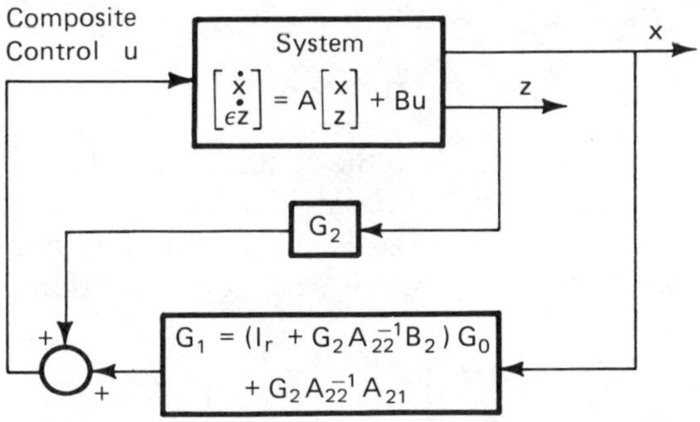

Fig. 3.1. Composite state feedback control.

Theorem 2.1

Let G_2 be designed such that $\text{Re}\,\lambda(A_{22} + B_2 G_2) < 0$. Then there exists an $\varepsilon^* > 0$ such that if the composite control

$$u = [(I_r + G_2 A_{22}^{-1} B_2) G_0 + G_2 A_{22}^{-1} A_{21}]x + G_2 z \tag{2.11}$$

is applied to the system (2.1), (2.2), the state and control of the resulting closed-loop system, starting from any bounded initial condition x^0 and z^0, are approximated according to

$$x(t) = x_s(t) + O(\varepsilon), \tag{2.12}$$

$$z(t) = -A_{22}^{-1}(A_{21} + B_2 G_0)x_s(t) + z_f(t) + O(\varepsilon), \tag{2.13}$$

$$u(t) = u_s(t) + u_f(t) + O(\varepsilon) \tag{2.14}$$

for all finite $t \geq t_0$ and all $\varepsilon \in (0, \varepsilon^*]$. If in addition G_0 is designed such that $\operatorname{Re}\lambda(A_0 + B_0 G_0) < 0$, there exists an $\varepsilon^* > 0$ such that the resulting closed-loop system is asymptotically stable and (2.12)–(2.14) hold for all $\varepsilon \in (0, \varepsilon^*]$ and $t \in [t_0, \infty)$.

Proof Consider a feedback control of the form (2.9) for any G_1, and not only G_1 given by (2.10), applied to the system (2.1)–(2.3). The closed-loop system may be written as

$$\begin{bmatrix} \dot{x} \\ \varepsilon\dot{z} \end{bmatrix} = \begin{bmatrix} F_{11} & F_{12} \\ F_{21} & F_{22} \end{bmatrix}\begin{bmatrix} x \\ z \end{bmatrix}, \quad \begin{bmatrix} x(t_0) \\ z(t_0) \end{bmatrix} = \begin{bmatrix} x^0 \\ z^0 \end{bmatrix}, \tag{2.15}$$

where

$$F_{11} = A_{11} + B_1 G_1, \quad F_{12} = A_{12} + B_1 G_2, \tag{2.16a}$$

$$F_{21} = A_{21} + B_2 G_1, \quad F_{22} = A_{22} + B_2 G_2. \tag{2.16b}$$

The closed-loop system (2.15) is in the standard singularly perturbed form where, as a Hurwitz matrix, F_{22} is nonsingular. Hence a complete separation of (2.15) into slow and fast systems is achieved by the nonsingular transformation into block-diagonal form introduced in Section 2.4:

$$\begin{bmatrix} \zeta \\ \eta \end{bmatrix} = \begin{bmatrix} I_n - \varepsilon HL & -\varepsilon H \\ L & I_m \end{bmatrix}\begin{bmatrix} x \\ z \end{bmatrix}, \quad \begin{bmatrix} x \\ z \end{bmatrix} = \begin{bmatrix} I_n & \varepsilon H \\ -L & I_m - \varepsilon LH \end{bmatrix}\begin{bmatrix} \zeta \\ \eta \end{bmatrix}, \tag{2.17}$$

where for $\varepsilon \in (0, \varepsilon^*]$, $\varepsilon^* > 0$, the matrices $L = L(\varepsilon)$ and $H = H(\varepsilon)$ respectively satisfy

$$0 = F_{22}L - \varepsilon LF_{11} + \varepsilon LF_{12}L - F_{21} = 0 \tag{2.18}$$

and

$$\varepsilon(F_{11} - F_{12}L)H - H(F_{22} + \varepsilon LF_{12}) + F_{12} = 0. \tag{2.19}$$

In the new coordinates, the exact system is

$$\begin{bmatrix} \dot{\zeta} \\ \varepsilon\dot{\eta} \end{bmatrix} = \begin{bmatrix} F_{11} - F_{12}L & 0 \\ 0 & F_{22} + \varepsilon L F_{12} \end{bmatrix} \begin{bmatrix} \zeta \\ \eta \end{bmatrix}, \tag{2.20}$$

with the initial conditions

$$\zeta(t_0) = x^0 - \varepsilon H \eta^0,$$
$$\eta(t_0) = z^0 + L x^0.$$

Instead of using G_1 and G_2 as design parameters, we use G_s and G_2 as design parameters with G_1 given by

$$G_1 = G_s + G_2 L. \tag{2.21}$$

Then substitution of (2.21) and (2.16) into the quadratic equation (2.18) results in the equation

$$A_{22}L - \varepsilon L(A_{11} + B_1 G_s) + \varepsilon L A_{12} L - (A_{21} + B_2 G_s) = 0. \tag{2.22}$$

It is observed that (2.22) is *independent* of the fast gain matrix G_2. Rearranging (2.22) as

$$L = A_{22}^{-1}(A_{21} + B_2 G_s) + \varepsilon A_{22}^{-1} L[(A_{11} + B_1 G_s) - A_{12} L] \tag{2.23}$$

and substituting (2.23) into (2.20), using (2.16) and (2.21), results in the exact closed-loop system

$$\begin{bmatrix} \dot{\zeta} \\ \varepsilon\dot{\eta} \end{bmatrix} = \begin{bmatrix} A_s + B_s G_s & 0 \\ 0 & A_f + B_f G_2 \end{bmatrix} \begin{bmatrix} \zeta \\ \eta \end{bmatrix}, \tag{2.24}$$

with the same initial conditions as (2.20), where

$$A_s \triangleq A_0 - \varepsilon A_{12} A_{22}^{-1} L(A_{11} - A_{12} L), \tag{2.25}$$

$$B_s \triangleq B_0 - \varepsilon A_{12} A_{22}^{-1} L B_1, \tag{2.26}$$

$$A_f \triangleq A_{22} + \varepsilon L A_{12}, \quad B_f \triangleq B_2 + \varepsilon L B_1. \tag{2.27}$$

Since L in (2.23)–(2.26) is independent of G_2, the matrix $A_s + B_s G_s$ is independent of G_2. The design parameters G_s and G_2 are chosen to shape the slow and fast models to the desired accuracy. For $O(\varepsilon)$ approximations let $G_s = G_0$ and approximate A_s and B_s by A_0 and B_0. Then G_0 is chosen to control the closed-loop matrix $A_0 + B_0 G_0$. Similarly, A_f and B_f are approximated by A_{22} and B_2, and G_2 is chosen to control the closed-loop matrix $A_{22} + B_2 G_2$. In calculating G_1 using (2.21), it suffices to use an $O(\varepsilon)$

approximation of G_1. Hence, by (2.23) G_1 is approximated by $G_0 + G_2 L_0$, where

$$L_0 \triangleq A_{22}^{-1}(A_{21} + B_2 G_0), \tag{2.28}$$

and this results in the expression (2.10) for G_1. When this value of G_1 is used and the analysis leading to (2.24) is applied to the closed-loop system, the resulting closed-loop slow model will be $O(\varepsilon)$ close to $A_0 + B_0 G_0$, and the resulting closed-loop fast model will be $O(\varepsilon)$ close to $A_{22} + B_2 G_2$. The remainder of the proof then follows from the stability of $F_{22} = A_{22} + B_2 G_2$, in complete analogy with that of Theorem 5.1 of Chapter 2. □

It is noted that an upper bound ε^* for ε can be obtained by applying Lemma 2.2 of Section 2.2 to the closed-loop system (2.15), where for A_{11} we write F_{11}, etc. The preceding two-time-scale design procedure has two important features: first, a reduction in computational requirements is achieved through solving two lower-order control problems in separate time scales; second, the resulting composite feedback control (2.11) does not require knowledge of the singular perturbation parameter ε, which may represent small uncertain parameters. Since the control of the fast system (2.6) occurs in an $1/\varepsilon$ faster time scale than that of the slow system (2.3), the response of the full system (2.1), (2.2) will be dominated by the reduced system state (2.3) after the decay of fast-system transients; that is, after an initial short boundary layer interval $[0, t_1]$, $z_f(t) \sim 0$ and $u_f(t) \sim 0$ in (2.12)–(2.14) for $t \geq t_1$.

Also, if A_{22} is a Hurwitz matrix, one need only stabilize the reduced system (2.3). This is simply effected by setting $G_2 = 0$ in (2.11) to obtain the following corollary to Theorem 2.1.

Corollary 2.1

If A_{22} is a Hurwitz matrix and the control

$$u = G_0 x \tag{2.29}$$

is applied to the system (2.1), (2.2), then there exists an $\varepsilon^* > 0$ such that the relations (2.12)–(2.14) hold for all finite $t \geq t_0$ and all $\varepsilon \in (0, \varepsilon^*]$. If G_0 is designed such that $A_0 + B_0 G_0$ is also Hurwitz, then there exists an $\varepsilon^* > 0$ such that the closed-loop system is asymptotically stable for all $\varepsilon \in (0, \varepsilon^*]$.

Corollary 2.1 is a *fundamental robustness result* in that it states that the system zero-order approximations of Theorem 2.1 remain valid for a control using feedback of the slow state $x(t)$ only, provided that the neglected fast modes are asymptotically stable. The corollary provides a theoretical justification for disregarding the parasitic elements of the full-system model

(2.1), (2.2) in the design of the control (2.29) by showing that, for sufficiently small ε, the actual performance of the closed-loop system is arbitrarily close to that predicted by the lower-order model (2.3)–(2.5): that is, the control strategy is robust with respect to the neglect of parasitic elements represented by the small scalar $\varepsilon > 0$. This is in contrast with the output-feedback control problem of Section 3.7, where we shall see that such a strategy may be *destabilizing* even when the neglected system modes are asymptotically stable.

In the same way, and for similar reasons, as in Section 2.5, *corrected* or $O(\varepsilon^2)$ approximations of the state trajectories for the system (2.1), (2.2) under linear state-feedback control (2.9) can be achieved through the use of separate corrected slow and fast designs. The key to this development, as in the uncorrected state-feedback design of Theorem 2.1, is the exact two-time-scale decomposition of the closed-loop system in (2.24).

For the corrected design, we choose

$$G_s = G_0 + \varepsilon G_0^1 \triangleq G_{0c}. \tag{2.30}$$

Then, by (2.30) and (2.23),

$$L = L_0 + O(\varepsilon) = L_1 + O(\varepsilon^2), \tag{2.31}$$

where

$$L_1 \triangleq A_{22}^{-1}(A_{21} + B_2 G_{0c}) + \varepsilon A_{22}^{-1} L_0 (A_0 + B_0 G_0). \tag{2.32}$$

Also, using (2.19) and (2.16), let us define

$$H_0 = (A_{12} + B_1 G_2)(A_{22} + B_2 G_2)^{-1}. \tag{2.33}$$

In analogy with (5.36) of Chapter 2, the corrected slow model is (cf. (2.25) and (2.26))

$$\dot{x}_{sc}(t) = A_{0c} x_{sc}(t) + B_{0c} u_{sc}(t), \tag{2.34}$$

$$A_{0c} \triangleq A_0 - \varepsilon A_{12} A_{22}^{-1} L_0 (A_{11} - A_{12} L_0), \tag{2.35}$$

$$B_{0c} \triangleq B_0 - \varepsilon A_{12} A_{22}^{-1} L_0 B_1, \tag{2.36}$$

with the initial condition

$$x_{sc}(t_0) = x^0 - \varepsilon H_0 (z^0 + L_0 x^0). \tag{2.37}$$

Similarly, in analogy with (5.37) of Chapter 2, the corrected fast model in the $\tau = (t - t_0)/\varepsilon$ scale is (cf. (2.27))

$$\frac{dz_{fc}(\tau)}{d\tau} = A_{22c} z_{fc}(\tau) + B_{2c} u_{fc}(\tau), \tag{2.38}$$

$$A_{22c} \triangleq A_{22} + \varepsilon L_0 A_{12}, \tag{2.39}$$

$$B_{2c} \triangleq B_2 + \varepsilon L_0 B_1, \tag{2.40}$$

with the initial condition

$$z_{fc}(0) = z^0 + L_1 x^0. \tag{2.41}$$

In analogy with the $O(\varepsilon)$ approximations of Theorem 2.1, $O(\varepsilon^2)$ approximations are obtained by way of a composite state-feedback controller based on separate lower-order corrected feedback designs

$$u_{sc} = G_{0c} x_{sc}, \quad u_{fc} = G_2 z_{fc} \tag{2.42}$$

for (2.34) and (2.38) in different time scales.

Theorem 2.2

If the composite control

$$u = [G_{0c} + G_2 L_1]x + G_2 z \tag{2.43}$$

is applied to the full system (2.1), (2.2) and if $A_{22} + B_2 G_2$ is a Hurwitz matrix, then there exists an $\varepsilon^* > 0$ such that the state and control of the resulting closed-loop system, starting from any bounded initial conditions x^0 and z^0 are approximated according to

$$x(t) = x_{sc}(t) + \varepsilon H_0 z_{fc}(\tau) + O(\varepsilon^2), \tag{2.44}$$

$$z(t) = -L_1 x_{sc}(t) + (I_m - \varepsilon L_0 H_0) z_{fc}(\tau) + O(\varepsilon^2), \tag{2.45}$$

$$u(t) = u_{sc}(t) + u_{fc}(\tau) + O(\varepsilon^2) \tag{2.46}$$

for all finite $t \geq t_0$ and all $\varepsilon \in (0, \varepsilon^*]$, where L_0, L_1 and H_0 are defined by (2.28), (2.32) and (2.33) respectively. If $A_0 + B_0 G_0$ is also Hurwitz then there exists an $\varepsilon^* > 0$ such that the closed-loop system is asymptotically stable and (2.44) –(2.46) hold for all $\varepsilon \in (0, \varepsilon^*)$ and $t \in [t_0, \infty)$.

Proof It is immediate from (2.25), (2.26) and (2.35), (2.36) that A_s and B_s are approximated to within $O(\varepsilon^2)$ according to

$$A_s = A_{0c} + O(\varepsilon^2), \quad B_s = B_{0c} + O(\varepsilon^2). \tag{2.47}$$

Similarly, by (2.27) and (2.39), (2.40),

$$A_f = A_{22c} + O(\varepsilon^2), \quad B_f = B_{2c} + O(\varepsilon^2). \tag{2.48}$$

The gain matrices G_{0c} and G_2 are designed to control $A_{0c} + B_{0c} G_{0c}$ and $A_{22c} + B_{2c} G_2$ respectively, where, by (2.30), (2.47) and (2.48),

$$A_s + B_s G_s = A_{0c} + B_{0c} G_{0c} + O(\varepsilon^2) \tag{2.49}$$

and

$$A_{\mathrm{f}} + B_{\mathrm{f}}G_2 = A_{22\mathrm{c}} + B_{2\mathrm{c}}G_2 + O(\varepsilon^2). \tag{2.50}$$

Also, substituting (2.30) in (2.21), we have the composite control (2.43), where

$$G_1 = G_{0\mathrm{c}} + G_2 L_1 + O(\varepsilon^2). \tag{2.51}$$

Since $A_{22} + B_2 G_2$ is Hurwitz, the remainder of the proof proceeds in analogy with that of Theorem 5.2 of Chapter 2. □

Similar to Corollary 2.1, one has the following useful robustness result, this time using a corrected reduced control.

Corollary 2.2

If A_{22} is a Hurwitz matrix, and the control

$$u = G_{0\mathrm{c}} x \tag{2.52}$$

is applied to the full system (2.1), (2.2), then there exists an $\varepsilon^* > 0$ such that the relations (2.44)–(2.46) hold for all finite $t \geq t_0$ and all $\varepsilon \in (0, \varepsilon^*]$. If $A + B_0 G_0$ is also Hurwitz, then there exists an $\varepsilon^* > 0$ such that the closed-loop system is asymptotically stable and (2.44)–(2.46) hold for all $\varepsilon \in (0, \varepsilon^*]$ and $t \in [t_0, \infty)$.

Another design possibility is a composite control law based upon a corrected slow design and an uncorrected fast design. All feedback control strategies, involving corrected slow designs, yield the same $O(\varepsilon^2)$ approximation after the decay of fast-system transients in the initial short boundary layer interval.

It is observed in Theorem 2.2 and Corollary 2.2 that the corrected system models (2.34) and (2.38) depend on the uncorrected slow gain matrix G_0. This disadvantage is not as grave as it might appear, since it is reasonable to first assess the uncorrected state approximations of Theorem 2.1 or Corollary 2.1 before proceeding to Theorem 2.2 or Corollary 2.2. By construction in (2.30), the corrected slow gain $G_{0\mathrm{c}}$ is related to the uncorrected one G_0 by $G_{0\mathrm{c}}(\varepsilon) = G_0 + \varepsilon G_0'$ for some matrix G_0' that is $O(1)$. When the design problem has more than one solution, as in the multi-input eigenvalue assignment problem of Section 3.3, a solution for the uncorrected design G_0 may not be the limit of the solution for the corrected design $G_{0\mathrm{c}}$ unless the designer carefully picks $G_{0\mathrm{c}}$ such that $\lim_{\varepsilon \to 0} G_{0\mathrm{c}}(\varepsilon) = G_0$.

Corollary 2.3

If A_{22} is nonsingular, the exact slow model (A_s, B_s) of (2.24) is invariant with respect to the class of fast feedback controls $u = v + G_2 z$.

Proof It is sufficient to note that for the exact slow model pair (A_s, B_s) in (2.24) the solution $L(\varepsilon)$ of the quadratic equation (2.22) or (2.23) does not depend on the fast feedback gain matrix G_2 of the original linear feedback control law (2.9), applied prior to L-H transformation. □

It follows from Corollary 2.3 that not only are the slow designs independent of fast feedback design, but the controllability properties of the uncorrected and corrected slow models are unaffected by fast state feedback.

3.3 Eigenvalue Assignment

We observed in Theorem 6.1 of Chapter 2 that if the eigenvalues of the slow system (2.3) and the eigenvalues of the fast system (2.6) are controllable, then the corresponding eigenvalues of the original system (2.1), (2.2) are controllable for ε sufficiently small. Given the well known fact (Kwakernaak and Sivan 1972) that under linear state feedback, controllability is equivalent to the arbitrary assignability of the eigenvalues of the closed-loop system matrix, it follows that the separate stabilization of the systems (2.3) and (2.6) implies the stabilization of (2.1), (2.2).

Suppose it is required to assign $n + m$ closed-loop eigenvalues of the system (2.1), (2.2) to arbitrary positions $\{\lambda_1^c, \lambda_2^c, \ldots, \lambda_{n+m}^c\}$ in the open left-half complex plane by way of the linear feedback control (2.9). Given the separation in the open-loop eigenvalue spectrum of (2.1), (2.2), it is natural to assign n small eigenvalues to the slow system and to assign m large eigenvalues to the fast system.

Theorem 3.1

If A_{22}^{-1} exists and if the slow system pair (A_0, B_0) and the fast system pair (A_{22}, B_2) are each controllable, and G_0 and G_2 are designed to assign distinct eigenvalues λ_i, $i = 1, \ldots, n$ and λ_j, $j = 1, \ldots, m$, to the matrices $A_0 + B_0 G_0$ and $A_{22} + B_2 G_2$ respectively, then there exists an $\varepsilon^* > 0$ such that for all $\varepsilon \in (0, \varepsilon^*]$ the application of the composite feedback control

$$u = [(I_r + G_2 A_{22}^{-1} B_2) G_0 + G_2 A_{22}^{-1} A_{21}] x + G_2 z \tag{3.1}$$

to the system (2.1), (2.2) results in a closed-loop system containing n small eigenvalues $\{\lambda_1^c, \lambda_2^c, \ldots, \lambda_n^c\}$ and m large eigenvalues $\lambda_{n+1}^c, \ldots, \lambda_{n+m}^c\}$, which are approximated by

$$\lambda_i^c = \lambda_i(A_0 + B_0 G_0) + O(\varepsilon), \quad i = 1, \ldots, n, \tag{3.2}$$

$$\lambda_i^c = [\lambda_j(A_{22} + B_2 G_2) + O(\varepsilon)]/\varepsilon, \quad i = n + j, \quad j = 1, \ldots, m. \tag{3.3}$$

Proof Complete controllability of the pairs (A_0, B_0) and (A_{22}, B_2) implies the existence of gain matrices G_0 and G_2 which arbitrarily assign corresponding eigenvalues to the matrices $A_0 + B_0G_0$ and $A_{22} + B_2G_2$ respectively (Kwakernaak and Sivan 1972). Since the closed-loop eigenvalues λ_i^c, $i = 1, \ldots, n + m$, are precisely those of (2.24), the proof follows from the fact that $A_0 + B_0G_0$ and $A_{22} + B_2G_2$ are $O(\varepsilon)$ regular perturbations of those of (2.24) and the application of Theorem 3.1 of Chapter 2. □

The above separation property suggests the following two-time-scale design procedure for an $O(\varepsilon)$ approximate eigenvalue assignment of the singularly perturbed linear system (2.1), (2.2): construct G_0 so as to place the n small eigenvalues of $A_0 + B_0G_0$; separately construct G_2 so as to place the m large eigenvalues of $A_{22} + B_2G_2$; then form the composite feedback control (3.1). It is important to observe that the success of the composite control idea, here applied to eigenvalue assignment, hinges on the implicit requirement that the design objective be compatible with the slow–fast nature of the open-loop system (2.1), (2.2). In other words, the design objective should allow $O(1)$ feedback to be applied in such a way as to keep what is slow slow and likewise what is fast fast.

In the light of Theorem 3.1 of Chapter 2, should one remove the assumption that the assigned eigenvalues $\lambda_i(A_0 + B_0G_0)$ are distinct and the assigned eigenvalues $\lambda_j(A_{22} + B_2G_2)$ are distinct, the result (3.2), (3.3) of Theorem 3.1 is replaced by the weaker one: namely, as $\varepsilon \to 0$, the first n eigenvalues of the closed-loop system tend to fixed positions in the complex plane defined by $\lambda_i(A_0 + B_0G_0)$ while the remaining m eigenvalues of the closed-loop system tend to infinity with the rate $1/\varepsilon$ along asymptotes defined by $\lambda_j(A_{22} + B_2G_2)$, that is to $[\lambda_j(A_{22} + B_2G_2)]/\varepsilon$.

As an illustration of the preceding two-time-scale design procedure, consider the following example.

Example 3.1

Recall for the DC motor of Case Study 7.1 in Chapter 1 that the normalized motor state equations are

$$\begin{bmatrix} \dot{x} \\ \varepsilon\dot{z} \end{bmatrix} = \begin{bmatrix} 0 & 1 \\ -1 & -1 \end{bmatrix} \begin{bmatrix} x \\ z \end{bmatrix} + \begin{bmatrix} 0 \\ 1 \end{bmatrix} u, \tag{3.4}$$

where x, z and u are the deviations in angular velocity, armature current and armature voltage from their respective nominal values. Suppose that the desired eigenvalue spectrum to be assigned is

$$\{-3, -2/\varepsilon\}. \tag{3.5}$$

We consider the two-time-scale feedback design of Theorem 3.1. Since the slow system pair $(A_0, B_0) = (-1, 1)$ and the fast system pair $(A_{22}, B_2) = (-1, 1)$ are each completely controllable, the choice of $G_0 = -2$ places the eigenvalue of $A_0 + B_0 G_0$ at -3, while the choice of $G_2 = -1$ places the eigenvalue of $A_{22} + B_2 G_2$ at -2. The ensuing composite control, by (3.1), is

$$u_c = -5x - z, \tag{3.6}$$

and results in a closed-loop system with the characteristic equation

$$\lambda^2 + \frac{2}{\varepsilon}\lambda + \frac{6}{\varepsilon} = 0. \tag{3.7}$$

Using the frequency-scale analysis of polynomials introduced in Section 2.3, we have that, in the slow time-scale, the characteristic equation (3.17) may be rewritten as

$$\varepsilon\lambda^2 + 2\lambda + 6 = 0. \tag{3.8}$$

As $\varepsilon \to 0$ the characteristic equation (3.8) tends to $2\lambda + 6$, associated with the desired slow eigenvalue

$$\lambda = -3. \tag{3.9}$$

In the fast frequency-scale let $p = \varepsilon\lambda$. Then the characteristic equation (3.8) may be rewritten as

$$p^2 + 2p + 6\varepsilon = 0. \tag{3.10}$$

As $\varepsilon \to 0$, the characteristic equation (3.10) tends to $p^2 + 2p = p(p + 2) = 0$, the large root of which, $p = -2$, yields the desired fast eigenvalue

$$\lambda = -2/\varepsilon. \tag{3.11}$$

It is of interest to observe that the state feedback controller that exactly assigns the desired spectrum (3.5) or the characteristic equation

$$(\lambda + 3)\left(\lambda + \frac{2}{\varepsilon}\right) = \lambda^2 + \left(3 + \frac{2}{\varepsilon}\right)\lambda + \frac{6}{\varepsilon} = 0 \tag{3.12}$$

is given by the ε-dependent control law

$$u(\varepsilon) = -5x - (1 + 3\varepsilon)z. \tag{3.13}$$

Finally, noting that the fast mode is stable, and experiences an exponential decay of the order of $e^{-t/\varepsilon}$, the reduced control, based upon the slow model $(A_0, B_0) = (-1, 1)$, is robust and takes the form

$$u_r = -2x, \tag{3.14}$$

which results in a stable closed-loop system with the characteristic equation

$$\lambda^2 + \frac{1}{\varepsilon}\lambda + \frac{3}{\varepsilon} = 0. \tag{3.15}$$

As a check, let us perform a frequency analysis of the characteristic equation (3.15) along the lines of (3.8)–(3.11). In the slow frequency-scale, the characteristic equation (3.15) or $\varepsilon\lambda^2 + \lambda + 3 = 0$ tends to $\lambda + 3$, as $\varepsilon \to 0$, which is associated with the desired slow eigenvalue $\lambda = -3$. Also, in the fast frequency scale let $p = \varepsilon\lambda$, so that (3.15) may be rewritten as

$$p^2 + p + 3\varepsilon = 0. \tag{3.16}$$

As $\varepsilon \to 0$ the characteristic equation (3.16) tends to $p^2 + p = p(p + 1) = 0$, the nonzero root of which, $p = -1$, yields the desired fast eigenvalue $\lambda = -1/\varepsilon$. Observe that in the limit as $\varepsilon \to 0$ the fast eigenvalue is not shifted by the reduced control (3.14), but remains at its open-loop value of $\lambda = -1/\varepsilon$.

A closer placement of the eigenvalues of the system (2.1), (2.2), to within $O(\varepsilon^2)$ of desired closed-loop values, is achieved by way of the corrected composite control (2.43) based on eigenvalue placements for the corrected slow and fast models (2.34) and (2.38) in separate time scales.

Theorem 3.2

If A_{22}^{-1} exists and if the slow system pair (A_0, B_0) and the fast system pair (A_{22}, B_2) are each controllable then there exists an $\varepsilon^* > 0$ and gain matrices G_{0c} and G_2 which arbitrarily assign distinct eigenvalues λ_i, $i = 1, \ldots, n$ and λ_j, $j = 1, \ldots, m$ to the matrices $A_{0c} + B_{0c}G_{0c}$ and $A_{22c} + B_{2c}G_2$, respectively, such that for all $\varepsilon \in (0, \varepsilon^*]$ the application of the composite feedback control

$$u = [G_{0c} + G_2L_1]x + G_2z, \tag{3.17}$$

where

$$L_1 = A_{22}^{-1}(A_{21} + B_2 G_{0c}) + \varepsilon A_{22}^{-2}(A_{21} + B_2 G_0)(A_0 + B_0 G_0), \quad (3.18)$$

to the system (2.1), (2.2) results in a closed-loop system containing n small eigenvalues $\{\lambda_1^c, \lambda_2^c, \ldots, \lambda_n^c\}$ and m large eigenvalues $\{\lambda_{n+1}^c, \ldots, \lambda_{n+m}^c\}$, which are approximated by

$$\lambda_i^c = \lambda_i(A_{0c} + B_{0c} G_{0c}) + O(\varepsilon^2), \quad i = 1, \ldots, n \quad (3.19)$$

$$\lambda_i^c = [\lambda_j(A_{22c} + B_{2c} G_2) + O(\varepsilon^2)]/\varepsilon, \quad i = n + j, \quad j = 1, \ldots, m. \quad (3.20)$$

Proof The proof is similar to that of Theorem 3.1. As in the proof of Theorem 6.1 of Chapter 2, (A_0, B_0) and (A_{22}, B_2) controllable imply that (A_{0c}, B_{0c}) and (A_{22c}, B_{2c}) respectively are controllable for $\varepsilon \in [0, \varepsilon^*]$. Thus there exist gain matrices G_{0c} and G_{2c} which arbitrarily assign eigenvalues to $A_{0c} + B_{0c}G_{0c}$ and $A_{22c} + B_{2c}G_2$ respectively. The proof then follows from (2.24) and the application of equations (3.18) and (3.19) of Chapter 2. □

Again, as in Theorem 2.2, we note the dependence of the corrected slow and fast models (2.34) and (2.38) on the uncorrected slow gain matrix G_0. Thus, the price of achieving a closer eigenvalue assignment than in Theorem 3.1 is that it is necessary to first solve an uncorrected slow eigenvalue problem to obtain G_0. This disadvantage is mitigated by the fact that an uncorrected slow eigenvalue assignment will usually be attempted before proceeding to the corrected design. As remarked in Section 3.2, it is important to choose G_{0c} such that $\lim_{\varepsilon \to 0} G_{0c}(\varepsilon) = G_0$. An illustration of the approximation in eigenvalue assignment and the resulting state trajectories achieved using both uncorrected and corrected composite control is provided by the next example.

Example 3.2

Consider the two-tank chemical reactor system referred to in Example 6.1 of Chapter 2, where it is required to maintain the concentration of liquid in the second tank at a desired level, in spite of variation of inlet concentration to the first tank, by the addition of reactant through a control valve.

A linearized model of the control scheme is depicted in Fig. 3.2, where deviations from desired steady-state values are described by

$C_r = 0$ = deviation in desired concentration,
x_1 = deviation in concentration of second tank,

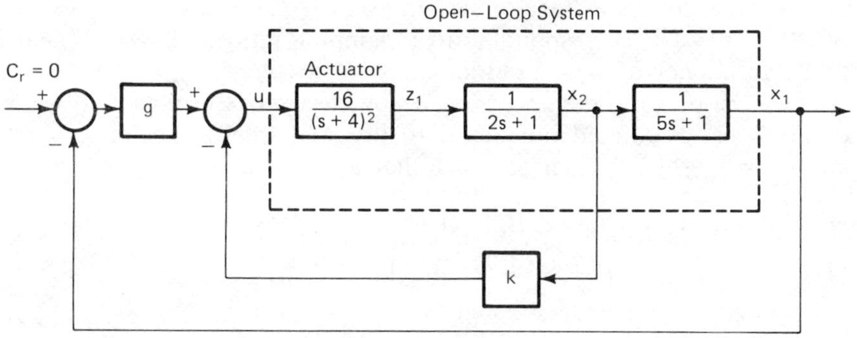

Fig. 3.2. Two-tank chemical reactor system.

x_2 = deviation in concentration of first tank,
u = actuator control value signal.

By inspection, the open-loop system has four poles at $\{-\frac{1}{5}, -\frac{1}{2}, -4, -4\}$, of which the last two are associated with the actuator dynamics. The actuator block in Fig. 3.2 may be represented, through scaling the complex frequency s by $\varepsilon = \frac{1}{4}$, as

$$\frac{1}{(\varepsilon s)^2 + 2(\varepsilon s) + 1},$$

so that the complete system model takes the singularly perturbed form

$$\begin{bmatrix} \dot{x}(t) \\ \varepsilon\dot{z}(t) \end{bmatrix} = \left[\begin{array}{cc|cc} -\frac{1}{5} & \frac{1}{5} & 0 & 0 \\ 0 & -\frac{1}{2} & \frac{1}{2} & 0 \\ \hline 0 & 0 & 0 & 1 \\ 0 & 0 & -1 & -2 \end{array}\right] \begin{bmatrix} x(t) \\ z(t) \end{bmatrix} + \left[\begin{array}{c} 0 \\ 0 \\ \hline 0 \\ 1 \end{array}\right] u, \tag{3.21}$$

with the initial condition

$$\begin{bmatrix} x(0) \\ z(0) \end{bmatrix} = \begin{bmatrix} x_1(0) \\ x_2(0) \\ z_1(0) \\ z_2(0) \end{bmatrix} = \begin{bmatrix} 2.5 \\ 2.0 \\ 0 \\ 0 \end{bmatrix}, \tag{3.22}$$

where $x \in R^2$ and $z \in R^2$ respectively describe the slow reactor and fast actuator dynamics. Since the actuator poles $\{-4, -4\}$ are stable, or equivalently, from Example 6.1 of Chapter 2, A_{22} is Hurwitz, the control problem is approached as one of designing the slow state-feedback control ($G_2 = 0$)

$$u(t) = G_1 x(t), \quad G_1 = [-g \quad -k] \tag{3.23}$$

such that the dominant reactor system poles $\{-\frac{1}{5}, -\frac{1}{2}\}$ are shifted to near $\{-0.707 \pm j0.707\}$, corresponding to a damping ratio of $\zeta = 0.707$ and a natural frequency of $\omega_n = 1$ rad/s.

Let us first perform the uncorrected slow design of Corollary 2.1, where A_0 and B_0 are as defined in equation (6.18) of Chapter 2. The uncorrected gain, $G_0 = [-g_0 \quad -k_0]$, chosen such that

$$\begin{aligned}
&\det[sI_2 - (A_0 + B_0G_0)] \\
&= s^2 + [\tfrac{1}{5} + \tfrac{1}{2}(1 + k_0)]s + [\tfrac{1}{10}(1 + g_0 + k_0)] \\
&= s^2 + 2\zeta\omega_n s + \omega_n^2,
\end{aligned} \tag{3.24}$$

where $\zeta = 0.707$ and $\omega_n = 1$ rad/s, is given by

$$G_0 = [-7.572 \quad -1.428]. \tag{3.25}$$

The eigenvalues of the actual full system (3.21) under this feedback control (3.23), (3.25) are

$$\{-2.224, -5.366, -0.5551 \pm j1.0162\}, \tag{3.26}$$

of which the slow eigenvalues are observed to be within $O(\frac{1}{4})$ of the desired eigenvalues $\{-0.707 \pm j0.707\}$. The corresponding slow states x_1 and x_2 of the original system (3.21) under feedback control (3.23), (3.25) are compared with those of the slow closed-loop system (2.3), (3.25) or

$$\dot{x}_s = \begin{bmatrix} -0.2 & 0.2 \\ -3.786 & -1.214 \end{bmatrix} x_s, \quad x_s(0) = \begin{bmatrix} 2.5 \\ 2.0 \end{bmatrix} \tag{3.27}$$

in Fig. 3.3.

For a closer slow eigenvalue assignment and closed-loop system response, we proceed to the corrected slow design of Corollary 2.2 for the slow model pair (A_{0c}, B_{0c}) of (2.35), (2.36) or

$$A_{0c} = \begin{bmatrix} -0.2 & 0.2 \\ -1.7302 & -0.5548 \end{bmatrix}, \quad B_{0c} = \begin{bmatrix} 0 \\ 0.5 \end{bmatrix}. \tag{3.28}$$

The corrected gain G_{0c}, chosen such that

$$\det[sI_2 - (A_{0c} + B_{0c}G_{0c})] = s^2 + 2\zeta\omega_n s + \omega_n^2 = 0 \tag{3.29}$$

for the design values of $\zeta = 0.707$ and $\omega_n = 1$ rad/s, is

$$G_{0c} = [-4.116 \quad -1.3184]. \tag{3.30}$$

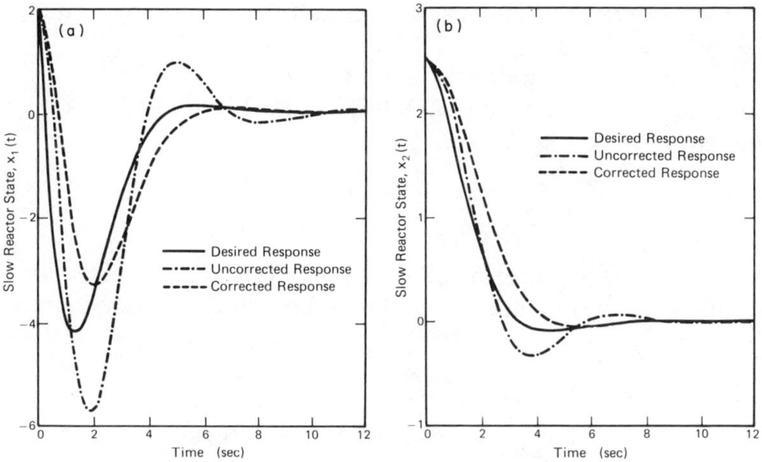

Fig. 3.3. (a), (b) Reactor system response.

The eigenvalues of the actual full system (3.21) under this feedback control (3.23), (3.30) are

$$\{-1.7576, -5.3789, -0.7817 \pm j0.6907\}, \tag{3.31}$$

of which the slow eigenvalues are this time within $O((\frac{1}{4})^2)$ of the desired eigenvalues $\{-0.707 \pm j0.707\}$. The corresponding responses of the slow states of the original system (3.21) under corrected feedback control (3.23), (3.30) are compared with those of the uncorrected design (3.21), (3.23), (3.25), and with those of the desired slow closed-loop system (3.27) in Fig. 3.3.

Example 3.3

As an illustration of weak controllability, consider again the state realization (b) of Example 6.2 of Chapter 2, where it is desired to shift the eigenvalues $\{-1, -1/\varepsilon\}$ of the open-loop system

$$\begin{bmatrix} \dot{x} \\ \varepsilon\dot{z} \end{bmatrix} = \begin{bmatrix} -1 & 0 \\ 1 & -1 \end{bmatrix} \begin{bmatrix} x \\ z \end{bmatrix} + \begin{bmatrix} 1 \\ 0 \end{bmatrix} u \tag{3.32}$$

to the positions $\{-2, -2/\varepsilon\}$ of the complex plane. Although the system (3.32) is completely controllable for $\varepsilon > 0$, its fast eigenvalue is but weakly controllable, as noted in Example 6.2. Consequently, although a reduced slow design is possible, the composite design procedure of Theorem 3.1

cannot apply because of our inability to shift the "fast" eigenvalue -1 in τ-scale.

On the other hand, the linear state feedback that assigns the desired characteristic polynomial $(\lambda + 2)(\lambda + 2/\varepsilon)$ exactly, for $\varepsilon > 0$, is given by the high-gain control

$$u(\varepsilon) = -\left(1 + \frac{1}{\varepsilon}\right)x + \left(-2 + \frac{1}{\varepsilon}\right)z. \tag{3.33}$$

The gains in (3.33) are large on account of the fact that the control u needs to force its effect on the fast transient of z through the predominantly slow state x of (3.32).

3.4 Near-Optimal Regulators

The infinite-time optimal linear regulator problem for the linear time-invariant system

$$\begin{bmatrix} \dot{x} \\ \varepsilon\dot{z} \end{bmatrix} = \begin{bmatrix} A_{11} & A_{12} \\ A_{21} & A_{22} \end{bmatrix}\begin{bmatrix} x \\ z \end{bmatrix} + \begin{bmatrix} B_1 \\ B_2 \end{bmatrix}u, \quad \begin{bmatrix} x(0) \\ z(0) \end{bmatrix} = \begin{bmatrix} x^0 \\ z^0 \end{bmatrix}, \tag{4.1}$$

$$y = [M_1 \quad M_2]\begin{bmatrix} x \\ z \end{bmatrix}, \tag{4.2}$$

where $x \in R^n$, $z \in R^m$, $u \in R^r$ and $y \in R^p$, is defined as the problem of finding a control $u(t) \in R^r$, $t \in [0, \infty]$, so as to regulate the state $\left[\begin{smallmatrix} x(t) \\ z(t) \end{smallmatrix}\right]$ to the origin by way of minimizing the quadratic performance index

$$J = \frac{1}{2}\int_0^\infty (y^{\mathrm{T}}y + u^{\mathrm{T}}Ru)\,\mathrm{d}t, \quad R > 0, \tag{4.3}$$

where the $r \times r$ weighting matrix R penalizes excessive values of control $u(t)$.

The solution of the linear quadratic regulator problem (4.1)–(4.3) is the optimal linear feedback control law (Kwakernaak and Sivan 1972)

$$u_{\mathrm{opt}} = -R^{-1}B^{\mathrm{T}}K(\varepsilon)\begin{bmatrix} x \\ z \end{bmatrix} = G(\varepsilon)\begin{bmatrix} x \\ z \end{bmatrix}, \tag{4.4}$$

where $K = K(\varepsilon)$ satisfies the algebraic matrix Riccati equation

$$0 = -KA - A^{\mathrm{T}}K + KBR^{-1}B^{\mathrm{T}}K - M^{\mathrm{T}}M \tag{4.5}$$

with

$$A = \begin{bmatrix} A_{11} & A_{12} \\ \dfrac{A_{21}}{\varepsilon} & \dfrac{A_{22}}{\varepsilon} \end{bmatrix}, \quad B = \begin{bmatrix} B_1 \\ \dfrac{B_2}{\varepsilon} \end{bmatrix}, \tag{4.6}$$

$$M = [M_1 \quad M_2]. \tag{4.7}$$

Moreover, the optimal cost is given by

$$J_{\text{opt}} = \tfrac{1}{2}[x^{\text{T}}(0) \quad z^{\text{T}}(0)]K \begin{bmatrix} x(0) \\ z(0) \end{bmatrix}. \tag{4.8}$$

Conditions providing for the existence of a solution (4.4)–(4.8) to the optimal linear regulator problem (4.1)–(4.3) are as follows.

Theorem 4.1

If the slow system triple (M_0, A_0, B_0) and the fast system triple (M_2, A_{22}, B_2) are each stabilizable and detectable [observable] then there exists an $\varepsilon^* > 0$ such that the original optimal linear regulator problem (4.1)–(4.3) has a solution (4.4)–(4.8) where the $(n + m) \times (n + m)$ matrix $K = K(\varepsilon)$ exists as the unique positive-semidefinite [positive-definite] stabilizing solution of the algebraic Riccati equation (4.5) for all $\varepsilon \in (0, \varepsilon^*]$.

Proof By Corollaries 6.1 and 6.2 of Chapter 2, the conditions of Theorem 4.1 imply that the full system triple (M, A, B) is stabilizable and detectable [observable]. The theorem then immediately follows from standard linear regulator theory (Kwakernaak and Sivan 1972). □

In Theorem 4.1 advantage has been taken of the singularly perturbed nature of the system (4.1), (4.2) to give existence conditions for a solution to the full regulator problem in terms of stabilizability–detectability conditions on lower-order systems, (M_0, A_0, B_0) and (M_2, A_{22}, B_2), in separate time-scales. It is natural to suspect that the original regulator problem (4.1)–(4.3) itself can be decomposed into two separate regulator problems: one for the slow system (M_0, A_0, B_0) of (2.3), and the other for the fast system (M_2, A_{22}, B_2) of (2.6). Indeed, Theorem 2.1 suggests that from the system optimal controls u_{s} and u_{f} a composite control

$$u_{\text{c}} = u_{\text{s}} + u_{\text{f}} \tag{4.9}$$

can be formed for implementation on the original system (4.1), (4.2).

First, let us see how the original regulator problem (4.1)–(4.3) may be decomposed into two regulator problems in separate time scales. Equation (4.2) may be decomposed, using (2.12) and (2.13), into a slow part and a fast part; that is,

$$\begin{aligned} y &= M_1 x + M_2 z \\ &= M_1[x_s + O(\varepsilon)] + M_2[-A_{22}^{-1}(A_{21}x_s + B_2 u_s) + z_f + O(\varepsilon)] \\ &= y_s(t) + y_f(t) + O(\varepsilon), \end{aligned} \tag{4.10}$$

where, separating slow and fast variables, the output

$$y_s = M_0 x_s + N_0 u_s, \tag{4.11}$$

with

$$M_0 \triangleq M_1 - M_2 A_{22}^{-1} A_{21}, \quad N_0 \triangleq -M_2 A_{22}^{-1} B_2, \tag{4.12}$$

is associated with the slow system (2.3), and the output

$$y_f(t) = M_2 z_f \tag{4.13}$$

is associated with the fast system (2.6).

Now, after the decay of fast-system transients in the initial short boundary layer interval $[0, t_1]$, we have from (2.14) and (4.10) that $u(t) \sim u_s(t)$ and $y(t) \sim y_s(t)$ for $t \geq t_1$. By the same token, for $t \geq t_1$ or in the slow time-scale, the original regulator problem (4.1)–(4.3) is described by the following one.

Slow regulator problem

Find u_s so as to minimize

$$J_s = \frac{1}{2}\int_0^\infty (y_s^T y_s + u_s^T R u_s)\, dt, \quad R > 0, \tag{4.14}$$

for the slow system (2.3), (4.11).

In terms of x_s and u_s, (4.14) becomes

$$J_s = \frac{1}{2}\int_0^\infty [x_s^T M_0^T M_0 x_s + 2u_s^T N_0^T x_s + u_s^T R_0 u_s]\, dt \tag{4.15}$$

where

$$R_0 = R + N_0^T N_0. \tag{4.16}$$

Lemma 4.1

If the triple (M_0, A_0, B_0) is stabilizable and detectable [observable] then there exists a unique positive-semidefinite [positive-definite] stabilizing $n \times n$ solution K_s of the algebraic matrix Riccati equation

$$0 = -K_s(A_0 - B_0R_0^{-1}N_0^TM_0) - (A_0 - B_0R_0^{-1}N_0^TM_0)^TK_s + K_sB_0R_0^{-1}B_0^TK_s - M_0^T(I - N_0R_0^{-1}N_0^T)M_0, \qquad (4.17)$$

and the optimal control for (2.3) and (4.14) is

$$u_s = -R_0^{-1}(N_0^TM_0 + B_0^TK_s)x_s = G_0x_s. \qquad (4.18)$$

Proof It is well known (Anderson and Moore 1971) that if the triple $(\hat{M}_0, A_0 - B_0R_0^{-1}N_0^TM_0, B_0)$ is stabilizable and detectable [observable], where $\hat{M}_0^T\hat{M}_0 = M_0^T(I - N_0^TR_0^{-1}N_0^T)M_0$, there exists a unique positive-semi-definite [positive-definite] stabilizing solution K. Since stabilizability is unaffected by linear state feedback, the pair $(A_0 - B_0R_0^{-1}N_0^TM_0, B_0)$ is stabilizable. From the matrix identity, proved by multiplying both sides by $I + N_0R^{-1}N_0^T$,

$$I - N_0R_0^{-1}N_0^T = (I + N_0R^{-1}N_0^T)^{-1} > 0 \qquad (4.19)$$

it follows that there exists a nonsingular Q_0 such that $Q_0^TQ_0 = I - N_0R_0^{-1}N_0^T$. Hence, as a dual of the preceding stabilizability result, the pair $(A_0 - B_0R_0^{-1}N_0^TM_0, Q_0M_0)$ is detectable if and only if the pair (M_0, A_0) is detectable. Thus the stabilizability and detectability [observability] of the triple $(\hat{M}_0, A_0 - B_0R_0^{-1}N_0^TM_0, B_0)$ is equivalent to that of the triple (M_0, A_0, B_0). □

In the fast time scale $\tau = t/\varepsilon$ only the fast variations $z_f(\tau)$, $u_f(\tau)$ and $y_f(\tau)$ of (2.13), (2.14) and (4.10) are active; x_s, u_s and y_s are constant during the boundary layer interval $[0, t_1]$. Thus, in the fast time scale, the original regulator problem (4.1), (4.3) is described by the following one.

Fast regulator problem

Find u_f so as to minimize

$$J_f = \frac{1}{2}\int_0^\infty (y_f^Ty_f + u_f^TRu_f)\,dt, \quad R > 0, \qquad (4.20)$$

for the fast system (2.6), (4.13).

Lemma 4.2

If the triple (M_2, A_{22}, B_2) is stabilizable and detectable [observable] then there exists a unique positive-semidefinite [positive-definite] stabilizing $m \times m$ solution K_f of the algebraic matrix Riccati equation

$$0 = -K_f A_{22} - A_{22}^T K_f + K_f B_2 R^{-1} B_2^T K_f - M_2^T M_2, \tag{4.21}$$

and the optimal control for (2.6), (4.13) and (4.20) is

$$u_f = -R^{-1} B_2^T K_f z_f = G_2^{(0)} z_f. \tag{4.22}$$

Proof The lemma is a special case of Lemma 4.1. □

The controls u_s and u_f, defined by (4.18) and (4.22), are optimal only for the slow and fast systems. According to (2.11), (4.18) and (4.22), a composite control u_c, operating upon the actual states x and z, is given by

$$\begin{aligned} u_c = -[(I_r - R^{-1} B_2^T K_f A_{22}^{-1} B_2) R_0^{-1} (N_0^T M_0 + B_0^T K_s) \\ + R^{-1} B_2^T K_f A_{22}^{-1} A_{21}] x - R^{-1} B_2^T K_f z. \end{aligned} \tag{4.23}$$

Again, the composite control (4.23) does not explicitly depend upon ε, which might represent small unknown parameters. Given that the gains G_0 and $G_2^{(0)}$ are only optimal for their respective slow and fast regulator problems, the question arises as to the optimality or near-optimality of the composite feedback control law (2.11) or (4.23).

3.4.1. Near-Optimality of the Composite Control

In order to compare the composite control u_c with the exact optimal control (4.4), it is convenient to express u_c in a more suggestive form.

Lemma 4.3

The composite control (4.23) is identical with the composite control

$$u_c = -R^{-1} B^T \begin{bmatrix} K_s & 0 \\ \varepsilon K_m^T & \varepsilon K_f \end{bmatrix} \begin{bmatrix} x \\ z \end{bmatrix} = -R^{-1} B^T M_c \begin{bmatrix} x \\ z \end{bmatrix} \tag{4.24}$$

where B is as in (4.6) and

$$\begin{aligned} K_m = [K_s(B_1 R^{-1} B_2^T K_f - A_{12}) - (A_{21}^T K_f + M_1^T M_2)] \\ \times (A_{22} - B_2 R^{-1} B_2^T K_f)^{-1}. \end{aligned} \tag{4.25}$$

Proof Consistent with the identity

$$I - G_2(A_{22} + B_2G_2)^{-1}B_2 = (I + G_2A_{22}^{-1}B_2)^{-1}$$

and (4.22), define

$$H = I + R^{-1}B_2^{\mathrm{T}}K_{\mathrm{f}}(A_{22} - B_2R^{-1}B_2^{\mathrm{T}}K_{\mathrm{f}})^{-1}B_2,$$
$$H^{-1} = I - R^{-1}B_2^{\mathrm{T}}K_{\mathrm{f}}A_{22}^{-1}B_2.$$

H and H^{-1} are well-defined provided A_{22} and $A_{22} - B_2R^{-1}B_2^{\mathrm{T}}K_{\mathrm{f}}$ are nonsingular. Using (4.21),

$$\begin{aligned} H^{-\mathrm{T}}RH^{-1} &= R + B_2^{\mathrm{T}}A_{22}^{-\mathrm{T}}(-K_{\mathrm{f}}A_{22} - A_{22}^{\mathrm{T}}K_{\mathrm{f}} + K_{\mathrm{f}}B_2R^{-1}B_2^{\mathrm{T}}K_{\mathrm{f}})A_{22}^{-1}B_2 \\ &= R + N_0^{\mathrm{T}}N_0 = R_0, \end{aligned}$$

which, upon inversion and rearrangement, yields the identity

$$R^{-1}H^{\mathrm{T}} = H^{-1}R_0^{-1}$$

or

$$\begin{aligned} &(I - R^{-1}B_2^{\mathrm{T}}K_{\mathrm{f}}A_{22}^{-1}B_2)R_0^{-1} \\ &\qquad = R^{-1}[I + B_2^{\mathrm{T}}(A_{22} - B_2R^{-1}B_2^{\mathrm{T}}K_{\mathrm{f}})^{-\mathrm{T}}K_{\mathrm{f}}B_2R^{-1}]. \end{aligned}$$

Hence, using (4.21), (4.23) can be rewritten as

$$u_{\mathrm{c}} = -R^{-1}(B_1^{\mathrm{T}}K_{\mathrm{s}}x + B_2^{\mathrm{T}}K_{\mathrm{m}}^{\mathrm{T}}x + B_2^{\mathrm{T}}K_{\mathrm{f}}z),$$

or (4.24), where K_{m} is given by (4.25). □

The composite control u_{c} in (4.24) is now in a form that invites direct comparison with the exact optimal control u_{opt} of (4.4). It is of interest to analyze the relationship between $K(\varepsilon)$ and M_{c} in (4.24) for ε small and positive. For this purpose, we construct a power series expansion of $K(\varepsilon)$, where $K(\varepsilon)$ is assumed to be scaled, as in equation (6.36) of Chapter 2 to avoid unboundedness of coefficients in the feedback gain matrix (4.4).

Theorem 4.2

Under the conditions of Theorem 4.1, the positive-semidefinite stabilizing solution $K = K(\varepsilon)$ of the algebraic Riccati equation (4.5) possesses a power series expansion at $\varepsilon = 0$, that is,

$$K = \begin{bmatrix} K_1^{(0)} & \varepsilon K_2^{(0)} \\ \varepsilon K_2^{(0)\mathrm{T}} & \varepsilon K_3^{(0)} \end{bmatrix} + \sum_{i=1}^{\infty} \frac{\varepsilon^i}{i!} \begin{bmatrix} K_1^{(i)} & \varepsilon K_2^{(i)} \\ \varepsilon K_2^{(i)\mathrm{T}} & \varepsilon K_3^{(i)} \end{bmatrix}. \tag{4.26}$$

Furthermore, the matrices $K_1^{(0)}, K_2^{(0)}, K_3^{(0)}$ satisfy the identities

$$K_1^{(0)} = K_s, \quad K_2^{(0)} = K_m, \quad K_3^{(0)} = K_f, \tag{4.27}$$

where K_s, K_f and K_m are defined by (4.17), (4.21) and (4.25) respectively.

Proof It is convenient to rewrite the original Riccati equation (4.5) in the Lyapunov form

$$0 = K(A + BG) + (A + BG)^T K + G^T RG + M^T M, \tag{4.28}$$

where G is as defined in (4.4). Then, the substitution of (4.26) into (4.28) yields at $\varepsilon = 0$ the equations

$$\begin{aligned} 0 = {} & K_1^{(0)} F_{11} + F_{11}^T K_1^{(0)} + K_2^{(0)} F_{21} + F_{21}^T K_2^{(0)T} + G_1^{(0)T} RG_1^{(0)} \\ & + M_1^T M_1, \end{aligned} \tag{4.29a}$$

$$0 = K_1^{(0)} F_{12} + K_2^{(0)} F_{22} + F_{21}^T K_3^{(0)} + G_1^{(0)T} RG_2^{(0)} + M_1^T M_2, \tag{4.29b}$$

$$0 = K_3^{(0)} F_{22} + F_{22}^T K_3^{(0)} + G_2^{(0)T} RG_2^{(0)} + M_2^T M_2, \tag{4.29c}$$

where $G(0) \triangleq [G_1^{(0)} \quad G_2^{(0)}]$ and the partitions of the closed-loop system matrix $F_{11} \triangleq A_{11} + B_1 G_1^{(0)}$, etc., are defined in analogy with (2.16). Under the conditions of Lemma 4.2, (4.21), (4.22) and (4.29c) imply that $K_3^{(0)} = K_f$. Also, by Lemma 4.2, $F_{22} \triangleq A_{22} + B_2 G_2^{(0)}$ is a Hurwitz matrix and so F_{22}^{-1} exists.

Therefore, (4.29b) can be rewritten as

$$K_2 = -[K_1^{(0)} F_{12} + F_{21}^T K_3^{(0)} + G_1^{(0)T} RG_2^{(0)} + M_1^T M_2] F_{22}^{-1}, \tag{4.30}$$

whereupon substitution of (4.30) in (4.29a) yields

$$\begin{aligned} 0 = {} & K_1^{(0)} [F_{11} - F_{12} F_{22}^{-1} F_{21}] + [F_{11} - F_{12} F_{22}^{-1} F_{21}]^T K_1^{(0)} \\ & - F_{21}^T K_3^{(0)} F_{22}^{-1} F_{21} - F_{21}^T F_{22}^{-T} K_3^{(0)} F_{21} - G_1^{(0)T} RG_2^{(0)} F_{22}^{-1} F_{21} \\ & - F_{21}^T F_{22}^{-T} G_2^{(0)T} RG_1^{(0)} - M_1^T M_2 F_{22}^{-1} F_{21} - F_{21}^T F_{22}^{-T} M_2^T M_1 \\ & + G_1^{(0)T} RG_1^{(0)} + M_1^T M_1. \end{aligned} \tag{4.31}$$

The remainder of the proof is devoted to showing that (4.31) is *independent* of the fast Riccati solution $K_3^{(0)} = K_f$ (and $G_2^{(0)}$) and is quite the same as (4.17). Now, equations (2.18) and (2.23) of Section 3.2 imply the following identity:

$$F_{22}^{-1} F_{21} = A_{22}^{-1} (A_{21} + B_2 G_0). \tag{4.32}$$

(Alternatively, as a simple exercise the reader may verify (4.32) by pre-multiplication of (4.32) by F_{22}.) Also, in view of (4.32) and (2.10), we have

$$F_{11} - F_{12}F_{22}^{-1}F_{21} = A_0 + B_0G_0. \tag{4.33}$$

Substituting (4.33), (4.29c) and (2.10) in (4.31) yields

$$\begin{aligned}
0 = {} & K_1^{(0)}(A_0 + B_0G_0) + (A_0 + B_0G_0)^{\mathrm{T}}K_1^{(0)} \\
& + F_{21}^{\mathrm{T}}F_{22}^{-\mathrm{T}}[G_2^{(0)\mathrm{T}}RG_2^{(0)} + M_2^{\mathrm{T}}M_2]F_{22}^{-1}F_{21} \\
& - [G_0 + G_2^{(0)}A_{22}^{-1}(A_{21} + B_2G_0)]^{\mathrm{T}}RG_2^{(0)}F_{22}^{-1}F_{21} \\
& - F_{21}^{\mathrm{T}}F_{22}^{-\mathrm{T}}G_2^{(0)\mathrm{T}}R[G_0 + G_2^{(0)}A_{22}^{-1}(A_{21} + B_2G_0)] \\
& - M_1^{\mathrm{T}}M_2F_{22}^{-1}F_{21} - F_{21}^{\mathrm{T}}F_{22}^{-\mathrm{T}}M_2^{\mathrm{T}}M_1 + M_1^{\mathrm{T}}M_1 \\
& + [G_0 + G_2^{(0)}A_{22}^{-1}(A_{21} + B_2G_0)]^{\mathrm{T}} \\
& \times R[G_0 + G_2^{(0)}A_{22}^{-1}(A_{21} + B_2G_0)].
\end{aligned} \tag{4.34}$$

Noting (4.32) and performing the indicated cancellations in (4.34), we have

$$\begin{aligned}
0 = {} & K_1^{(0)}(A_0 + B_0G_0) + (A_0 + B_0G_0)^{\mathrm{T}}K_1^{(0)} + G_0^{\mathrm{T}}RG_0 \\
& + [M_1 - M_2F_{22}^{-1}F_{21}]^{\mathrm{T}}[M_1 - M_2F_{22}^{-1}F_{21}].
\end{aligned} \tag{4.35}$$

Since, by (4.32),

$$M_1 - M_2F_{22}^{-1}F_{21} = M_0 + N_0G_0,$$

we finally arrive, after a rearrangement of terms in (4.35), at the expression

$$\begin{aligned}
0 = {} & K_1^{(0)}(A_0 + B_0G_0) + (A_0 + B_0G_0)^{\mathrm{T}}K_1^{(0)} + M_0^{\mathrm{T}}M_0 \\
& + M_0^{\mathrm{T}}N_0G_0 + G_0^{\mathrm{T}}N_0^{\mathrm{T}}M_0 + G_0^{\mathrm{T}}(R + N_0^{\mathrm{T}}N_0)G_0,
\end{aligned} \tag{4.36}$$

which is the Lyapunov form of the Riccati equation (4.17), where G_0 is defined by (4.18). Thus, under the conditions of Lemma 4.1, (4.17) and (4.36) imply that $K_1^{(0)} = K_s$. Also, substituting (2.16) and (4.22) in (4.30), we have, upon comparison with (4.25), that $K_2^{(0)} = K_m$. The existence of the series (4.26) then follows from the implicit function theorem (Dieudonné, 1982). □

Theorem 4.2 establishes that the composite feedback control u_c in (4.23) is stabilizing and is $O(\varepsilon)$ close to u_{opt} of (4.4). This implies that J_c, the value of the performance index J of the system (4.1), (4.2) with u_c, is at least $O(\varepsilon)$ close to the value of the optimal performance index J_{opt}. The following analysis of J_c and J_{opt} reveals that u_c in fact yields an $O(\varepsilon^2)$ approximation of J_{opt}.

Since

$$J_{\text{opt}} = \tfrac{1}{2}[x^{0\text{T}} \quad z^{0\text{T}}]K\begin{bmatrix} x^0 \\ z^0 \end{bmatrix} \quad \text{and} \quad J_{\text{c}} = \tfrac{1}{2}[x^{0\text{T}} \quad z^{0\text{T}}]P_{\text{c}}\begin{bmatrix} x^0 \\ z^0 \end{bmatrix},$$

where P_{c} is the positive-semidefinite solution of the Lyapunov equation

$$P_{\text{c}}(A - SM_{\text{c}}) + (A - SM_{\text{c}})^{\text{T}}P_{\text{c}} = -M_{\text{c}}^{\text{T}}SM_{\text{c}} - M^{\text{T}}M \qquad (4.37)$$

with $S \triangleq BR^{-1}B^{\text{T}}$, the following theorem holds.

Theorem 4.3

Under the conditions of Theorem 4.1, the first two terms of the power series of J_{c} and J_{opt} at $\varepsilon = 0$ are the same, that is,

$$J_{\text{c}} = J_{\text{opt}} + O(\varepsilon^2), \qquad (4.38)$$

and hence the composite feedback control (4.23) is an $O(\varepsilon^2)$ near-optimal solution to the complete regulator problem (4.1)–(4.3).

Proof Adding (4.5) to (4.37) and rearranging, we obtain a Lyapunov equation for $P_{\text{c}} - K = W$:

$$W(A - SM_{\text{c}}) + (A - SM_{\text{c}})^{\text{T}}W + (K - M_{\text{c}}^{\text{T}})S(K - M_{\text{c}}) = 0. \quad (4.39)$$

By an application of the implicit function theorem to (4.37), it can be shown that P_{c} possesses a power series at $\varepsilon = 0$. Thus W can also be expanded as follows:

$$W = \sum_{i=0}^{\infty} \frac{\varepsilon^i}{i!}\begin{bmatrix} W_1^{(i)} & W_2^{(i)} \\ \varepsilon W_2^{(i)\text{T}} & \varepsilon W_3^{(i)} \end{bmatrix}. \qquad (4.40)$$

From (4.24), (4.26) and (4.27) we have

$$(K - M_{\text{c}}^{\text{T}})S(K - M_{\text{c}}) = O(\varepsilon^2),$$

and, since the matrices $A_0 - B_0R_0^{-1}(N_0^{\text{T}}M_0 + B_0^{\text{T}}K_{\text{s}})$ and $(A_{22} - B_2R^{-1}B_2^TK_f)$ are Hurwitz, the substitution of (4.40) into (4.39) yields $W_j^{(0)} = 0$ and $W_j^{(1)} = 0$, $j = 1, 2, 3$. Hence $W = O(\varepsilon^2)$, which proves (4.38). □

It has been shown that the composite control (4.23), even though it does not contain ε explicitly, guarantees an $O(\varepsilon^2)$ approximation of the optimal performance. Knowledge of ε would, however, be needed for a higher-order approximation. This is considered in Section 3.5.

3.4.2 Near-Optimal Reduced Control

Similar to the stabilizing reduced control of Corollary 2.1, let us now consider the approximation achieved by optimizing only the slow system. If A_{22} is a Hurwitz matrix and $G_2 = 0$, then (4.23) reduces to

$$u_r = -R_0^{-1}(N_0^T M_0 + B_0^T K_s)x = -F\begin{bmatrix} x \\ z \end{bmatrix}. \tag{4.41}$$

The value J_r of the performance index J with u_r in (4.41) applied to the system (4.1), (4.2) is

$$J_r = \tfrac{1}{2}[x^{0T} \quad z^{0T}]P_r\begin{bmatrix} x^0 \\ z^0 \end{bmatrix}, \tag{4.42}$$

where P_r is the positive-semidefinite solution of the Lyapunov equation

$$P_r(A - BF) + (A - BF)^T P_r = -F^T RF - M^T M. \tag{4.43}$$

In contradistinction to the composite control u_c, the reduced control u_r of (4.41) is not $O(\varepsilon)$ close to u_{opt} owing to the absence of the boundary layer term u_f in (2.14). After the decay of the boundary layer term u_f away from $t = 0$, u_r is $O(\varepsilon)$ close to u_{opt}, so that its performance J_r does approximate J_{opt} to within $O(\varepsilon)$.

Theorem 4.4

If A_{22} is a Hurwitz matrix, then the constant terms of the power series of J_r and J_{opt} at $\varepsilon = 0$ are equal, that is,

$$J_r = J_{opt} + O(\varepsilon), \tag{4.44}$$

and hence the feedback control u_r in (4.41) is an $O(\varepsilon)$ near-optimal solution to the complete regulator problem (4.1)–(4.3).

Proof Using (4.21) and the identity

$$(I - R^{-1}B_2^T K_f A_{22}^{-1} B_2)R_0^{-1}$$
$$= R^{-1}[I + B_2^T(A_{22} - B_2 R^{-1} B_2^T K_f)^{-T} K_f B_2 R^{-1}],$$

u_r in (4.41) can be expressed as

$$u_r = -R^{-1}B^T\begin{bmatrix} K_s & 0 \\ \varepsilon K_r^T & 0 \end{bmatrix}x = -R^{-1}B^T M_r x, \tag{4.45}$$

where

$$K_r = [K_m + (K_s S_{12} - A_{21}^T) A_{22}^{-T} K_f][I + S_2(A_{22} - S_2 K_f)^{-T} K_f], \quad (4.46)$$

with $S_{12} \triangleq B_1 R^{-1} B_2^T$ and $S_2 \triangleq B_2 R^{-1} B_2^T$. Hence, (4.43) can be rewritten as

$$P_r(A - SM_r) + (A - SM_r)^T P_r = M_r^T S M_r - M^T M, \quad (4.47)$$

where $S \triangleq BR^{-1}B^T$ and P_r possesses a power series in ε. Adding (4.5) to (4.47) and rearranging, we obtain a Lyapunov equation for $P_r - K = V$:

$$V(A - SM_r) + (A - SM_r)^T V + (K - M_r^T) S (K - M_r) = 0. \quad (4.48)$$

Substituting the power series

$$V = \begin{bmatrix} V_1^{(0)} & \varepsilon V_2^{(0)} \\ \varepsilon V_2^{(0)T} & \varepsilon V_3^{(0)} \end{bmatrix} + \sum_{i=1}^{\infty} \frac{\varepsilon^i}{i!} \begin{bmatrix} V_1^{(i)} & \varepsilon V_2^{(i)} \\ \varepsilon V_2^{(i)T} & \varepsilon V_3^{(i)} \end{bmatrix} \quad (4.49)$$

into (4.48) and evaluating at $\varepsilon = 0$ yields

$$V_1^{(0)} \bar{A}_{11} + \bar{A}_{11}^T V_1^{(0)} + V_2^{(0)} \bar{A}_{21} + \bar{A}_{21}^T V_2^{(0)} + (K_m - K_r) S_2 (K_m - K_r)^T = 0, \quad (4.50)$$

$$V_1^{(0)} A_{12} + V_2^{(0)} A_{22} + A_{21}^{-T} V_3^{(0)} + (K_m - K_r) S_2 K_f = 0, \quad (4.51)$$

$$V_3^{(0)} A_{22} + A_{22}^T V_3^{(0)} + K_f S_2 K_f = 0, \quad (4.52)$$

where, with $S_1 \triangleq B_1 R^{-1} B_1^T$,

$$\bar{A}_{11} = A_{11} - S_1 K_s - S_{12} K_r^T, \quad (4.53)$$

$$\bar{A}_{21} = A_{21} - S_{12}^T K_s - S_2 K_r^T. \quad (4.54)$$

Since A_{22} is Hurwitz, there exists a unique positive-semidefinite solution $V_3^{(0)}$ of (4.52). Expressing $V_2^{(0)}$ in terms of $V_1^{(0)}$ and $V_3^{(0)}$ and substituting into (4.50), we obtain, using (4.52), that

$$V_1^{(0)}(\bar{A}_{11} - A_{12} A_{22}^{-1} \bar{A}_{21}) + (\bar{A}_{11} - A_{12} A_{22}^{-1} \bar{A}_{21})^T V_1^{(0)} = -(K_m - K_r - \bar{A}_{21}^T A_{22}^{-T} K_f) S_2 (K_m - K_r - \bar{A}_{21}^T A_{22}^{-T} K_f)^T. \quad (4.55)$$

Rearrangement of (4.46) yields

$$K_m - K_r - \bar{A}_{21}^T A_{22}^{-T} K_f = 0, \quad (4.56)$$

and hence the right-hand side of (4.55) is identically zero. Furthermore,

$\bar{A}_{11} - A_{12}A_{22}^{-1}\bar{A}_{21} = A_0 - B_0R_0^{-1}(N_0^TM_0 + B_0^TK_s)$, which is Hurwitz. Thus the solution of (4.55) is $V_1^{(0)} = 0$, implying $V = O(\varepsilon)$ and (4.44). □

Comparing Theorem 4.4 with Theorem 4.3, it is obvious that the major portion of the cost J is contributed by the slow system, while the contribution of the fast system is $O(\varepsilon)$.

3.4.3 Design Procedure

The preceding two-time-scale approach, applied to the optimal linear regulator problem (4.1)—(4.6), is summarized in the following design procedure.

For an $O(\varepsilon^2)$ near-optimal regulator, solve the separate lower-order algebraic Riccati equations (4.17) and (4.21), which are independent of ε, for K_s and K_f respectively. The composite feedback control to be implemented on the original system (4.1)–(4.6) is given by (4.23).

For an $O(\varepsilon)$ near-optimal regulator when A_{22} is Hurwitz, solve the $(n \times n)$-order Riccati equation (4.17) for K_s and implement the reduced control given by (4.41).

Example 4.1

Returning to the DC-motor control problem of Example 3.1, let us this time design near-optimal composite and reduced controls for the motor system

$$\begin{bmatrix} \dot{x} \\ \varepsilon\dot{z} \end{bmatrix} = \begin{bmatrix} 0 & 1 \\ -1 & -1 \end{bmatrix}\begin{bmatrix} x \\ z \end{bmatrix} + \begin{bmatrix} 0 \\ 1 \end{bmatrix}u, \quad \begin{bmatrix} x(0) \\ z(0) \end{bmatrix} = \begin{bmatrix} x^0 \\ z^0 \end{bmatrix}, \tag{4.57}$$

with the performance index

$$J = \frac{1}{2}\int_0^\infty (q_1x^2 + q_2z^2 + ru^2)\,\mathrm{d}t \tag{4.58}$$

where $q_1 > 0$, $q_2 > 0$ and $r > 0$.

The slow Riccati equation (4.17) is

$$0 = 2\left(\frac{r}{r+q_2}\right)K_s + \frac{K_s^2}{r+q_2} - \left(q_1 + \frac{rq_2}{r+q_2}\right) \tag{4.59}$$

of which the unique stabilizing positive root, under the conditions of Lemma 4.1, is

$$K_s = -r + [(r+q_1)(r+q_2)]^{1/2}, \tag{4.60}$$

resulting in the slow regulator gain given by (4.18):

$$G_0 = 1 - \left[\frac{r + q_1}{r + q_2}\right]^{1/2}. \tag{4.61}$$

The reduced control (4.41) is simply

$$u_r = \left(1 - \left[\frac{r + q_1}{r + q_2}\right]^{1/2}\right)x. \tag{4.62}$$

Similarly, the fast Riccati equation (4.21) is

$$0 = 2K_f + \frac{K_f^2}{r} - q_2, \tag{4.63}$$

of which the unique stabilizing positive root, under the conditions of Lemma 4.2, is

$$K_f = -r + [r(r + q_2)]^{1/2}, \tag{4.64}$$

resulting in the fast regulator gain matrix given by (4.22)

$$G_2 = 1 - \left[\frac{r + q_2}{r}\right]^{1/2}. \tag{4.65}$$

By (4.25), (4.60) and (4.64),

$$K_m = -r + [r(r + q_1)]^{1/2}, \tag{4.66}$$

while from (4.24) or (2.11), (4.61) and (4.65), the composite control is

$$u_c = \left(1 - \left[\frac{r + q_1}{r}\right]^{1/2}\right)x + \left(1 - \left[\frac{r + q_2}{r}\right]^{1/2}\right)z. \tag{4.67}$$

It is of interest to compare the composite control (4.67) with the optimal solution of the original regulator problem (4.57), (4.58). The ε-dependent Riccati equation (4.5) takes the partitioned form

$$0 = 2K_2(\varepsilon) + \frac{K_2^2(\varepsilon)}{r} - q_1, \tag{4.68a}$$

$$0 = -K_1(\varepsilon) + K_2(\varepsilon) + K_3(\varepsilon) + \frac{K_2(\varepsilon)K_3(\varepsilon)}{r}, \tag{4.68b}$$

$$0 = -2\varepsilon K_2(\varepsilon) + 2K_3(\varepsilon) + \frac{K_3^2(\varepsilon)}{r} - q_2. \tag{4.68c}$$

The positive root of (4.68a) is

$$K_2(\varepsilon) = -r + [r(r + q_1)]^{1/2}. \tag{4.69}$$

Given $K_2(\varepsilon)$ in (4.69), we solve (4.68c) for its positive root

$$K_3(\varepsilon) = -r + [r(r + q_2) + 2\varepsilon r[-r + [r(r + q_1)]^{1/2}]]^{1/2}. \tag{4.70}$$

Substituting (4.69) and (4.70) in (4.68b) and rearranging terms,

$$K_1(\varepsilon) = -r + [(r + q_1)(r + q_2 + 2\varepsilon[-r + [r(r + q_1)]^{1/2}])]^{1/2} \tag{4.71}$$

Upon performing Maclaurin series expansions of $K_1(\varepsilon)$ in (4.71) and $K_3(\varepsilon)$ in (4.70) at $\varepsilon = 0$, it is observed that

$$K_1(\varepsilon) = K_s + O(\varepsilon), \quad K_3(\varepsilon) = K_f + O(\varepsilon). \tag{4.72}$$

Also, from (4.66) and (4.69),

$$K_2(\varepsilon) = K_m. \tag{4.73}$$

Equations (4.72) and (4.73) mean that equations (4.26) and (4.27) of Theorem 4.2 are satisfied. Evaluating the optimal feedback control law (4.4), it is readily shown that

$$u_{\text{opt}}(t) = u_c(t) + O(\varepsilon). \tag{4.74}$$

It is left as an exercise to demonstrate that the composite cost

$$J_c = \tfrac{1}{2}[x^{0\text{T}} \quad z^{0\text{T}}]P_c\begin{bmatrix} x^0 \\ z^0 \end{bmatrix},$$

where P_c is the solution of (4.37), is near-optimal in the sense that

$$P_c = K + O(\varepsilon^2), \tag{4.75}$$

and that the reduced cost

$$J_r = \tfrac{1}{2}[x^{0\text{T}} \quad z^{0\text{T}}]P_r\begin{bmatrix} x^0 \\ z^0 \end{bmatrix},$$

where P_r is the solution of (4.43), is near-optimal in the sense that

$$P_r = K + O(\varepsilon). \tag{4.76}$$

Example 4.2

For the singularly perturbed system

$$\begin{bmatrix} \dot{x} \\ \varepsilon\dot{z} \end{bmatrix} = \begin{bmatrix} 0 & 0.4 & 0 & 0 \\ 0 & 0 & 0.345 & 0 \\ 0 & -0.524 & -0.465 & 0.262 \\ 0 & 0 & 0 & -1 \end{bmatrix} \begin{bmatrix} x \\ z \end{bmatrix} = \begin{bmatrix} 0 \\ 0 \\ 0 \\ 1 \end{bmatrix} u, \tag{4.77}$$

$$y = \begin{bmatrix} 1 & 0 & 0 & 0 \\ 0 & 0 & 1 & 0 \end{bmatrix} \begin{bmatrix} x \\ z \end{bmatrix} \tag{4.78}$$

it is desired to regulate the output y by way of minimizing the quadratic performance index

$$J = \frac{1}{2}\int_0^\infty (y^T y + u^T u)\,dt. \tag{4.79}$$

A near-optimal solution of the linear quadratic regulator problem (4.77)–(4.79) is sought through the independent solutions of lower-order slow and fast regulator problems in the manner of Lemma 4.1 and Lemma 4.2.

Slow problem: The triple

$$A_0 = \begin{bmatrix} 0 & 0.4 \\ 0 & -0.389 \end{bmatrix}, \quad B_0 = \begin{bmatrix} 0 \\ 0.194 \end{bmatrix}, \quad M_0 = \begin{bmatrix} 1 & 0 \\ 0 & -1.127 \end{bmatrix}$$

is completely controllable and completely observable in the sense of Lemma 6.1 of Chapter 2. Consequently, by Lemma 4.1 there exists a unique positive-definite stabilizing solution K_s of the algebraic matrix Riccati equation (4.17), which is

$$K_s = \begin{bmatrix} 7.39 & 5.91 \\ 5.91 & 7.15 \end{bmatrix}. \tag{4.80}$$

Substitution of (4.80), $R_0 = 1.316$ and $N_0 = \begin{bmatrix} 0 \\ 0.563 \end{bmatrix}$ in (4.18) gives the slow regulator gain

$$G_0 = [-0.87 \quad -0.57]. \tag{4.81}$$

Fast problem: The triple

$$A_{22} = \begin{bmatrix} -0.465 & 0.262 \\ 0 & -1 \end{bmatrix}, \quad B_2 = \begin{bmatrix} 0 \\ 1 \end{bmatrix}, \quad M_2 = \begin{bmatrix} 0 & 0 \\ 1 & 0 \end{bmatrix}$$

is completely controllable and completely observable. Consequently, by Lemma 4.2 there exists a unique positive-definite stabilizing solution of the algebraic Riccatic equation (4.21), which is

$$K_f = \begin{bmatrix} 1.04 & 0.18 \\ 0.18 & 0.05 \end{bmatrix} \tag{4.82}$$

By (4.82) and (4.22), the fast regulator gain is

$$G_2 = [-0.18 \quad -0.05]. \tag{4.83}$$

Composite control: By (4.23) or (2.11), (4.81) and (4.83) there results the composite control

$$u_c = [-1 \quad -0.89]x + [-0.18 \quad -0.05]z. \tag{4.84}$$

Reduced control: Since the eigenvalues of A_{22} are -0.47 and -1, A_{22} is a Hurwitz matrix and the two fast modes experience an exponential decay of the order of $e^{-0.47t/\varepsilon}$ and $e^{-t/\varepsilon}$ respectively without any fast regulator control ($G_2 = 0$). Accordingly, we would expect satisfactory state regulation to be achieved through the use of the reduced control (4.41), *viz*

$$u_r = [-0.87 \quad -0.57]x, \tag{4.85}$$

obtained from the solution of the slow Riccati equation (4.80).

Comparison with the optimal solution: For $\varepsilon = 0.1$, the optimal feedback solution to (4.77)–(4.79) is

$$u_{opt} = [-1 \quad -0.89]x + [-0.24 \quad -0.06]z, \tag{4.86}$$

obtained from solving a fourth-order ε-dependent Riccati equation (4.5) or ten scalar equations. For the initial condition $[x^{0T} \quad z^{0T}] = [1 \quad 0 \quad 1 \quad 0]$, the values of the performance index are

$$J_{opt} = 4.2406, \tag{4.87}$$

$$J_c = 4.2428, \tag{4.88}$$

$$J_r = 4.2506. \tag{4.89}$$

Hence the loss of performance compared with the optimal cost, with u_c less than 0.052% and u_r less than 0.24%, is practically negligible in both cases. By contrast with the optimal solution, the separate sets of three scalar equations associated with the "slow" and "fast" Riccati equations are independent of ε and are readily solved by hand.

Example 3.3 highlighted the fact that strong controllability [stabilizability] is an essential structural property without which no composite feedback design procedure is possible. If, instead of employing eigenvalue assignment for the weakly controllable singularly perturbed system (3.32), we were to consider the optimal linear regulator, similar difficulties would occur: namely, the exact optimal control (4.4) would be of the order of $1/\varepsilon$ large and the near-optimal composite control (4.23) would not exist. In the case of two-time-scale regulator design, it is not, however, necessary that unstable eigenvalues of the system (4.1), (4.2) be strongly observable (in the sense of Section 2.6) with respect to the performance index (4.3). The next example explores this issue.

Example 4.3

Consider the singularly perturbed optimal linear regulator problem

$$\begin{bmatrix} \dot{x} \\ \varepsilon\dot{z} \end{bmatrix} = \begin{bmatrix} 0 & 1 \\ 0 & 1 \end{bmatrix} \begin{bmatrix} x \\ z \end{bmatrix} + \begin{bmatrix} 0 \\ 1 \end{bmatrix} u, \tag{4.90}$$

$$J = \frac{1}{2}\int_0^\infty (x^2 + qz^2 + u^2)\,dt, \quad q > 0. \tag{4.91}$$

The system (4.90) is completely controllable for $\varepsilon > 0$, and as $\varepsilon \to 0$ the slow system $\dot{x}_s = -u_s$ and the fast system $\dot{z}_f = z_f + u_f$ are each completely controllable; by definition (Section 2.6), the system (4.90) is strongly controllable. Also, the system (4.90) is seen to be strongly observable with respect to the performance index (4.91). So far, all is well for a two-time-scale decomposition of regulator design along the lines of Lemmas 4.1 and 4.2.

Our attention is now focused on the detectability of the system (4.90), (4.91) as $q \to 0$, that is, on the detectability of the pair

$$\left([1, \;\; 0], \begin{bmatrix} 0 & 1 \\ 0 & 1/\varepsilon \end{bmatrix}\right).$$

This pair is detectable for $\varepsilon > 0$, but loses its detectability as $\varepsilon \to 0$ since the fast subsystem pair $(C_2, A_{22}) = (0, 1)$ is undetectable. In other words, the fast unstable eigenvalue $\lambda = 1/\varepsilon$ of the system (4.90) is weakly observable, as $q \to 0$, with respect to the performance index (4.91) and a two-time-scale separation of regulator design is inadmissible.

It is interesting, nonetheless, to examine the asymptotic behavior of the solution to the full regulator problem (4.90), (4.91) as q and ε both tend to zero. This we now do by solving the algebraic Riccati equation (4.5) for (4.90), (4.91) or the three scalar equations

$$0 = \frac{k_{12}^2}{\varepsilon^2} - 1, \tag{4.92a}$$

$$0 = -k_{11} - \frac{k_{12}}{\varepsilon} + \frac{k_{12}k_{22}}{\varepsilon^2}, \tag{4.92b}$$

$$0 = -2k_{12} - \frac{2k_{22}}{\varepsilon} + \frac{k_{22}^2}{\varepsilon^2} - q \tag{4.92c}$$

to obtain the unique solution

$$K(\varepsilon, q) = \begin{bmatrix} (1 + (2\varepsilon + q))^{1/2} & \varepsilon \\ \varepsilon & \varepsilon + \varepsilon(1 + (2\varepsilon + q))^{1/2} \end{bmatrix}. \tag{4.93}$$

The associated optimal linear feedback control (4.4) for (4.90), (4.91) is

$$u_{\text{opt}}(\varepsilon, q, t) = [-1, \quad -1 - (1 + 2\varepsilon + q)^{1/2}]\begin{bmatrix} x \\ z \end{bmatrix}, \tag{4.94}$$

which has the unique limit, as $q \to 0$ and $\varepsilon \to 0$, of

$$u_{\text{opt}} = [-1 \quad -2]\begin{bmatrix} x \\ z \end{bmatrix}. \tag{4.95}$$

Consider now the asymptotic behavior of the composite control (2.11) or (4.23). The Riccati equation (4.21) of the fast regulator is

$$0 = -2K_{\text{f}} + K_{\text{f}}^2 - q, \tag{4.96}$$

which, as $q \to 0$ thereby losing observability of the fast system eigenvalue, has not one but two positive-semidefinite solutions

$$K_{\text{f}} = 2 \quad \text{and} \quad K_{\text{f}} = 0. \tag{4.97}$$

Picking $K_{\text{f}} = 2$ results, by (4.22), in the stabilizing fast regulator gain $G_2^{(0)} = -2$, while it is left as a simple exercise to show that as $q \to 0$ the slow regulator gain is $G_0 = -1$ and the resulting composite control is identical with that of (4.95); that is, the composite control exists as $q \to 0$ and asymptotically approaches the optimal control as $\varepsilon \to 0$. However, if the other positive-semidefinite optimal solution $K_f = 0$ of (4.97) is picked then the fast regulator gain $G_2^{(0)}$ is not stabilizing and no stabilizing composite control exists in this case.

3.5 A Corrected Linear-Quadratic Design

The fact that the algebraic Riccati equation (4.5) possesses a power series expansion (4.26) at $\varepsilon = 0$ suggests that a closer approximation to the optimal performance index (4.8) can be obtained through evaluation of higher-order terms in the expansion. What we seek is a method that will result in a performance superior to the performance achievable by the uncorrected composite control (4.23).

The method adopted here is to approximate the optimal feedback gain matrix $G(\varepsilon)$ by

$$G(0) + \varepsilon\left[\frac{\mathrm{d}G(\varepsilon)}{\mathrm{d}\varepsilon}\right]_{\varepsilon=0} \triangleq \hat{G}(\varepsilon). \tag{5.1}$$

The analysis is facilitated by rewriting the exact solution (4.4), (4.5) in the following partitioned form:

$$G_1(\varepsilon) = -R^{-1}(B_1^{\mathrm{T}}K_1(\varepsilon) + B_2^{\mathrm{T}}K_2^{\mathrm{T}}(\varepsilon)), \tag{5.2}$$

$$G_2(\varepsilon) = -R^{-1}(\varepsilon B_1^{\mathrm{T}}K_2(\varepsilon) + B_2^{\mathrm{T}}K_3(\varepsilon)), \tag{5.3}$$

$$\begin{aligned} 0 = {} & -K_1(\varepsilon)(A_{11} - S_{12}K_2^{\mathrm{T}}(\varepsilon)) - (A_{11} - S_{12}K_2^{\mathrm{T}}(\varepsilon))^{\mathrm{T}}K_1(\varepsilon) \\ & + K_1(\varepsilon)S_1K_1(\varepsilon) - K_2(\varepsilon)A_{21} - A_{21}^{\mathrm{T}}K_2^{\mathrm{T}}(\varepsilon) \\ & + K_2(\varepsilon)S_2K_2^{\mathrm{T}}(\varepsilon) - M_1^{\mathrm{T}}M_1, \end{aligned} \tag{5.4a}$$

$$\begin{aligned} 0 = {} & K_2(\varepsilon)(S_2K_3(\varepsilon) - A_{22}) - K_1(\varepsilon)A_{12} \\ & - \varepsilon(A_{11}^{\mathrm{T}} - K_1(\varepsilon)S_1 - K_2(\varepsilon)S_{12}^{\mathrm{T}})K_2(\varepsilon) \\ & - A_{21}^{\mathrm{T}}K_3(\varepsilon) + K_1(\varepsilon)S_{12}K_3(\varepsilon) - M_1^{\mathrm{T}}M_2, \end{aligned} \tag{5.4b}$$

$$\begin{aligned} 0 = {} & -K_3(\varepsilon)A_{22} - A_{22}^{\mathrm{T}}K_3(\varepsilon) + K_3(\varepsilon)S_2K_3(\varepsilon) \\ & - \varepsilon K_2^{\mathrm{T}}(\varepsilon)(A_{12} - S_{12}K_3(\varepsilon)) - \varepsilon(A_{12} - S_{12}K_3(\varepsilon))^{\mathrm{T}}K_2(\varepsilon) \\ & + \varepsilon^2K_2^{\mathrm{T}}(\varepsilon)S_1K_2(\varepsilon) - M_2^{\mathrm{T}}M_2, \end{aligned} \tag{5.4c}$$

where $S_1 = B_1R^{-1}B_1^T$, $S_2 = B_2R^{-1}B_2^T$ and $S_{12} = B_1R^{-1}B_2^T$ as before, and the partitioned matrices $K_i(\varepsilon)$, $i = 1, 2, 3$ correspond to those of (4.26).

Setting $\varepsilon = 0$ in (5.4), we obtain, as in Subsection 3.4.1, the zeroth-order solution given by (4.29). In particular, by (5.2), (5.3), Lemma 4.3 and Theorem 4.2,

$$G_1(0) = (I_r + G_2^{(0)}A_{22}^{-1}B_2)G_0 + G_2^{(0)}A_{22}^{-1}A_{21}, \quad G_2(0) = G_2^{(0)}, \tag{5.5}$$

where G_0 and $G_2^{(0)}$ are the uncorrected slow and fast regulator gain matrices defined by (4.18) and (4.22) respectively.

Next, it is required to compute the gain correction $\varepsilon[\mathrm{d}G(\varepsilon)/\mathrm{d}\varepsilon]_{\varepsilon=0}$ in (5.1). Evaluating the derivatives of (5.2) and (5.3) with respect to ε, at $\varepsilon = 0$,

$$\left[\frac{\mathrm{d}G_1(\varepsilon)}{\mathrm{d}\varepsilon}\right]_{\varepsilon=0} = -R^{-1}(B_1^TK_1^{(1)} + B_2^TK_2^{(1)T}), \tag{5.6}$$

$$\left[\frac{\mathrm{d}G_2(\varepsilon)}{\mathrm{d}\varepsilon}\right]_{\varepsilon=0} = -R^{-1}(B_1^TK_2^{(0)} + B_2^TK_3^{(1)}), \tag{5.7}$$

where the derivatives $K_i^{(1)} \triangleq [\mathrm{d}K_i(\varepsilon)/\mathrm{d}\varepsilon]_{\varepsilon=0}$, $i = 1, 2, 3$, satisfy the following equations:

$$\begin{aligned}0 = {} & K_1^{(1)}(A_{11} - S_1K_1^{(0)} - S_{12}K_2^{(0)T}) + (A_{11} - S_1K_1^{(0)} - S_{12}K_2^{(0)T})^TK_1^{(1)} \\ & + K_2^{(1)}(A_{21} - S_{12}^TK_1^{(0)} - S_2K_2^{(0)T}) \\ & + (A_{21} - S_{12}^TK_1^{(0)} - S_2K_2^{(0)T})^TK_2^{(1)T},\end{aligned} \tag{5.8a}$$

$$\begin{aligned}0 = {} & K_1^{(1)}(A_{12} - S_{12}K_3^{(0)}) + K_2^{(1)}(A_{22} - S_2K_3^{(0)}) \\ & + (A_{21} - S_{12}^TK_1^{(0)} - S_2K_2^{(0)T})^TK_3^{(1)} \\ & + (A_{11}^T - K_1^{(0)}S_1 - K_2^{(0)}S_{12}^T)K_2^{(0)},\end{aligned} \tag{5.8b}$$

$$\begin{aligned}0 = {} & K_3^{(1)}(A_{22} - S_2K_2^{(0)}) + (A_{22} - S_2K_3^{(0)})^TK_3^{(1)} \\ & + K_2^{(0)T}(A_{12} - S_{12}K_3^{(0)}) + (A_{12} - S_{12}K_3^{(0)})^TK_2^{(0)}.\end{aligned} \tag{5.8c}$$

Observing that the corrected state-feedback gain matrix $\hat{G}$ in (5.1) approximates the optimal gain matrix G to within $O(\varepsilon^2)$, as the solution

$$\begin{bmatrix} K_1 & \varepsilon K_2 \\ \varepsilon K_2^T & \varepsilon K_3 \end{bmatrix} = \begin{bmatrix} K_1^{(0)} & \varepsilon K_2^{(0)} \\ \varepsilon K_2^{(0)T} & \varepsilon K_3^{(0)} \end{bmatrix} + \varepsilon \begin{bmatrix} K_1^{(1)} & \varepsilon K_2^{(1)} \\ \varepsilon K_2^{(1)T} & \varepsilon K_3^{(1)} \end{bmatrix} \tag{5.9}$$

approximates the Riccati equation solution in (4.26), it is clear that the corresponding cost J_c is at least within $O(\varepsilon^2)$ of the optimal cost J_{opt} of (4.8).

Theorem 5.1

Under the conditions of Theorem 4.1, the first four terms of the power series of J_c and J_{opt} at $\varepsilon = 0$ are the same, that is,

$$J_c = J_{opt} + O(\varepsilon^4), \tag{5.10}$$

and hence the corrected feedback control (5.1) is an $O(\varepsilon^4)$ near-optimal solution to the regulator problem (4.1)–(4.3).

Proof The proof (Sannuti and Kokotović, 1969a) is a lengthier version of that of Theorem 4.3 and is omitted.

It should be noted (and checked) that (5.8) in the corrected solution partitions $K_i^{(1)}$, $i = 1, 2, 3$, is also obtained by formal substitution of (5.9) in the original matrix Riccati equation (5.4) and equating coefficients of ε.

In order to solve the *linear* equations (5.8) for $K_1^{(1)}$, $K_2^{(1)}$ and $K_3^{(1)}$, it is useful to first rewrite (5.8c) as

$$0 = K_3^{(1)} F_{22} + F_{22}^{\mathrm{T}} K_3^{(1)} + K_2^{(0)\mathrm{T}} F_{12} + F_{12}^{\mathrm{T}} K_2^{(0)}, \tag{5.11}$$

where

$$F_{11} \triangleq A_{11} + B_1 G_1^{(0)}, \quad F_{12} \triangleq A_{12} + B_1 G_2^{(0)}, \tag{5.12a}$$

$$F_{21} \triangleq A_{21} + B_2 G_1^{(0)}, \quad F_{22} \triangleq A_{22} + B_2 G_2^{(0)} \tag{5.12b}$$

are the partitions of the uncorrected closed-loop system matrix, as in (4.29). Rearrangement of (5.8b) using (5.12), where under the conditions of Lemma 4.2 F_{22} Hurwitz implies F_{22}^{-1} exists, yields

$$K_2^{(1)} = -[K_1^{(1)} F_{12} + F_{21}^{\mathrm{T}} K_3^{(1)} + F_{11}^{\mathrm{T}} K_2^{(0)}] F_{22}^{-1}. \tag{5.13}$$

Similarly, (5.8a) may be rewritten as

$$0 = K_1^{(1)} F_{11} + F_{11}^{\mathrm{T}} K_1^{(1)} + K_2^{(1)} F_{21} + F_{21}^{\mathrm{T}} K_2^{(1)\mathrm{T}}, \tag{5.14}$$

whereupon substitution of $K_2^{(1)}$ from (5.13) in (5.14) gives

$$\begin{aligned} 0 = {} & K_1^{(1)} [F_{11} - F_{12} F_{22}^{-1} F_{21}] + [F_{11} - F_{12} F_{22}^{-1} F_{21}]^{\mathrm{T}} K_1^{(1)} \\ & - F_{21}^{\mathrm{T}} F_{22}^{-\mathrm{T}} [F_{22}^{\mathrm{T}} K_3^{(1)} + K_3^{(1)} F_{22}] F_{22}^{-1} F_{21} \\ & - F_{11}^{\mathrm{T}} K_2^{(0)} F_{22}^{-1} F_{21} - F_{21}^{\mathrm{T}} F_{22}^{-\mathrm{T}} K_2^{(0)\mathrm{T}} F_{11}. \end{aligned} \tag{5.15}$$

Using (5.11), (5.15) becomes

$$
\begin{aligned}
0 = {} & K_1^{(1)}[F_{11} - F_{12}F_{22}^{-1}F_{21}] + [F_{11} - F_{12}F_{22}^{-1}F_{21}]^{\mathrm{T}}K_1^{(1)} \\
& - (F_{22}^{-1}F_{21})^{\mathrm{T}}K_2^{(0)\mathrm{T}}[F_{11} - F_{12}F_{22}^{-1}F_{21}] \\
& - [F_{11} - F_{12}F_{22}^{-1}F_{21}]^{\mathrm{T}}K_2^{(0)}F_{22}^{-1}F_{21}.
\end{aligned} \tag{5.16}
$$

Using the identities (4.32) and (4.33), (5.16) finally assumes the form, linear in $K_1^{(1)}$,

$$
\begin{aligned}
0 = {} & K_1^{(1)}(A_0 + B_0G_0) + (A_0 + B_0G_0)^{\mathrm{T}}K_1^{(1)} \\
& - (A_{21} + B_2G_0)^{\mathrm{T}}A_{22}^{-\mathrm{T}}K_2^{(0)\mathrm{T}}(A_0 + B_0G_0) \\
& - (A_0 + B_0G_0)^{\mathrm{T}}K_2^{(0)}A_{22}^{-1}(A_{21} + B_2G_0).
\end{aligned} \tag{5.17}
$$

In summary, the corrected near-optimal design consists of the following steps.

1. Solve the slow and fast regulator equations of Lemmas 4.1 and 4.2, with $K_{\mathrm{s}} = K_1^{(0)}$, $K_{\mathrm{f}} = K_3^{(0)}$ and $K_{\mathrm{m}} = K_2^{(0)}$ in (4.25), to obtain the uncorrected gain matrix $G(0) = [G_1(0) \quad G_2(0)]$ of (5.5).
2. Solve $\frac{1}{2}n(n+1)$ linear scalar equations (5.17) for $K_1^{(1)}$.
3. Solve $\frac{1}{2}m(m+1)$ linear scalar equations (5.11) for $K_3^{(1)}$.
4. Obtain $K_2^{(1)}$ from (5.13).
5. Compute the corrected gain $\hat{G}(\varepsilon)$ in (5.1) from $G(0)$ (step 1), (5.6) and (5.7).

The next example indicates the significant improvement that the corrected design can have over the uncorrected design of Section 3.4 when ε is known.

Example 5.1

In a voltage regulator problem, the plant is defined by the block diagram of Fig. 3.4, where

$$T_1 = 5, \quad T_2 = 2, T_3 = 0.07, \quad T_4 = 0.04, \quad T_5 = 0.1, \tag{5.18}$$

$$a_1 = 2.5, \quad a_2 = 3.2, \quad a_3 = 6, \quad a_4 = 3, \quad a_5 = 3. \tag{5.19}$$

Now, the small time constants can be expressed as multiples of a single small parameter $\varepsilon = 0.1$:

$$T_3 = 0.7\varepsilon, \quad T_4 = 0.4\varepsilon, \quad T_5 = \varepsilon. \tag{5.20}$$

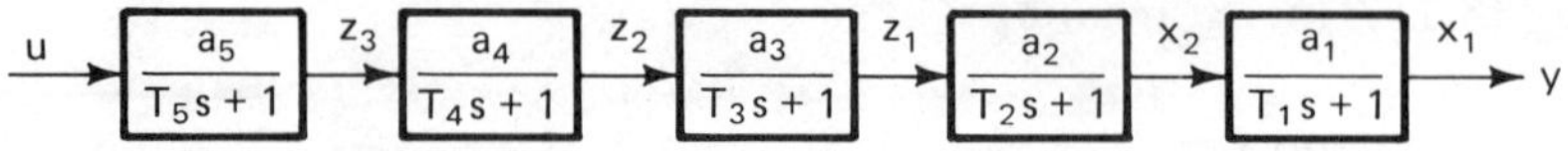

Fig. 3.4. Block diagram of a voltage regulator.

The associated state model accordingly assumes the explicit singularly perturbed form

$$\dot{x} = \begin{bmatrix} -0.2 & 0.5 \\ 0 & -0.5 \end{bmatrix} x + \begin{bmatrix} 0 & 0 & 0 \\ 1.6 & 0 & 0 \end{bmatrix} z, \tag{5.21}$$

$$\varepsilon \dot{z} = \begin{bmatrix} \frac{-10}{7} & \frac{60}{7} & 0 \\ 0 & -2.5 & 7.5 \\ 0 & 0 & -1 \end{bmatrix} z + \begin{bmatrix} 0 \\ 0 \\ 3 \end{bmatrix} u. \tag{5.22}$$

Taking into account the fact that A_{22} is a Hurwitz matrix, it is desired to regulate the output (voltage) $y = x_1$ by way of minimizing the quadratic performance index

$$J = \frac{1}{2} \int_0^\infty (x_1^2 + u^2)\, \mathrm{d}t, \tag{5.23}$$

where it is observed that (5.23) is of the form (4.2), (4.3) with $M_2 = 0$. Consequently, by (4.21) $K_f = K_3^{(0)} = 0$, and by (4.22) $G_2^{(0)} = 0$. Combining $G_2(0) = 0$ with the solution $G_1(0) = G_0$ of (4.17) and (4.18), we have, by (5.5), that

$$G(0) = [0.958 \quad 0.1 \quad 0 \quad 0 \quad 0]. \tag{5.24}$$

Also, solving (5.17), (5.11) and (5.13) for $K_1^{(1)}$, $K_3^{(1)}$ and $K_2^{(1)}$ respectively, and substituting in (5.6) and (5.7),

$$\left[\frac{\mathrm{d}G(\varepsilon)}{\mathrm{d}\varepsilon}\right]_{\varepsilon=0} = [-0.402 \quad 0.901 \quad 0.112 \quad 0.383 \quad 2.870]. \tag{5.25}$$

For the values of T_3, T_4 and T_5 given in (5.18), $\varepsilon = 0.1$, and in view of (5.24) and (5.25), the corrected near-optimal gain matrix in (5.1) is given by

$$\begin{aligned} \hat{G}(0.1) &= G(0) + 0.1\left[\frac{\mathrm{d}G(\varepsilon)}{\mathrm{d}\varepsilon}\right]_{\varepsilon=0} \\ &= [0.918 \quad 0.19 \quad 0.011 \quad 0.038 \quad 0.287]. \end{aligned} \tag{5.26}$$

It is of interest to compare the gain matrix $\hat{G}(0.1)$ with the optimal gain matrix $G(0.1)$ obtained from solving the full-order problem (5.21)—(5.23) using (4.4) and (4.5):

$$G(0.1) = [0.924 \quad 0.171 \quad 0.016 \quad 0.039 \quad 0.264]. \tag{5.27}$$

The comparison of the responses of the optimal (solid line) and the corrected near-optimal (dotted line) systems in Fig. 3.5 indicate the closeness of the approximation.

It should be pointed out that the corrected near-optimal design may serve

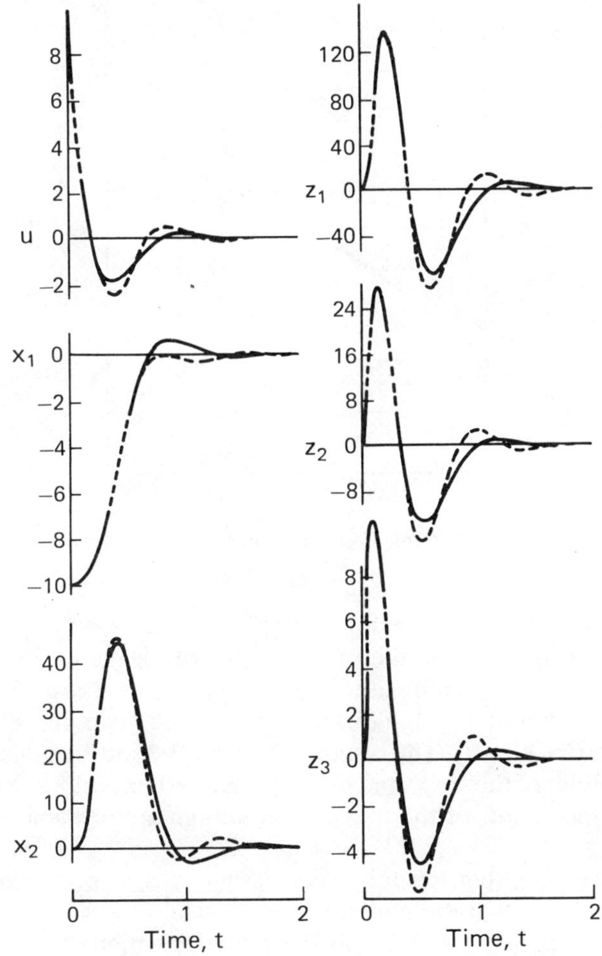

Fig. 3.5. Responses of optimal (solid line) and corrected near-optimal (dotted line) systems.

not only for $\varepsilon = 0.1$, as required in this example, but also for a range of values of ε. To see how large this range is, the corrected near-optimal design (5.1) is compared with the optimal design (4.4), (4.5) and with the uncorrected near-optimal design (4.18). Figure 3.6 gives the values of the performance index J obtained by these three designs, with the initial condition $x_1(0) = -10$, $x_2(0) = z_1(0) = z_2(0) = z_3(0) = 0$, for a range of values of ε (similar results are obtained for other initial conditions). For

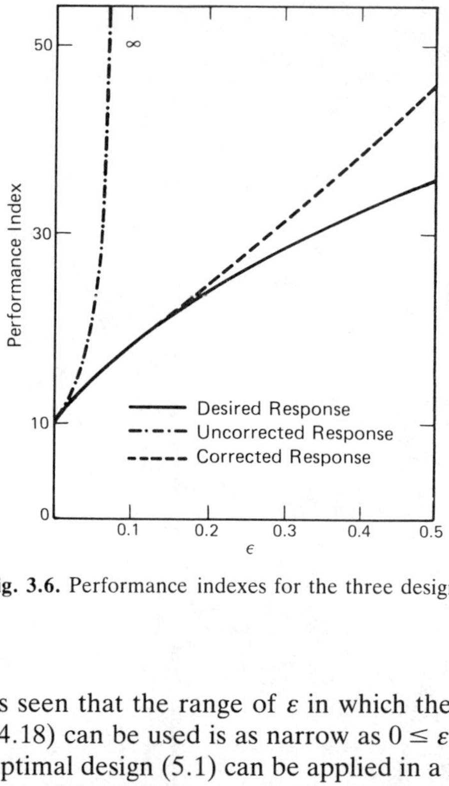

Fig. 3.6. Performance indexes for the three designs.

this example it is seen that the range of ε in which the uncorrected near-optimal design (4.18) can be used is as narrow as $0 \le \varepsilon \le 0.025$, while the corrected near-optimal design (5.1) can be applied in a range about twelve times larger, $0 \le \varepsilon \le 0.3$. The responses in Figs. 3.7 and 3.8 indicate that the (voltage) output of the corrected near-optimal system is close to the optimal one, while the results of the uncorrected design are unstable even at $\varepsilon = 0.07$.

Finally, we note that the full-order design consists of solving a system of $\frac{5}{2}(5 + 1) = 15$ quadratic equations (4.5), while the $\frac{3}{2}(3 + 1) = 6$ linear equations (5.8c) and the $\frac{2}{2}(2 + 1) = 3$ linear equations (5.8a) are easily solvable by hand.

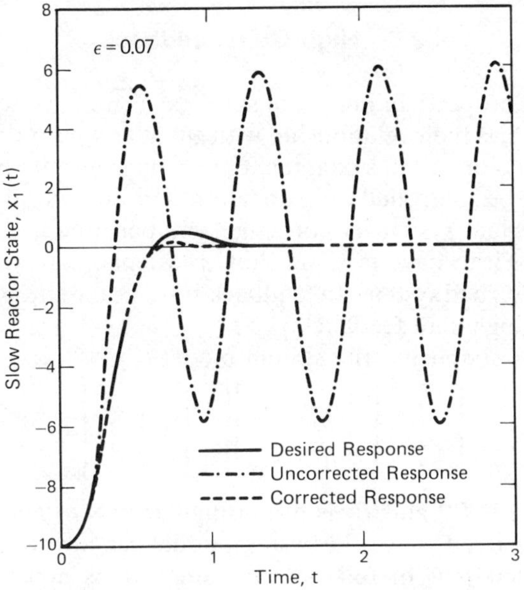

Fig. 3.7. State $x_1(t)$ for the three designs when $\varepsilon = 0.07$.

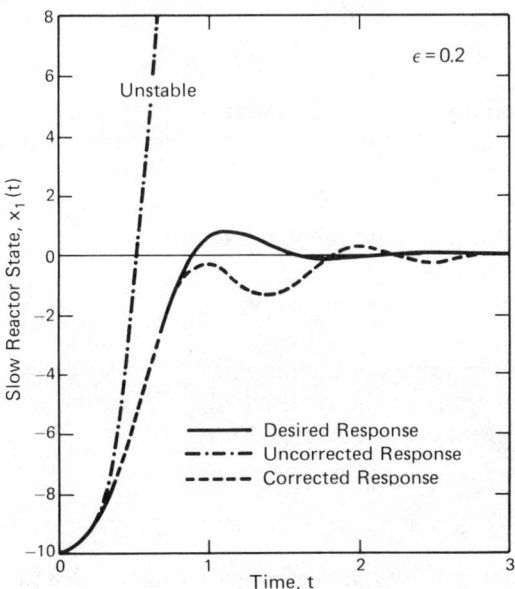

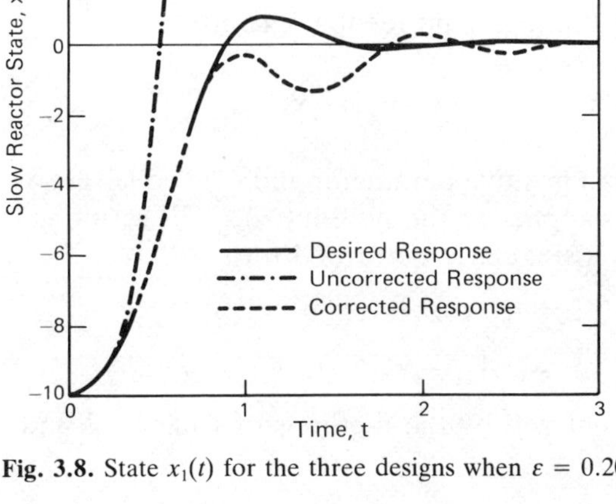

Fig. 3.8. State $x_1(t)$ for the three designs when $\varepsilon = 0.20$.

3.6 High-Gain Feedback

A fundamental property of high-gain systems, as we have seen in Example 6.2 of Chapter 1, is their relationship with singularly perturbed systems. So far in our discussions in this chapter, the singular perturbation properties of the system to be controlled were not altered by the control law. However, even if the original system is not singularly perturbed, a strong control action can force it to have fast and slow transients, that is, to behave like a singularly perturbed system. In feedback systems the strong control action is achieved by high-gain feedback.

Consider, for simplicity, the system model

$$\begin{bmatrix} \dot{x} \\ \dot{z} \end{bmatrix} = \begin{bmatrix} A_{11} & A_{12} \\ A_{21} & A_{22} \end{bmatrix} \begin{bmatrix} x \\ z \end{bmatrix} + \begin{bmatrix} 0 \\ B_2 \end{bmatrix} u, \tag{6.1}$$

where $x \in R^n$, $z \in R^m$ and B_2 is a nonsingular $m \times m$ matrix. No loss of generality is incurred, since the system model can always be transformed to (6.1) provided B is of full rank m. Since B_2 is nonsingular the pair (A_{22}, B_2) is completely controllable for any A_{22}. Also, the next lemma is immediate from the application of Lemma 6.1 of Chapter 2 to the system (6.1), where B_2 is nonsingular.

Lemma 6.1

If the system pair (A, B) of the system (6.1) is completely controllable [stabilizable], then the pair (A_{11}, A_{12}) is completely controllable [stabilizable].

Suppose that the high-gain feedback control

$$u = \frac{1}{\varepsilon}[G_1 x + G_2 z], \tag{6.2}$$

where ε is a small positive parameter and G_1 and G_2 are as yet unspecified gain matrices, is applied to the system (6.1). The resulting feedback system (6.1), (6.2) has a singularly perturbed form

$$\dot{x} = A_{11}x + A_{12}z, \tag{6.3a}$$

$$\varepsilon\dot{z} = (\varepsilon A_{21} + B_2 G_1)x + (\varepsilon A_{22} + B_2 G_2)z, \tag{6.3b}$$

which can be analyzed by the methods of Chapter 2 under the following assumption.

Assumption 6.1

B_2G_2 is a nonsingular matrix.

If, under Assumption 6.1, G_2 is chosen such that

$$\operatorname{Re}\lambda[B_2G_2] < 0 \tag{6.4}$$

then Theorem 5.1 of Chapter 2 holds and a two-time-scale design is possible by designing the slow system, obtained by setting $\varepsilon = 0$ in (6.3),

$$\dot{x}_s = [A_{11} - A_{12}(B_2G_2)^{-1}B_2G_1]x_s \tag{6.5}$$

and the fast system

$$\frac{dz_f(\tau)}{d\tau} = B_2G_2z_f(\tau). \tag{6.6}$$

Taking $G_1 = G_2G_0$, (6.5) becomes

$$\dot{x}_s = (A_{11} - A_{12}G_0)x_s. \tag{6.7}$$

Thus, invoking Theorem 3.1 of Chapter 2 and Lemma 6.1, the two-time-scale high-gain design method is encapsulated by the following eigenvalue assignment theorem.

Theorem 6.1

Let the system (6.1) be completely controllable, and let the $m \times n$ matrix G_0 and the $m \times m$ matrix G_2 be chosen such that

$$\lambda_i(A_{11} - A_{12}G_0) = p_i, \quad \operatorname{Re} p_i < 0, \quad i = 1, \ldots, n, \tag{6.8}$$

$$\lambda_j(B_2G_2) = q_j, \quad \operatorname{Re} q_j < 0, \quad j = 1, \ldots, m, \tag{6.9}$$

where p_i and q_j/ε are the prescribed locations of the distinct slow and distinct fast eigenvalues respectively. Then, there exists an $\varepsilon^* > 0$ such that for all $\varepsilon \in (0, \varepsilon^*]$ the application of the high-gain feedback control

$$u = \frac{1}{\varepsilon}[G_2G_0x + G_2z] \tag{6.10}$$

assigns the eigenvalues of the system (6.1) to $p_i + O(\varepsilon)$ and $[q_j + O(\varepsilon)]/\varepsilon$.

The foregoing theorem, in analogy with Theorem 3.1, suggests a two-time-scale design procedure for the assignment of the closed-loop system eigenvalues using high-gain feedback: construct G_0 so as to place the $O(1)$

eigenvalues of $A_{11} - A_{12}G_0$; separately construct G_2 so as to place the $O(1/\varepsilon)$ large eigenvalues of B_2G_2; then form the composite feedback control (6.10) for application to the system (6.1). Although the two separate lower-order eigenvalue assignment problems (6.8) and (6.9) do not depend on ε, the composite control (6.10), in contradistinction to (3.1), does depend on ε. The singularly perturbed form of the closed-loop system (6.3) corresponds to the frequently observed fact that systems (6.1) under high-gain feedback control (6.10) exhibit a two-time-scale response in that the initial fast transient (boundary layer) is followed by a slow time evolution. The fast transient and the slow evolution are due respectively to the eigenvalues $\frac{\lambda_j}{\varepsilon}(B_2G_2)$ and $\lambda_i(A_{11} - A_{12}G_0)$ specified in Theorem 6.1.

Corollary 6.1

If B_2G_2 and $A_{11} - A_{12}G_0$ are Hurwitz matrices, and the high-gain feedback control (6.10) is applied to the system (6.1), then there exists an $\varepsilon^* > 0$ such that the resulting closed-loop system (6.1), (6.10) is asymptotically stable for all $\varepsilon \in (0, \varepsilon^*]$.

Proof Although the proof is immediate from Theorem 6.1, the following alternative proof is of independent interest. Any high-gain control of the form

$$u = \frac{1}{\varepsilon} v, \quad v = O(1) \tag{6.11}$$

when applied to the open-loop system (6.1) induces it to assume the singularly perturbed form

$$\begin{bmatrix} \dot{x} \\ \varepsilon \dot{z} \end{bmatrix} = \begin{bmatrix} A_{11} & A_{12} \\ \varepsilon A_{21} & \varepsilon A_{22} \end{bmatrix} \begin{bmatrix} x \\ z \end{bmatrix} + \begin{bmatrix} 0 \\ B_2 \end{bmatrix} v. \tag{6.12}$$

Since εA_{22} is singular at $\varepsilon = 0$, application of the fast feedback control

$$v = G_2 z + \tilde{v} \tag{6.13}$$

to (6.12) is required to convert (6.12) to the standard singularly perturbed form

$$\begin{bmatrix} \dot{x} \\ \varepsilon \dot{z} \end{bmatrix} = \begin{bmatrix} A_{11} & A_{12} \\ \varepsilon A_{21} & \varepsilon A_{22} + B_2 G_2 \end{bmatrix} \begin{bmatrix} x \\ z \end{bmatrix} + \begin{bmatrix} 0 \\ B_2 \end{bmatrix} \tilde{v}. \tag{6.14}$$

To stabilize this system we apply the composite feedback control of Section 3.2. Upon setting $\varepsilon = 0$ in (6.14), the slow system of (6.14), described by the pair$(A_{11}, -A_{12}G_2^{-1})$, is stabilized by the control

$$\tilde{v} = G_2 G_0 x, \tag{6.15}$$

since $A_{11} - A_{12}G_0$ is Hurwitz. With B_2G_2 also Hurwitz, it is enough to use the reduced control (6.15). The closed-loop system (6.14) is asymptotically stable by Corollary 2.1, while (6.15), (6.13) and (6.11) yield the overall stabilizing control (6.10). □

Example 6.1

Consider a system in the form of (6.1):

$$\dot{x} = \begin{bmatrix} 3 & 1 & 1 \\ -6 & 1 & 0 \\ 0 & 0 & 3 \end{bmatrix} x + \begin{bmatrix} 0 & 0 \\ 1 & 1 \\ 0 & 1 \end{bmatrix} u. \tag{6.16}$$

First, suppose that a high-gain feedback control is to be found to place the slow eigenvalue near $\lambda_1 = -3$ and the fast eigenvalues near $\lambda_{2,3} = (-1 \pm j1)/\varepsilon$. We solve the two lower-order eigenvalue assignment problems (6.8) and (6.9) and obtain

$$G_0 = \begin{bmatrix} 6 \\ 0 \end{bmatrix}, \quad G_2 = \begin{bmatrix} 0 & 1 \\ -2 & 0 \end{bmatrix}. \tag{6.17}$$

The high-gain composite feedback (6.10) is

$$u = \frac{1}{\varepsilon} \begin{bmatrix} 0 & 0 & 1 \\ -12 & -2 & 0 \end{bmatrix} x. \tag{6.18}$$

The exact eigenvalues are computed for comparison purposes and the results for values of ε of 0.01 and 0.001 are summarized in Table 3.1. It is

TABLE 3.1

Gain $1/\varepsilon$	Exact eigenvalues	Specified eigenvalues	Percentage error
	−2.69	−3.0	14
100	−95.15		
	±j107.96	−100±j100	
	−2.97	−3.0	
1000	−995.02		~1
	±j1008.0	−1000 ± j1000	

observed that as $\varepsilon \to 0$, $\lambda_{1,2,3}$ tend to the specified values. For design purposes the gain $1/\varepsilon$ does not have to be very high, since with $1/\varepsilon = 100$ the accuracy of 14% in eigenvalue location is often acceptable. Finally, should a closer eigenvalue assignment be desired, a corrected high-gain feedback design based upon Theorem 3.2 may be pursued (see Exercise 3.9).

3.6.1 The Cheap Control Problem

High-gain feedback systems also result from the optimization of the system (6.1) with respect to a quadratic performance index having a small penalty on u:

$$J = \frac{1}{2}\int_0^\infty \left\{ [x^T \quad z^T] \begin{bmatrix} Q_{11} & Q_{12} \\ Q_{12}^T & Q_{22} \end{bmatrix} \begin{bmatrix} x \\ z \end{bmatrix} + \varepsilon^2 u^T R u \right\} dt, \tag{6.19}$$

where $Q \geq 0$, $R > 0$ and ε is a small positive scalar.

Using the high-gain control (6.11), the so-called "cheap" control problem (6.19), (6.1) is transformed to the equivalent linear–quadratic control problem

$$J = \frac{1}{2}\int_0^\infty \left\{ [x^T \quad z^T] \begin{bmatrix} Q_{11} & Q_{12} \\ Q_{12}^T & Q_{22} \end{bmatrix} \begin{bmatrix} x \\ z \end{bmatrix} + v^T R v \right\} dt \tag{6.20}$$

for the system (6.12) where it is observed that the singularity at $\varepsilon = 0$ has been transferred from the performance index J to the dynamic system model (6.12). Under the conditions of Theorem 4.1, the solution to this equivalent optimal linear regulator problem (6.20), (6.12) for any given $\varepsilon > 0$ is of the form (4.4), (4.5). Although the system model (6.12) is a nonstandard singularly perturbed model, the corresponding algebraic Riccati equation (4.5) nonetheless possesses a power series expansion at $\varepsilon = 0$ of the form of (4.26) where $K_1^{(0)}$, $K_2^{(0)}$ and $K_3^{(0)}$ satisfy (4.28)–(4.30) with $A_{21} = 0$, $A_{22} = 0$, $S_1 = 0$ and $S_{12} = 0$:

$$0 = -K_1^{(0)} A_{11} - A_{11}^T K_1^{(0)} + K_2^{(0)} S_2 K_2^{(0)T} - Q_{11}, \tag{6.21}$$

$$0 = K_2^{(0)} S_2 K_3^{(0)} - K_1^{(0)} A_{12} - Q_{12}, \tag{6.22}$$

$$0 = K_3^{(0)} S_2 K_3^{(0)} - Q_{22}. \tag{6.23}$$

Assumption 6.2

Q_{22} is a positive-definite matrix.

Under this assumption, a two-time-scale decomposition of the problem (6.20), (6.12) proceeds in the manner of Section 3.4. Rewriting (6.23) as

$$S_2^{1/2}K_3^{(0)}S_2^{1/2}S_2^{1/2}K_3^{(0)}S_2^{1/2} = (S_2^{1/2}K_3^{(0)}S_2^{1/2})^2 = S_2^{1/2}Q_{22}S_2^{1/2}, \qquad (6.24)$$

and taking the square root of both sides of (6.24), we have, under Assumption 6.2 where B_2 is nonsingular, that the unique solution of (6.23) is

$$K_3^{(0)} = S_2^{-1/2}(S_2^{1/2}Q_{22}S^{1/2})^{1/2}S^{-1/2} > 0. \qquad (6.25)$$

Rearranging (6.22) as

$$K_2^{(0)} = (Q_{12} + K_1^{(0)}A_{12})(S_2K_3^{(0)})^{-1} \qquad (6.26)$$

and substituting (6.26) in (6.21), noting (6.23), results in the nth-order algebraic Riccati equation

$$\begin{aligned} 0 = &-K_1^{(0)}(A_{11} - A_{12}Q_{22}^{-1}Q_{12}^{\mathrm{T}}) - (A_{11} - A_{12}Q_{22}^{-1}Q_{12}^{\mathrm{T}})^{\mathrm{T}}K_1^{(0)} \\ &+ K_1^{(0)}A_{12}Q_{22}^{-1}A_{12}^{\mathrm{T}}K_1^{(0)} - (Q_{11} - Q_{12}Q_{22}^{-1}Q_{12}^{\mathrm{T}}). \end{aligned} \qquad (6.27)$$

Lemma 6.2

If the pair (A, B) is stabilizable and the pair (D, A_{11}) is detectable, where

$$D^{\mathrm{T}}D \triangleq Q_{11} - Q_{12}Q_{22}^{-1}Q_{12}^{\mathrm{T}}, \qquad (6.28)$$

then there exists a unique positive-semidefinite stabilizing solution of the algebraic Riccati equation (6.27).

Proof The proof follows from Lemma 6.1 and Lemma 4.1. □

Consider the control

$$v = G_1x + G_2z, \qquad (6.29)$$

$$G_1 \triangleq -R^{-1}B_2^{\mathrm{T}}K_2^{(0)\mathrm{T}}, \quad G_2 = -R^{-1}B_2^{\mathrm{T}}K_3^{(0)}. \qquad (6.30)$$

Substituting (6.26) in (6.30), and using (6.23), we have

$$G_1 = -R^{-1}B_2^{\mathrm{T}}K_3^{(0)}Q_{22}^{-1}(Q_{12}^{\mathrm{T}} + A_{12}^{\mathrm{T}}K_1^{(0)}) \qquad (6.31)$$

or

$$G_1 = G_2G_0, \qquad (6.32)$$

where

$$G_0 = Q_{22}^{-1}(Q_{12}^{\mathrm{T}} + A_{12}^{\mathrm{T}}K_1^{(0)}) \qquad (6.33)$$

and G_2 is as defined in (6.30). By (6.30)–(6.33), and Theorem 4.2, the original optimal control (4.4) for the regulator problem (6.20), (6.12) is approximated to within $O(\varepsilon)$ by the composite control

$$u = \frac{1}{\varepsilon} v = \frac{1}{\varepsilon}[G_2 G_0 x + G_2 z]. \tag{6.34}$$

It is observed that (6.34) is of the *same* structure as the high-gain composite control (6.10). This time, G_0, given by (6.33) and (6.27), is the solution of a reduced-order regulator problem for the slow system (6.7), while G_2, given by (6.30) and (6.25), is the solution of a reduced-order regulator problem for the fast system (6.6).

Beyond the fact that the composite control (6.34) is within $O(\varepsilon)$ of the optimal one, it can be proved using Theorem 4.3 that the composite control is *near-optimal* in the sense that the performance J of the feedback system (6.1), (6.34) is $O(\varepsilon^2)$ close to its optimum performance.

To regulate the output

$$y = M\begin{bmatrix} x \\ z \end{bmatrix} = M_1 x + M_2 z \tag{6.35}$$

we set $Q = M^{\mathrm{T}} M$ as in Section 3.4.

Assumption 6.3

$M_2 B_2$ is a nonsingular matrix.

By assumption 6.3 and the nonsingularity of B_2, M_2 is nonsingular, and an $O(\varepsilon^2)$ near-optimal control can be found by the preceding decomposition procedure. Since $Q_{12} = M_1^{\mathrm{T}} M_1$, $Q_{12} = M_1^{\mathrm{T}} M_2$, $Q_{22} = M_2^{\mathrm{T}} M_2$ and M_2 is nonsingular, then $D = 0$ in (6.28), and the detectability condition of Lemma 6.2 is replaced by

$$\operatorname{Re} \lambda(A_{11} - A_{12} M_2^{-1} M_1) < 0. \tag{6.36}$$

Also, with $D = 0$ the solution of (6.27) is

$$K_1 = 0. \tag{6.37}$$

By (6.37), $\lambda(A_{11} - A_{12} M_2^{-1} M_1)$ are the eigenvalues of the slow system (6.7) under the slow control (6.33). In fact, $\lambda(A_{11} - A_{12} M_2^{-1} M_1)$ are the *transmission zeros* of the open-loop system (6.1) with output (6.35). Condition (6.36) requires that the transmission zeros of the system (6.1), (6.35) should be in the open left-half plane (Young, Kokotović and Utkin, 1977).

3.7 Robust Output-Feedback Design

It was established in Corollary 2.1 that a state-feedback control law, designed to stabilize the reduced-order system model is robust in the sense that it will stabilize the actual full-order system model for sufficiently small ε provided the neglected fast modes are asymptotically stable. Unfortunately, a similar conclusion for output-feedback control does not hold in general.

In this section, we discuss the robustness of output-feedback control design for linear time-invariant models to a common class of *structured uncertainties*, that of *high-frequency parasitics*. As before, the structuring of high-frequency uncertainties is achieved through the use of singular perturbation methods. The actual system model P_ε is represented as

$$\left.\begin{aligned} \dot{x} &= A_{11}x + A_{12}z + B_1u, && (7.1a)\\ \varepsilon\dot{z} &= A_{21}x + A_{22}z + B_2u, && (7.1b)\\ y &= C_1x + C_2z, && (7.1c) \end{aligned}\right\} P_\varepsilon$$

where $x \in R^n$ is the dominant state vector, $z \in R^m$ is the state vector of the parasitic dynamics, $y \in R^p$ is the output vector, and $\varepsilon > 0$ is a small positive scalar representing the parasitics (e.g. small inductances, capacitances, inertias). Neglect of the parasitic elements by setting $\varepsilon = 0$ in (7.1) results in the reduced-order design model P_0, assumed to be stabilizable and detectable:

$$\left.\begin{aligned} \dot{x} &= A_0x + B_0u, && (7.2a)\\ y &= C_0x + D_0u, && (7.2b) \end{aligned}\right\} P_0$$

where

$$A_0 = A_{11} - A_{12}A_{22}^{-1}A_{21}, \quad B_0 = B_1 - A_{12}A_{22}^{-1}B_2, \tag{7.3}$$

$$C_0 = C_1 - C_2A_{22}^{-1}A_{21}, \quad D_0 = -C_2A_{22}^{-1}B_2, \tag{7.4}$$

and where A_{22} is assumed to be nonsingular.

The robustness issue under discussion is whether an output-feedback controller, designed to stabilize the nominal system model in (7.2), will in fact stabilize the actual system model P_ε in (7.1) for ε sufficiently small.

A fairly general output-feedback control scheme for the design model (7.2), depicted in Fig. 3.9, is

$$\left.\begin{aligned} \dot{v} &= G_1v + G_2y + G_3u, && (7.5a)\\ u &= F_1v + F_2y, && (7.5b) \end{aligned}\right\} C$$

where (7.5a) is a dynamic compensator of order l; $0 \le l \le n$. The closed-

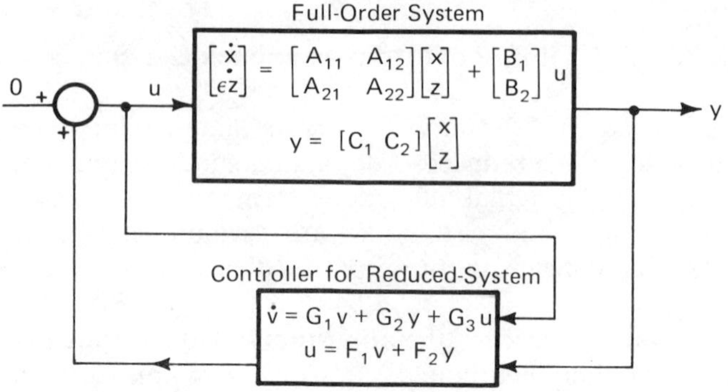

Fig. 3.9. Closed-loop output feedback control system.

loop design model (7.2), (7.5) is given by

$$\begin{bmatrix} \dot{x} \\ \dot{v} \end{bmatrix} = W \begin{bmatrix} x \\ v \end{bmatrix} \triangleq \begin{bmatrix} W_{11} & W_{12} \\ W_{21} & W_{22} \end{bmatrix} \begin{bmatrix} x \\ v \end{bmatrix}, \tag{7.6}$$

where, assuming F_2 is chosen such that $I - F_2 D_0$ is nonsingular,

$$\begin{aligned} W_{11} &= A_0 + B_0(I - F_2 D_0)^{-1} F_2 C_0, \\ W_{12} &= B_0(I - F_2 D_0)^{-1} F_1, \\ W_{21} &= [G_2 + (G_3 + G_2 D_0)(I - F_2 D_0)^{-1} F_2] C_0, \\ W_{22} &= G_1 + (G_3 + G_2 D_0)(I - F_2 D_0)^{-1} F_1. \end{aligned}$$

The actual closed-loop model (7.1), (7.5) is given by

$$\begin{bmatrix} \dot{x} \\ \dot{v} \\ \hline \varepsilon \dot{z} \end{bmatrix} = \left[\begin{array}{c|c} \Gamma_1 & \Gamma_2 \\ \hline \Gamma_3 & \Gamma_4 \end{array}\right] \begin{bmatrix} x \\ v \\ \hline z \end{bmatrix}, \tag{7.7}$$

where $[x^{\mathrm{T}} \quad v^{\mathrm{T}}]^{\mathrm{T}}$ is the dominant state, z is the fast state and

$$\Gamma_1 = \begin{bmatrix} A_{11} + B_1 F_2 C_1 & B_1 F_1 \\ (G_2 + G_3 F_2) C_1 & G_1 + G_3 F_1 \end{bmatrix},$$

$$\Gamma_2 = \begin{bmatrix} A_{12} + B_1 F_2 C_2 \\ (G_2 + G_3 F_2) C_2 \end{bmatrix},$$

$$\Gamma_3 = [A_{21} + B_2 F_2 C_1 \quad B_2 F_1],$$

$$\Gamma_4 = A_{22} + B_2 F_2 C_2.$$

Performing a standard two-time-scale decomposition in the manner of Section 2.4, the closed-loop singularly perturbed model (7.7) assumes, in transformed coordinates, the decoupled form

$$\begin{bmatrix} \dot{\xi} \\ \varepsilon\dot{\eta} \end{bmatrix} = \begin{bmatrix} \Gamma_0 + 0(\varepsilon) & 0 \\ 0 & \Gamma_4 + O(\varepsilon) \end{bmatrix} \begin{bmatrix} \xi \\ \eta \end{bmatrix}, \tag{7.8}$$

provided Γ_4 is nonsingular to ensure

$$\Gamma_0 \triangleq \Gamma_1 - \Gamma_2\Gamma_4^{-1}\Gamma_3 \tag{7.9}$$

is well defined.

Let us verify that Γ_4 is in fact nonsingular. Use of the identity

$$(A_{22} + B_2F_2C_2)\Gamma_4^{-1} = I_m \tag{7.10}$$

yields

$$\Gamma_4^{-1} = A_{22}^{-1}[I - B_2F_2C_2\Gamma_4^{-1}]. \tag{7.11}$$

Rearranging (7.11), we have

$$A_{22}^{-1} = (I + A_{22}^{-1}B_2F_2C_2)\Gamma_4^{-1}, \tag{7.12}$$

so that

$$\begin{aligned} F_2C_2A_{22}^{-1} &= F_2C_2(I + A_{22}^{-1}B_2F_2C_2)\Gamma_4^{-1} \\ &= (I + F_2C_2A_{22}^{-1}B_2)F_2C_2\Gamma_4^{-1}. \end{aligned} \tag{7.13}$$

Premultiplication of both sides of (7.13) by $(I - F_2D_0)^{-1}$ results in

$$(I - F_2D_0)^{-1}F_2C_2A_{22}^{-1} = F_2C_2\Gamma_4^{-1}. \tag{7.14}$$

Hence, substituting (7.14) into (7.11), we have

$$\Gamma_4^{-1} = A_{22}^{-1}[I - B_2(I - F_2D_0)^{-1}F_2C_2A_{22}^{-1}], \tag{7.15}$$

and so Γ_4 is nonsingular.

Similarly, it can be shown after lengthy manipulation that

$$\Gamma_0 \triangleq \Gamma_1 - \Gamma_2\Gamma_4^{-1}\Gamma_3 = W. \tag{7.16}$$

Equation (7.16) is an intuitively pleasing result in that it implies that the slow system associated with the full singularly perturbed closed-loop system (7.7) is identical with the system (7.6) obtained from applying output-feedback control to the open-loop slow system (7.2).

By design, the controller parameters l, G_1, G_2, G_3, F_1 and F_2 are chosen such that the closed-loop design model (7.6) is asymptotically stable, that is

$$\operatorname{Re}\lambda(W) < 0. \tag{7.17}$$

However, even if $\operatorname{Re}\lambda(A_{22}) < 0$, there is no guarantee that the second condition

$$\operatorname{Re}\lambda(A_{22} + B_2F_2C_2) < 0 \tag{7.18}$$

for overall closed-loop system stability in (7.8) is observed. What is provided in the next theorem are conditions that ensure (7.18) is satisfied.

Theorem 7.1

If $\operatorname{Re}\lambda(A_{22}) < 0$, and any one of the following three conditions is satisfied:

(a) (A_{22}, B_2) is uncontrollable: weak controllability of high-frequency parasitics;
(b) (C_2, A_{22}) is unobservable: weak observability of high-frequency parasitics;
(c) $F_2 = 0$: no static feedback from output to input;

then there exists an $\varepsilon^* > 0$ such that the feedback controller (7.5) stabilizes the actual system (7.1) for all $\varepsilon \in (0, \varepsilon^*]$.

Remark 7.1: If none of the conditions (a), (b) and (c) are observed, stability of the actual closed-loop model (7.7) or (7.8) *cannot* be guaranteed.

Remark 7.2: It is recalled from Section 2.6 that weak controllability and weak observability of the fast eigenvalues (high-frequency parasitics) means that they will not be affected by more than $O(\varepsilon)$ through output feedback.

Remark 7.3: For a system model (7.1) with high-frequency parasitics, conditions (a) and (b) may be replaced by

(a) $B_2 = 0$: weak control of parasitics;
(b) $C_2 = 0$: weak observation of parasitics.

Remark 7.4: The state-feedback control $u = G_0x$ of Corollary 2.1 is *always* robust since $C_1 = I_n$ and $C_2 = 0$.

A General Robust Design Rule

Avoid static output feedback for sytems with unmodeled high-frequency dynamics.

In the light of Theorem 7.1 and the above design rule, one may expect an instability problem in the case of static output feedback, as shown by the following simple example.

Example 7.1

Consider the second-order system

$$\left.\begin{aligned} \dot{x} &= z, \\ \varepsilon\dot{z} &= -x - z + u, \\ y &= 2x + z, \end{aligned}\right\} \tag{7.19}$$

for which $\operatorname{Re}\lambda(A_{22}) = -1 < 0$. Upon neglecting the parasitic element by setting $\varepsilon = 0$, one obtains the reduced-order slow model

$$\left.\begin{aligned} \dot{x} &= -x + u, \\ y &= x + u. \end{aligned}\right\} \tag{7.20}$$

Using this first-order model as the design model, a designer has chosen the output-feedback control

$$u = 2y \tag{7.21}$$

so as to place the eigenvalue at -3. When the same feedback control (7.21) is applied to the actual system (7.19), the resulting closed-loop system is

$$\left.\begin{aligned} \dot{x} &= z, \\ \varepsilon\dot{z} &= 3x + z, \end{aligned}\right\} \tag{7.22}$$

which indeed has a slow eigenvalue $-3 + O(\varepsilon)$, as desired. However, it also has an unstable fast eigenvalue $(1 + O(\varepsilon))/\varepsilon$; that is, the design is nonrobust. The source of instability is the presence of z fed from the input ($B_2 \neq 0$: strong controllability of parasitics) *and* the presence of z in the output equation of (7.19) ($C_2 \neq 0$: strong observability of parasitics) and static feedback from output to input ($F_2 \neq 0$).

3.7.1 Frequency-Domain Robustness Design

In order to further assess the robustness of output feedback using the frequency-domain models of Section 2.7, the actual open-loop system model P_ε of (7.1) is described in the block diagonal form of Section 2.4 by

$$\begin{bmatrix} \dot{\zeta} \\ \varepsilon\dot{\eta} \end{bmatrix} = \begin{bmatrix} A_0 + O(\varepsilon) & 0 \\ 0 & A_{22} + O(\varepsilon) \end{bmatrix} \begin{bmatrix} \zeta \\ \eta \end{bmatrix} + \begin{bmatrix} B_0 + O(\varepsilon) \\ B_2 + O(\varepsilon) \end{bmatrix} u, \tag{7.23a}$$

$$y = [C_0 + O(\varepsilon) \quad C_2 + O(\varepsilon)] \begin{bmatrix} \zeta \\ \eta \end{bmatrix}, \tag{7.23b}$$

where the matrix triple $\{A_0, B_0, C_0\}$ is as defined in (7.3) and (7.4). If A_0 and A_{22} have no eigenvalues on the imaginary axis, the open-loop system model P_ε has a transfer-function matrix $G(s, \varepsilon)$ which is approximated for *all* frequencies $s = j\omega$ according to

$$G(s, \varepsilon) = G_s(s) + G_f(\varepsilon s) - D_0 + O(\varepsilon), \tag{7.24}$$

where

$$G_s(s) \triangleq C_0(sI_n - A_0)^{-1}B_0 + D_0 \tag{7.25}$$

is the transfer-function matrix of the design model (7.2) and is the asymptotic form of $G(s, \varepsilon)$ in the low-frequency scale $s = O(1)$, and

$$G_f(\varepsilon s) \triangleq C_2(\varepsilon s I_m - A_{22})^{-1}B_2 = C_2(pI_m - A_{22})^{-1}B_2 \tag{7.26}$$

is the transfer-function matrix of the parasitic block and is the asymptotic form of $G(s, \varepsilon)$ in the high-frequency scale $s = p/\varepsilon$, $p = O(1)$.

Consider the application of a controller $C(s)$, designed on the basis of the nominal model $G_s(s)$, to the actual system $G(s)$ as depicted in Fig. 3.10. Given that the controller (7.5) has the transfer-function matrix

$$C(s) = -F_1(sI_l - G_1 - G_3F_1)^{-1}(G_2 + G_3F_2) - F_2, \tag{7.27}$$

the robustness condition $F_2 = 0$ of Theorem 7.1 requires that the controller transfer-function matrix $C(s)$ in (7.27) be *strictly proper*. In other words, the robustness condition (c) of Theorem 7.1 is equivalent to requiring that the controller $C(s)$ have the *low-pass property*

$$C(s) = O(\varepsilon), \quad s \in [j\omega_1/\varepsilon, \infty) \tag{7.28}$$

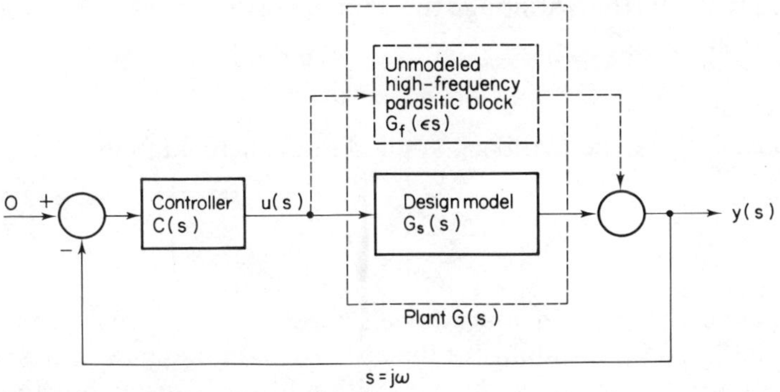

Fig. 3.10. Robust output feedback control.

Example 7.2

The robustness of output-feedback design based upon the full-order state observer (O'Reilly, 1983b) for (7.2),

$$\dot{\hat{x}} = A_0\hat{x} + B_0u + K_0[y - D_0u - C_0\hat{x}], \tag{7.29a}$$

$$u = F_0\hat{x}, \tag{7.29b}$$

follows from Theorem 7.1, since in this case $F_2 = 0$. In a frequency-domain context, the observer-based controller design (7.29) is robust by virtue of the fact that its transfer-function matrix

$$C(s) = -F_0(sI_n - A_0 - (B_0 - K_0D_0)F_0 + K_0C_0)^{-1}K_0 \tag{7.30}$$

has the low-pass property (7.28).

Consider again another example of static output feedback.

Example 7.3

The actual model (7.1) is given by

$$\dot{x} = x + [1 \quad 0]z + 2u, \tag{7.31a}$$

$$\varepsilon\dot{z} = \begin{bmatrix} 0 \\ 1 \end{bmatrix} x + \begin{bmatrix} -1 & 0 \\ -1 & -1 \end{bmatrix} z + \begin{bmatrix} -1 \\ -1 \end{bmatrix} u, \tag{7.31b}$$

$$y = x + [0 \quad 1]z. \tag{7.31c}$$

The nominal design model (7.2) is given by

$$\dot{x} = x + u, \tag{7.32a}$$

$$y = 2x. \tag{7.32b}$$

It is readily shown that, although the nominal plant model (7.32) is stabilized by

$$u = -ku \tag{7.33}$$

for all $k > \frac{1}{2}$, the same feedback applied to the actual model (7.31) leads to

$$A_{22} + B_2F_2C_2 = \begin{bmatrix} -1 & k \\ -1 & -1+k \end{bmatrix}, \tag{7.34}$$

which has unstable eigenvalues for all $k > 2$.

This potential for instability of the fast system modes may also be inferred from a two-frequency-scale decomposition of the transfer-function matrix of the open-loop system (7.31) along the lines of (7.24):

$$G(s, \varepsilon) \triangleq \frac{2}{s-1} - \frac{\varepsilon s}{(\varepsilon s + 1)^2} + O(\varepsilon). \tag{7.35}$$

The source of instability is signal transmission through the high-frequency block as in Fig. 3.10; that is,

$$G_{\mathrm{f}}(\varepsilon s) \triangleq \frac{-\varepsilon s}{(\varepsilon s + 1)^2} \quad \text{is } O(1) \text{ for } s \in [\mathrm{j}\omega_1/\varepsilon, \infty), \tag{7.36}$$

or in p-scale, $p = \varepsilon s$,

$$G_{\mathrm{f}}(p) = \frac{-p}{(p+1)^2} \quad \text{is } O(1) \text{ for } p = O(1).$$

Example 7.4

As a means of overcoming the problem of instability in Example 7.3, consider instead of the static feedback controller (7.33), the class of controller C in Figure 3.10 described by

$$C(s) = \frac{-k}{Ts + 1}, \quad T > 0. \tag{7.37}$$

This first-order lag controller is robust by virtue of the fact that it *blocks* high-frequency signal transmission in the sense of (7.28).

Another interpretation of the robustness property of the controller (7.37) is as follows. For a fixed $T > 0$, the low-pass filter $1/(Ts + 1)$ is described by

$$\dot{\psi} = -\frac{1}{T}\psi + \frac{1}{T}y, \quad y_{\text{filt}} = \psi. \tag{7.38}$$

Augmenting the actual model (7.31) with (7.38), one has

$$\begin{bmatrix} \dot{\psi} \\ \dot{x} \end{bmatrix} = \begin{bmatrix} \frac{-1}{T} & \frac{1}{T} \\ 0 & 1 \end{bmatrix} \begin{bmatrix} \psi \\ x \end{bmatrix} + \begin{bmatrix} 0 & \frac{1}{T} \\ 1 & 0 \end{bmatrix} z + \begin{bmatrix} 0 \\ 2 \end{bmatrix} u, \tag{7.39a}$$

$$\varepsilon \dot{z} = \begin{bmatrix} 0 & 0 \\ 0 & 1 \end{bmatrix} \begin{bmatrix} \psi \\ x \end{bmatrix} + \begin{bmatrix} -1 & 0 \\ -1 & -1 \end{bmatrix} z + \begin{bmatrix} -1 \\ -1 \end{bmatrix} u, \tag{7.39b}$$

$$y_{\text{filt}} = [1 \quad 0] \begin{bmatrix} \psi \\ x \end{bmatrix}. \tag{7.39c}$$

In the filtered outputs y_{filt} of (7.39c) the system parasitics are now weakly observable ($C_2 = 0$). Hence, in view of condition (b) of Theorem 7.1, stability of the actual closed-loop system is ensured by the application of any feedback loop

$$u = k y_{\text{filt}} \tag{7.40}$$

for any k that stabilizes the nominal design model (7.2) obtained from (7.39) by setting $\varepsilon = 0$.

In conclusion, a frequency-domain interpretation of our earlier robust design rule is as follows.

Frequency-Domain Robust Design Rule

Avoid closing high-frequency plant loops by using strictly proper controllers or controllers with the low-pass property (7.28).

3.8 Exercises

Exercise 3.1

Using the state transition matrix method or any other method, solve the normalized motor equations (3.4) of Example 3.1 under (a) the exact control $u(\varepsilon)$ of (3.13), (b) the composite control u_c of (3.6) and (c) the reduced control u_r of (3.14). Hence, for the system initial condition $x(0) = 4$, $z(0) = 3$, compare graphically the closed-loop responses $x(t)$ and $z(t)$ under controls (a), (b) and (c) for $\varepsilon = 0.1$ and $\varepsilon = 0.01$. Comment upon the accuracy of the state trajectory approximations. Also, comment on the possible significance of the singular perturbation parameter ε being unknown except that it is in the range $0.01 \leq \varepsilon \leq 0.1$.

Exercise 3.2

Deduce an upper bound on ε for Example 3.1 such that the closed-loop stability of the motor system is maintained (a) under the composite control u_c of (3.6) and (b) under the reduced control u_r of (3.14).

Hint: Use Lemma 2.2 of Chapter 2.

Would such a value of ε result in an acceptable approximation to the desired eigenvalue placement of $\{-3, -2/\varepsilon\}$?

Exercise 3.3

Derive discrete-time analogues of Theorem 2.1 and Theorem 3.1 for the linear system model

$$x(n+1) = (I + \varepsilon A_{11})x(n) + \varepsilon A_{12}z(n) + \varepsilon B_1 u(n), \quad x(0) = x^0,$$

$$z(n+1) = A_{21}x(n) + A_{22}z(n) + B_2 u(n), \quad z(0) = z^0,$$

where $I_m - A_{22}$ is nonsingular.

Hint: The slow system, evolving in the slow time scale εn, is

$$X(\varepsilon n + \varepsilon) - X(\varepsilon n) = \varepsilon A_{11}X(\varepsilon n) + \varepsilon A_{12}Z(\varepsilon n) + \varepsilon B_1 U(\varepsilon n),$$

$$Z(\varepsilon n) = A_{21}X(\varepsilon n) + A_{22}Z(\varepsilon n) + B_2 U(\varepsilon n),$$

which upon division by ε, letting $\varepsilon \to 0$, and eliminating Z, is of the *continuous-time* form (2.3) in the slow variables $x_s(t)$ and $u_s(t)$. The fast system is defined by the variables $x_f(n) = x(n) - x_s(t)$, $z_f(n) = z(n) - z_s(t)$ and $u_f(n) = u(n) - u_s(t)$. See Litkouhi and Khalil (1984) for more details.

Exercise 3.4

The linear composite control design (2.11) presumes A_{22}^{-1} exists. Construct a composite feedback control for the system (2.1), (2.2) when A_{22} is singular.

Hint: Use a preliminary fast feedback control, recalling Corollary 2.3, $u = Kz + v$, where $A_{22} + B_2K$ is nonsingular. (See Sannuti (1977).)

Exercise 3.5

Consider the weakly controllable system of Exercise 2.7:

$$\begin{bmatrix} \dot{x} \\ \varepsilon \dot{z} \end{bmatrix} = \begin{bmatrix} -1 & 1 \\ 0 & -1 \end{bmatrix} \begin{bmatrix} x \\ z \end{bmatrix} + \begin{bmatrix} -1 \\ 1 \end{bmatrix} u.$$

Why is it neither possible to design a reduced feedback control nor a composite feedback control? Calculate and comment on the exact control necessary to place the system eigenvalues at $\{-3, -1/\varepsilon\}$.

Exercise 3.6

Construct a corrected linear quadratic design for the DC-motor system (3.4) with $\varepsilon = 0.2$ and the performance index

$$J = \frac{1}{2}\int_0^\infty (2x^2 + u^2)\,\mathrm{d}t.$$

Compare the design and the associated performance index with the uncorrected near-optimal design.

Exercise 3.7

Derive a two-time-scale optimal linear regulator design for the discrete-time system model of Exercise 3.3 with the performance index

$$J = \varepsilon \sum_{n=0}^{\infty} [y^{\mathrm{T}}(n)y(n) + u^{\mathrm{T}}(n)Ru(n)], \quad R = R^{\mathrm{T}} > 0.$$

Prove its near-optimality along the lines of Theorem 4.2 and Theorem 4.3.
Hint Scale the exact Riccati equation solution as

$$\begin{bmatrix} \dfrac{P_1}{\varepsilon} & P_2 \\ P_2^{\mathrm{T}} & P_3 \end{bmatrix}.$$

(See Litkouhi and Khalil (1984) for more detail.)

Exercise 3.8

Design a high-gain composite control to assign the eigenvalue spectrum

$\{-3, -100, -150\}$ to the system

$$\dot{x} = \begin{bmatrix} 0 & 1 & 2 \\ -2 & 3 & 0 \\ -2 & -1 & 0 \end{bmatrix} x + \begin{bmatrix} 1 & 2 \\ 1 & 0 \\ 0 & 0 \end{bmatrix} u.$$

Comment on the accuracy of the scalar gain factor chosen.

Exercise 3.9

Derive a corrected high-gain feedback design, corresponding to the uncorrected one of Theorem 6.1, so as to assign the system eigenvalues to $p_i + O(\varepsilon^2)$, $i = 1, \ldots, n$, and $[q_j + O(\varepsilon^2)]/\varepsilon$, $j = 1, \ldots, m$.

Hint: Determine slow and fast corrected models so as to apply the eigenvalue assignment procedure of Theorem 3.2.

Exercise 3.10

Consider the second-order system

$$\dot{x} = x + z + u,$$

$$\varepsilon \dot{z} = x - z + u.$$

By selective use of a low-pass filter, a designer has the choice of measurements $y = x + z$ or $y = x$. Which measurement should he use if he is to employ a robust output feedback control $u = ky$, and why? Having made his choice, what design value of k is required in order to shift the dominant eigenvalue of the system to -6?

Exercise 3.11

Deduce conditions similar to those of Theorem 7.1 such that the linear system

$$\dot{x} = A_{11}x + A_{12}z + B_1 u,$$

$$\varepsilon \dot{z} = A_{21}x + A_{22}z + B_2 u$$

with direct input–output transmission

$$y = C_1 x + C_2 z + Du$$

is robust with respect to neglect of parasitics when under the static output feedback control $u = Ky$.

3.9 Notes and References

The complete decomposition of linear state-feedback design according to slow mode and fast mode performance specifications, and the subsequent recombination in the shape of an "order of ε" accurate composite control, is due to Chow and Kokotović (1976a). The composite state feedback controllers for corrected models in Theorem 2.2 and Corollary 2.2 are new, as is the invariance of the slow model to fast feedback in Corollary 2.3. The use of fast feedback in overcoming possible singularity of the matrix A_{22} was first noted by Kokotović and Haddad (1975a) and is further discussed in Sannuti (1977) and Porter (1977).

The eigenvalue assignment results of Theorem 3.1, implicit in Chow and Kokotović (1976a), are developed by Porter (1974, 1977), and independently by Suzuki and Miura (1976), through the application of a stability theorem of Klimushchev and Krasovskii (1962) to the closed-loop system. Theorem 3.2 on eigenvalue assignment for corrected models is new. An early general stabilizability condition for linear time-varying systems was formulated by Wilde (1972). Extensions of these eigenvalue assignment and stabilization procedures to time-varying systems, general two-time-scale systems, discrete-time systems, eigenstructure assignment and multivariable tracking are surveyed in Kokotović (1984) and Saksensa *et al.* (1984). More recently, composite control strategies for discrete-time systems have been developed by Litkouhi and Khalil (1985) for the cases of single-rate sampling and multirate sampling.

Since the publication by Sannuti and Kokotović (1969a), one of the most actively investigated singularly perturbed control problems has been the linear–quadratic (LQ) regulator problem. Approaches, in the main, have been either via the singularly perturbed Riccati equation or boundary-value problems (Kokotović *et al.* 1976). The particular scaling of the regulator Riccati equation was first proposed by Sannuti and Kokotović (1969a), while the first complete decomposition of a more general linear regulator problem into separate slow and fast regulator problems, upon which Section 3.4 is based, is achieved in Chow and Kokotović (1976a). Two-time-scale decomposition of near-optimal regulators for discrete-time linear systems in Litkouhi and Khalil (1984) parallels the earlier continuous-time results of Chow and Kokotović (1976a). A practical application of the near-optimal control to a drum boiler power plant is presented in Cori and Maffezzoni (1984).

The corrected linear quadratic design of Section 3.5 is an extension of the original paper by Sannuti and Kokotović (1969a) to the general case where $M_2 \neq 0$ and A_{22} may not be Hurwitz. Higher-order approximations to the optimal performance are possible through matched asymptotic expansions. Numerous other references on asymptotic expansions, multiple time scales, distributed parameter systems, quasi-conservative oscillatory systems, and large space structures for the optimal linear problem are given in Kokotović (1984) and Saksensa *et al.* (1984). The discussion in Sections 3.3 and 3.4 of the deleterious effect of weak controllability on linear state feedback control is based on Chow (1978b). Dual conclusions obtain for weak observability in the state observer design briefly introduced in Section 3.7.

High-gain feedback control problems have been investigated from many different standpoints: chiefly, cheap control, multivariable root loci, almost-singular state estimation, variable structure systems and the theory of almost invariant subspaces. See Saksensa *et al.* (1984), Kokotović (1984) and Saberi and Sannuti (1985, 1986) for more detailed discussions and numerous references therein. The unifying framework provided by the singular perturbation methods of Section 3.6 is based in part upon the exposition of Young *et al.* (1977) under the assumption that B_2G_2 is nonsingular.

Our detailed examination of the robustness of general output feedback control strategies in Section 3.7 is based on Khalil (1981b); see also Khalil (1984a), Kokotović (1984) and O'Reilly (1985). Earlier conditions for the cases of static output feedback and reduced-order observer design also take the form of "weakly observable parasitics" (Kokotović, 1984; Saksensa *et al.* 1984). Robust output feedback control strategies, based upon the full-order state observer O'Reilly (1983b), date from Porter (1974); see also Balas (1978), O'Reilly (1980) and Javid (1982). An analysis of the interaction of actuator and sensor parasitics—frequently neglected in feedback design—with the fast system modes is undertaken in Young and Kokotović (1982). A recent complementary result of Vidyasagar (1984, 1985) shows that the actual system model P_ε approaches the design model P_0 in the graph topology as the parasitic elements ε tend to zero if the high-frequency block $G_f(\varepsilon s) \equiv 0$ in Fig. 3.10. The design principle of avoiding closed-loop high-frequency signal transmission for robust feedback control carries over to adaptive systems and nonlinear systems. For instance, Ioannou and Kokotović (1983) establish the robustness of an adaptive observer through the use of a first-order filter. The two-frequency-scale description of general transfer-function matrix models by Luse and Khalil (1983) provides a starting point for more general studies of robustness within the feedback configuration of Fig. 3.6 as, for instance, described in Khalil (1985).

4 STOCHASTIC LINEAR FILTERING AND CONTROL

4.1 Introduction

It is to be expected that the linear–quadratic (LQ) designs of Sections 3.4 and 3.5 have their stochastic counterparts for systems where model uncertainties are by no means negligible and partial measurements of states are accompanied by measurement noise. As with the deterministic problems of Chapter 3, advantage may be taken of the singularly perturbed nature of the system model to decompose the stochastic filtering and control problems into well-conditioned subproblems in separate time scales. Additional care must, however, be exercised in the slow–fast decomposition of systems driven by white noise, since, as discussed in Section 4.2, the white noise process "fluctuates" faster than the fast states no matter how small the singular perturbation parameter $\varepsilon > 0$ is. Taking the variance of the fast states into account, state approximations are achieved which are of the order of $\varepsilon^{1/2}$ rather than of the order of ε previously obtained for deterministic systems in Section 2.5.

This two-time-scale decomposition for stochastic system models is extended in Section 4.3 to that of the Kalman–Bucy filtering problem. It is shown that the estimates of the slow and fast states, provided by Kalman–Bucy filters in separate time scales, are within an order of $\varepsilon^{1/2}$ of those provided by a Kalman–Bucy filter for the original system. The associated slow and fast filter gain matrices are independently determined by well-conditioned lower-order Riccati equations which are dual to those of the near-optimal regulator of Section 3.4.

Section 4.4 proceeds with the stochastic version of the near-optimal regulator solution of Section 3.4 for systems with incomplete state information. In the spirit of Chapter 3, the essence of the approach is the construction of a near-optimal composite control from the solutions of lower-order stochastic control problems in separate time scales. By the well known separation theorem of stochastic control (Kwakernaak and Sivan, 1972), these slow and fast solutions are provided by the slow and fast regulators of Section 3.4 in tandem with the corresponding slow and fast Kalman–Bucy filters of Section 4.3. The near-optimality of the composite stochastic control is illustrated in Section 4.5 by way of application to the aircraft autopilot Case Study 7.2 of Chapter 1.

As we have seen in Chapter 3, an asset of the singular perturbation approach to control system design is the ability to construct corrected designs should the performance of the initial uncorrected design prove inadequate. A higher-order approximation to the optimal strategy is achieved in Section 4.6 through the use of a corrected linear–quadratic Gaussian (LQG) design that parallels the corrected near-optimal regulator design of Section 3.5. As before, the key to the corrected design is the parameterization of the problem solution in ε.

Finally, the case of scaling the contribution of the white noise disturbances by a fractional power of ε is discussed in Section 4.7. It is shown that when all white noise inputs are scaled by the same factor, the approximation and design schemes of Sections 4.2–4.6 remain valid.

4.2 Slow–Fast Decomposition in the Presence of White-Noise Inputs

The stochastic counterpart of the singularly perturbed system model studied in Chapter 2 is

$$\dot{x}(t) = A_{11}x(t) + A_{12}z(t) + B_1 w(t), \quad x(0) = x_0, \tag{2.1}$$

$$\varepsilon\dot{z}(t) = A_{21}x(t) + A_{22}z(t) + B_2 w(t), \quad z(0) = z_0, \tag{2.2}$$

where x, z and w are n-, m- and s-dimensional vectors respectively. The input $w(t)$ is zero-mean stationary Gaussian white noise with intensity matrix $V > 0$, i.e. $\mathrm{E}\{w(t)w^{\mathrm{T}}(s)\} = V\delta(t-s)$. The initial conditions x_0, z_0 are jointly Gaussian random vectors independent of w. It is assumed, as usual, that A_{22} is nonsingular.

Our objective is to define a slow subsystem that describes the slow dynamics and a fast subsystem that describes the fast dynamics; then use their solutions to approximate x and z. We start by formally decomposing the system by extending the deterministic ideas of Chapters 1 and 2 to the stochastic case. To define the slow subsystem, we neglect the fast dynamics, which is equivalent to setting $\varepsilon = 0$ in (2.2), and dropping the initial condition z_0. The resulting equations are

$$\dot{\bar{x}}(t) = A_{11}\bar{x}(t) + A_{12}\bar{z}(t) + B_2 w(t), \quad \bar{x}(0) = x_0, \tag{2.3}$$

$$\bar{z}(t) = -A_{22}^{-1}(A_{21}\bar{x}(t) + B_2 w(t)). \tag{2.4}$$

Unlike deterministic systems with slow driving inputs, $\bar{z}$ defined by (2.4) is not a valid approximation of z as $\varepsilon \to 0$. In fact, the variance of $z - \bar{z}$ is infinite for all ε because of the white noise component of $\bar{z}$. However, $\bar{z}$ can still be used to approximate z as an input to a slow system. This is an important point that is worth more explanation. Let $\eta(t)$ satisfy

$$\varepsilon\dot{\eta}(t) = \Gamma\eta(t) + w(t), \quad \eta(0) = 0,$$

where Γ is a Hurwitz matrix and $w(t)$ is white noise. As we explained above, $-\Gamma^{-1}w(t)$ is not a valid approximation of $\eta(t)$ as $\varepsilon \to 0$.

Let $\xi(t) = \int_0^t \eta(\tau)\,\mathrm{d}\tau$, which represents the dynamics of a slow system (an integrator in this case). As $\varepsilon \to 0$, $\xi(t)$ can be approximated by replacing $\eta(t)$ by $-\Gamma^{-1}w(t)$. To see this, define $\bar{\xi}(t) = -\int_0^t \Gamma^{-1}w(\tau)\,\mathrm{d}\tau$ and calculate the variance of $\xi(t) - \bar{\xi}(t)$:

$$\begin{aligned} e(t) = \xi(t) - \bar{\xi}(t) &= \int_0^t \eta(\sigma)\,\mathrm{d}\sigma + \int_0^t \Gamma^{-1}w(\tau)\,\mathrm{d}\tau \\ &= \int_0^t \frac{1}{\varepsilon}\int_0^\sigma \mathrm{e}^{\Gamma(\sigma-\tau)/\varepsilon} w(\tau)\,\mathrm{d}\tau\,\mathrm{d}\sigma + \int_0^t \Gamma^{-1}w(\tau)\,\mathrm{d}\tau. \end{aligned}$$

Reversing the order of integration in the first term, we obtain

$$\begin{aligned} e(t) &= \int_0^t \int_\tau^t \frac{1}{\varepsilon}\mathrm{e}^{\Gamma(\sigma-\tau)/\varepsilon}\,\mathrm{d}\sigma\, w(\tau)\,\mathrm{d}\tau + \int_0^t \Gamma^{-1}w(\tau)\,\mathrm{d}\tau \\ &= \int_0^t \Gamma^{-1}[\mathrm{e}^{\Gamma(t-\tau)/\varepsilon} - I]w(\tau)\,\mathrm{d}\tau + \int_0^t \Gamma^{-1}w(\tau)\,\mathrm{d}\tau \\ &= \int_0^t \Gamma^{-1}\,\mathrm{e}^{\Gamma(t-\tau)/\varepsilon} w(\tau)\,\mathrm{d}\tau. \end{aligned}$$

Now

$$\begin{aligned} \mathrm{E}\{e(t)e^{\mathrm{T}}(t)\} &= \int_0^t \int_0^t \Gamma^{-1}\,\mathrm{e}^{\Gamma(t-\tau)/\varepsilon} V\delta(\tau-\sigma)\,\mathrm{e}^{\Gamma^{\mathrm{T}}(t-\sigma)/\varepsilon}\Gamma^{-\mathrm{T}}\,\mathrm{d}\tau\,\mathrm{d}\sigma \\ &= \int_0^t \Gamma^{-1}\,\mathrm{e}^{\Gamma(t-\tau)/\varepsilon} V\,\mathrm{e}^{\Gamma^{\mathrm{T}}(t-\tau)/\varepsilon}\Gamma^{-\mathrm{T}}\,\mathrm{d}\tau. \end{aligned}$$

Therefore

$$\|\mathrm{E}\{e(t)e^{\mathrm{T}}(t)\}\| \leq \int_0^t K\,\mathrm{e}^{-\alpha(t-\tau)/\varepsilon}\,\mathrm{d}\tau \leq \frac{\varepsilon K}{\alpha}$$

for some positive constants α and K.

Thus $\bar{\xi}(t)$ is a valid approximation of $\xi(t)$ as $\varepsilon \to 0$ in the mean-square sense. Actually, $\bar{\xi}(t)$ is an $O(\varepsilon^{1/2})$ approximation of $\xi(t)$ according to the following definition, which will be adopted throughout the chapter.

Definition 4.1

A random variable $f(\varepsilon)$ is said to converge to f in a mean-square sense (m.s.s.) if $\mathrm{E}(f(\varepsilon)-f)^2 \to 0$ as $\varepsilon \to 0$. Furthermore, $f(\varepsilon)-f$ is said to be $O(\varepsilon^\gamma)$ for some $\gamma > 0$ if there exist $\varepsilon^* > 0$ and $K > 0$ such that for all $\varepsilon \in (0, \varepsilon^*)$, $[\mathrm{E}(f(\varepsilon)-f)^2]^{1/2} \leq K\varepsilon^\gamma$.

Coming back to (2.3), (2.4), we see that as far as $\bar{x}$ is concerned (2.4) can be used to approximate $\bar{z}$ in (2.3). Substituting (2.4) in (2.3) and defining $x_s(t) = \bar{x}(t)$, we obtain the slow subsystem

$$\dot{x}_s(t) = A_0 x_s(t) + B_0 w(t), \quad x_s(0) = x_0, \tag{2.5}$$

where

$$A_0 = A_{11} - A_{12}A_{22}^{-1}A_{21}, \quad B_0 = B_1 - A_{12}A_{22}^{-1}B_2. \tag{2.6}$$

The slow state $x_s(t)$ is a candidate for approximating $x(t)$.

To define the fast subsystem, we introduce the stretching time scale $\tau = t/\varepsilon$ and the notation $X(\tau) = x(\varepsilon\tau)$, $Z(\tau) = z(\varepsilon\tau)$ and $W(\tau) = \varepsilon^{1/2}w(\varepsilon\tau)$, where the $\varepsilon^{1/2}$ factor is used so that $W(\tau)$ and $w(t)$ have the same intensity matrix. In the τ-time scale (2.1) and (2.2) are given by

$$\frac{\mathrm{d}X}{\mathrm{d}\tau} = \varepsilon A_{11}X(\tau) + \varepsilon A_{12}Z(\tau) + \varepsilon^{1/2}B_1 W(\tau), \quad X(0) = x_0, \tag{2.7}$$

$$\frac{\mathrm{d}Z}{\mathrm{d}\tau} = A_{21}X(\tau) + A_{22}Z(\tau) + \varepsilon^{-1/2}B_2 W(\tau), \qquad Z(0) = z_0. \tag{2.8}$$

For sufficiently small ε, $X(\tau)$ is much slower than $Z(\tau)$, and the $A_{21}X(\tau)$ term on the right-hand side of (2.8) can be viewed as a constant input whose presence results in a bias in $Z(\tau)$. To remove the bias we define

$$Z_f(\tau) = Z(\tau) + A_{22}^{-1}A_{21}X(\tau). \tag{2.9}$$

Taking the derivative of $Z_f(\tau)$ with respect to τ, and using (2.7) and (2.8), it follows that the new variable $Z_f(\tau)$ satisfies the differential equation

$$\begin{aligned}\frac{\mathrm{d}Z_f}{\mathrm{d}\tau} = {} & \varepsilon A_{22}^{-1}A_{21}A_0X(\tau) + (A_{22} + \varepsilon A_{22}^{-1}A_{21}A_{12})Z_f(\tau) \\ & + \varepsilon^{-1/2}(B_2 + \varepsilon A_{22}^{-1}A_{21}B_1)W(\tau),\end{aligned} \tag{2.10}$$

with initial condition

$$Z_f(0) = z_0 + A_{22}^{-1}A_{21}x_0.$$

Neglecting coefficients that are $O(\varepsilon^{1/2})$ or smaller, we get

$$\frac{dZ_f}{d\tau} = A_{22}Z_f(\tau) + \varepsilon^{-1/2}B_2W(\tau), \quad Z_f(0) = z_0 + A_{22}^{-1}A_{21}x_0. \tag{2.11}$$

Equation (2.11) is the desired fast subsystem. Setting $z_f(t) = Z_f(t/\varepsilon)$, the fast subsystem can be expressed in the t-time scale as

$$\varepsilon\dot{z}_f(t) = A_{22}z_f(t) + B_2w(t), \quad x_f(0) = z_0 + A_{22}^{-1}A_{21}x_0. \tag{2.12}$$

Now, based on (2.9), $z_f(t) - A_{22}^{-1}A_{21}x_s(t)$ is taken as a candidate for approximating $z(t)$.

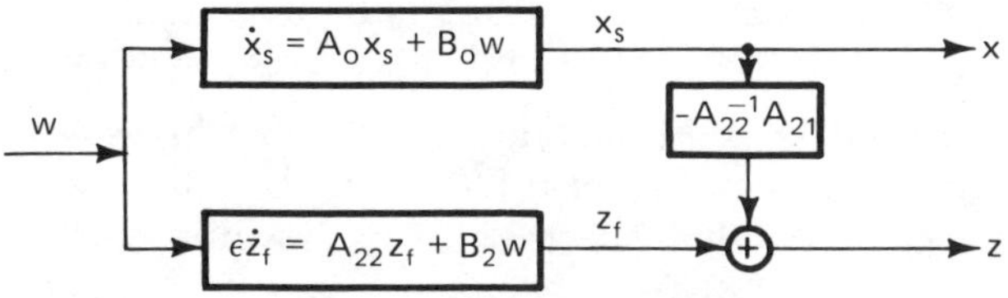

Fig. 4.1. Slow–fast decomposition of a singularly perturbed system driven by white noise.

In summary, the candidate decomposition, shown in Fig. 4.1, is given by

$$\dot{x}_s(t) = A_0x_s(t) + B_0w(t), \quad x_s(0) = x_0, \tag{2.13a}$$

$$\varepsilon\dot{z}_f(t) = A_{22}z_f(t) + B_2w(t), \quad z_f(0) = z_0 + A_{22}^{-1}A_{21}x_0, \tag{2.13b}$$

$$x(t) \sim x_s(t), \tag{2.13c}$$

$$z(t) \sim -A_{22}^{-1}A_{21}x_s(t) + z_f(t). \tag{2.13d}$$

It is important to notice that in the proposed decomposition the slow component of z is its component due to $A_{21}x$ and not its component due to the input w, as it is customary in singularly perturbed systems driven by slow inputs. No matter how small ε is, the fluctuations of the white noise input are faster than the fast dynamics. Therefore, the fast subsystem accounts not only for boundary layer transients but also for the component of z due to the white noise input. The following example illustrates the decomposition (2.13).

Example 2.1

Consider the second-order system

$$\left.\begin{aligned} \dot{x} &= -x + z, \\ \varepsilon\dot{z} &= -z + w, \end{aligned}\right\} \tag{2.14}$$

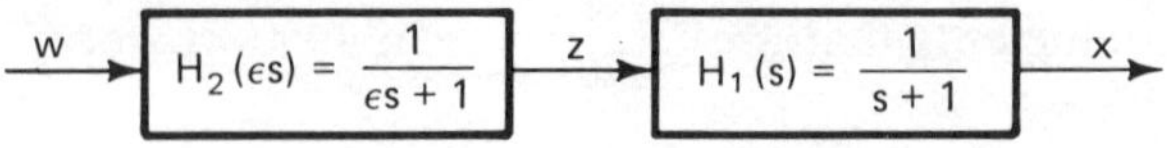

Fig. 4.2 Second-order stochastic system.

where w is white noise with intensity σ^2. A block diagram representation of this system is shown in Fig. 4.2.

Let us consider only the steady-state behavior of x and z. From Fig. 4.2, we see that the power spectra of x and z are given by

$$S_x(\omega) = \frac{\sigma^2}{(\omega^2 + 1)(\varepsilon^2\omega^2 + 1)}, \quad S_z(\omega) = \frac{\sigma^2}{\varepsilon^2\omega^2 + 1},$$

which are sketched in solid lines in Fig. 4.3. Setting $\varepsilon = 0$ in (2.14) is equivalent to replacing $H_2(\varepsilon s) = 1/(1 + \varepsilon s)$ by the identity transfer function. The corresponding spectrum of z, shown in dotted lines, is flat

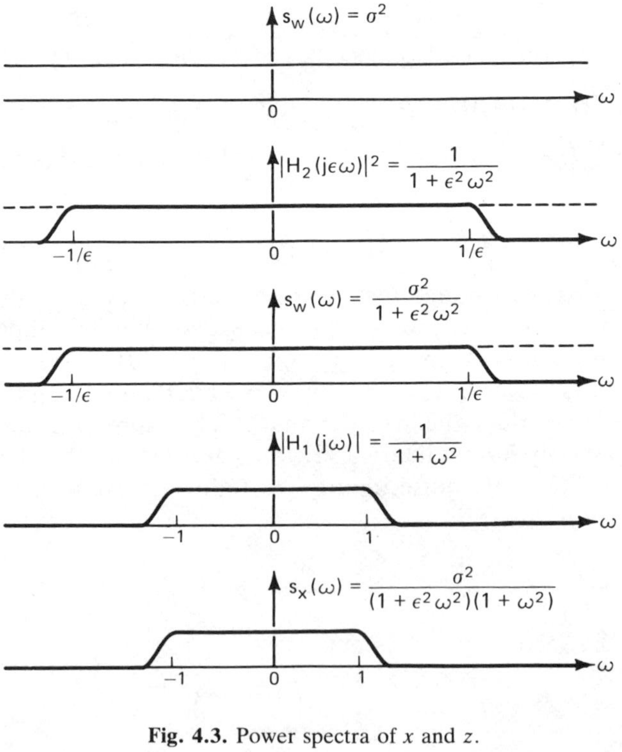

Fig. 4.3. Power spectra of x and z.

for all ω, i.e. white noise. It does not make sense to approximate z by white noise, because no matter how small ε is, the difference between the spectrum of z and the white noise spectrum will yield infinity when integrated over all ω. However, Fig. 4.3 shows that whether we use the exact spectrum $S_z(\omega)$ or its idealized white version, we still end up with the same spectrum $S_x(\omega)$. In other words, the error in approximating z by white noise is filtered out by the slow subsystem. That is why in defining the slow subsystem we take $z = w$ and $\dot{x}_s = -x_s + w$. On the other hand, to determine the wideband spectrum of z we have to use the equation $\varepsilon\dot{z} = -z + w$, which is the fast subsystem called for in the decomposition (2.13).

Asympototic validity of the decomposition (2.13) is established by the following theorem, which is the stochastic counterpart of the initial value theorem, Theorem 5.1, of Chapter 2. Notice, however, the order of the state approximation is $O(\varepsilon^{1/2})$ rather than $O(\varepsilon)$.

Theorem 2.1

Suppose that A_0 and A_{22} are Hurwitz matrices, then there exists an $\varepsilon^* > 0$ such that for all $\varepsilon \in (0, \varepsilon^*)$ (2.15) is satisfied for all $t \geq 0$:

$$\left.\begin{aligned} x(t) - x_s(t) &= O(\varepsilon^{1/2}), \\ z(t) + A_{22}^{-1}A_{21}x_s(t) - z_f(t) &= O(\varepsilon^{1/2}). \end{aligned}\right\} \tag{2.15}$$

Proof The proof is complicated by the fact that the variances of both z_f and z are of order $1/\varepsilon$. The order of the variance of z_f can be easily seen from (2.12), while the order of the variance of z follows from showing (Exercise 4.3) that the variance matrix of (x^T, z^T) takes the form

$$Q = \begin{bmatrix} Q_{11} & Q_{12} \\ Q_{12}^T & \frac{1}{\varepsilon}Q_{22} \end{bmatrix}.$$

The proof comprises three steps. First, the decoupling transformation of Section 2.4 is applied to the system (2.1), (2.2) to decouple the slow and fast dynamics. It is verified that the decomposition (2.13) is obtained from the decoupled system as perturbations of right-hand-side coefficients and initial conditions. Second, we apply an approximation result given in Appendix A for singularly perturbed systems driven by white noise. Finally, the inverse of the decoupling transformation is employed to show (2.15).

Application of the decoupling transformation (see Section 2.4) to (2.1),

(2.2) yields

$$\left.\begin{aligned}\dot{\xi}(t) &= [A_{11} - A_{12}L(\varepsilon)]\xi(t) + [B_1 - H(\varepsilon)B_2 - \varepsilon H(\varepsilon)L(\varepsilon)B_1]w(t),\\ \xi(0) &= [I_n - \varepsilon H(\varepsilon)L(\varepsilon)]x_0 - \varepsilon H(\varepsilon)z_0,\end{aligned}\right\} \tag{2.16}$$

$$\left.\begin{aligned}\varepsilon\dot{\eta}(t) &= [A_{22} + \varepsilon L(\varepsilon)A_{12}]\eta(t) + [B_2 + \varepsilon L(\varepsilon)B_1]w(t),\\ \eta(0) &= L(\varepsilon)x_0 + z_0.\end{aligned}\right\} \tag{2.17}$$

Using the fact that $L(\varepsilon)$ and $H(\varepsilon)$ are analytic at $\varepsilon = 0$ and $L(0) = A_{22}^{-1}A_{21}$, it can be easily seen that the right-hand-side coefficients and initial conditions of (2.5) and (2.12) are $O(\varepsilon)$ perturbations of (2.16) and (2.17), respectively. Hence, by Theorem A.1 of Appendix A,

$$\xi(t) - x_s(t) = O(\varepsilon),$$

$$\eta(t) - z_f(t) = O(\varepsilon^{1/2}).$$

Now, using the inverse of the decoupling transformation, we have

$$x(t) - x_s(t) = \xi(t) - x_s(t) + \varepsilon H(\varepsilon)\eta(t)$$

$$\begin{aligned}\mathrm{E}\|x(t) - x_s(t)\|^2 &\le 2\mathrm{E}\|\xi(t) - x_s(t)\|^2 + 2\varepsilon^2\|H\|^2\mathrm{E}\|\eta(t)\|^2\\ &\le C_1\varepsilon^2 + C_2\varepsilon \le C\varepsilon,\end{aligned}$$

where we have used the fact that $\operatorname{var}(\eta(t)) = O(1/\varepsilon)$.

Notice that the error in approximating the slow variable ξ by x_s is $O(\varepsilon)$, but the error in approximating the slow variable x by x_s is $O(\varepsilon^{1/2})$ owing to the $O(1/\varepsilon)$ order of variance of η.

Similarly,

$$\begin{aligned}z(t) + A_{22}^{-1}A_{21}x_s(t) - z_f(t) = &- L(0)[\xi(t) - x_s(t)]\\ &- [L(\varepsilon) - L(0)]\xi(t)\\ &+ [\eta(t) - z_f(t)]\\ &- \varepsilon L(\varepsilon)H(\varepsilon)\eta(t).\end{aligned}$$

Hence

$$\mathrm{E}\|z(t) + A_{22}^{-1}A_{21}x_s(t) - z_f(t)\|^2 \le C\varepsilon. \qquad \square$$

The use of the decoupling transformation in the proof of Theorem 2.1 provides a basis for obtaining higher-order approximations. Equations (2.16), (2.17), together with

$$x(t) = \xi(t) + \varepsilon H(\varepsilon)\eta(t) \tag{2.18}$$

and

$$z(t) = -L(\varepsilon)\xi(t) + [I_m - \varepsilon L(\varepsilon)H(\varepsilon)]\eta(t) \tag{2.19}$$

comprise an exact slow–fast decomposition of (2.1), (2.2). The decomposition (2.13), which yields $O(\varepsilon^{1/2})$ approximation, is obtained from (2.16)–(2.19) by $O(\varepsilon)$ perturbations of coefficients and initial conditions. Higher-order approximations can be obtained from (2.16)–(2.19) by using higher-order approximations of coefficients and initial conditions. In this respect the reader may consult our model building discussions of Chapter 2. Exercise 4.4 gives a slow–fast decomposition for an $O(\varepsilon^{3/2})$ approximation.

Example 2.2 Highly damped Langevin equation

The Langevin equation

$$m\ddot{y}(t) + f\dot{y}(t) + ky(t) = (2KTf)^{1/2}w(t) \tag{2.20}$$

describes the motion of a particle in a liquid (Papoulis, 1965). The process $y(t)$ is the position of the particle in one of its rectangular coordinates. The particle is subject to three forces: a collision force represented by the white noise term $(2KTf)^{1/2}w(t)$, a friction force represented by $-f\dot{y}$ and an external force proportional to its displacement represented by $-ky$. The constants m, f, K and T are mass, friction coefficient, Boltzmann constant and absolute temperature respectively, and $w(t)$ has intensity one (i.e. $w(t)$ is the derivative of a standard Wiener process). If $f^2 \gg km$ the poles of (2.20) are given by

$$\lambda_1 = -\frac{f}{2m} + \frac{f}{2m}\left(1 - \frac{4km}{f^2}\right)^{1/2} \sim -\frac{k}{f},$$

$$\lambda_2 = -\frac{f}{2m} - \frac{f}{2m}\left(1 - \frac{4km}{f^2}\right)^{1/2} \sim -\frac{f}{m}.$$

Let us suppose that the smallness of km/f^2 is due to the smallness of m or the largeness of f, but not due to the smallness o: k (as $k \to 0$ the external force $-ky$ disappears and the particle moves freely rather than harmonically bound; this case is not considered here). Rescaling time by setting $t = (f/k)s$ and setting $x(s) = y(fs/k)$ and $\tilde{w}(s) = (f/k)^{1/2}w(fs/k)$, (2.20) can be rewritten as

$$\frac{mk}{f^2}\frac{\mathrm{d}^2x}{\mathrm{d}s^2} + \frac{\mathrm{d}x}{\mathrm{d}s} + x = \left(\frac{2KT}{k}\right)^{1/2}\tilde{w}(s).$$

Taking $\varepsilon = mk/f^2 = |\lambda_1|/|\lambda_2|$ and $z = \mathrm{d}x/\mathrm{d}s$, we get the singularly perturbed system (2.21), which is of the form (2.1), (2.2):

$$\left.\begin{aligned}\frac{dx}{ds} &= z,\\ \varepsilon\frac{dz}{ds} &= -x - z + \left(\frac{2KT}{k}\right)^{1/2}\tilde{w}(s).\end{aligned}\right\}\tag{2.21}$$

Here $A_{22} = -1$ and $A_0 = A_{11} - A_{12}A_{22}^{-1}A_{21} = -1$, so that the stability conditions of Theorem 2.1 are satisfied. Therefore, by Theorem 2.1, x and z can be approximated for sufficiently small ε by x_s and $z_f - x_s$, respectively, where x_s satisfies (2.22) and z_f satisfies (2.23):

$$\frac{dx_s}{ds} = -x_s + \left(\frac{2KT}{k}\right)^{1/2}\tilde{w}(s), \quad x_s(0) = x(0), \tag{2.22}$$

$$\varepsilon\frac{dz_f}{ds} = -z_f + \left(\frac{2KT}{k}\right)^{1/2}\tilde{w}(s), \quad z_f(0) = x(0) + z(0). \tag{2.23}$$

4.3 The Steady-State Kalman–Bucy Filter

In Section 4.2 we have seen how to obtain slow–fast decompositions of singularly perturbed systems driven by white noise, and how to analyze the asymptotic behavior of such systems as $\varepsilon \to 0$. This analysis is now extended to an examination of the time-scale properties of Kalman–Bucy filters (Kwakernaak and Sivan, 1972) for stochastic singularly perturbed systems. To avoid the complication of having time-varying filter gains in the case of finite-time estimation, we limit our study to the steady-state estimation problem. This problem is one of generating an estimate

$$\begin{bmatrix}\hat{x}(t)\\ \hat{z}(t)\end{bmatrix} \text{ of the state } \begin{bmatrix}x(t)\\ z(t)\end{bmatrix}$$

in a least-square sense (Kwakernaak and Sivan, 1972) for the linear time-invariant system

$$\dot{x}(t) = A_{11}x(t) + A_{12}z(t) + F_1w(t), \quad x \in R^n, \quad w \in R^s, \tag{3.1}$$

$$\varepsilon\dot{z}(t) = A_{21}x(t) + A_{22}z(t) + F_2w(t), \quad z \in R^m, \tag{3.2}$$

on the basis of the observed output $\{y(\sigma); \sigma \leq t\}$, given by the output equation

$$y(t) = C_1x(t) + C_2z(t) + v(t), \quad y \in R^l, \tag{3.3}$$

where w and v are independent, zero-mean, stationary Gaussian white noise processes with intensities I_s (identity) and $V > 0$ respectively. Since the estimation problem will be studied at steady state (as $t \to \infty$), it is assumed that

$$\operatorname{Re}\lambda(A_{11} - A_{12}A_{22}^{-1}A_{21}) < 0, \quad \operatorname{Re}\lambda(A_{22}) < 0, \tag{3.4}$$

so that, by Corollary 3.1 of Chapter 2, the system is asymptotically stable for sufficiently small ε. The solution of this problem is the well-known steady-state Kalman–Bucy filter (Kwakernaak and Sivan, 1972) given by

$$\begin{bmatrix} \dot{\hat{x}}(t) \\ \dot{\hat{z}}(t) \end{bmatrix} = A \begin{bmatrix} \hat{x}(t) \\ \hat{z}(t) \end{bmatrix} + Q\left[y(t) - C \begin{bmatrix} \hat{x}(t) \\ \hat{z}(t) \end{bmatrix} \right], \tag{3.5}$$

where the filter gain matrix Q is specified by†

$$Q = PC^{\mathrm{T}}V^{-1}, \tag{3.6}$$

whereas $P = P(\varepsilon) \geq 0$ is the solution of the algebraic Riccati equation

$$0 = AP + PA^{\mathrm{T}} + FF^{\mathrm{T}} - PC^{\mathrm{T}}V^{-1}CP. \tag{3.7}$$

Here A, F and C stand for overall system matrices, i.e.

$$A = \begin{bmatrix} A_{11} & A_{12} \\ \dfrac{1}{\varepsilon}A_{21} & \dfrac{1}{\varepsilon}A_{22} \end{bmatrix}, \quad F = \begin{bmatrix} F_1 \\ \dfrac{1}{\varepsilon}F_2 \end{bmatrix}, \quad C = [C_1 \quad C_2].$$

Because of the $1/\varepsilon$ terms in A and F, the solution of the Riccati equation (3.7) takes the form

$$P(\varepsilon) = \begin{bmatrix} P_1(\varepsilon) & P_2(\varepsilon) \\ P_2^{\mathrm{T}}(\varepsilon) & \dfrac{1}{\varepsilon}P_3(\varepsilon) \end{bmatrix}, \tag{3.8}$$

which is similar to the solution of the Lyapunov equation (6.38) of Chapter 2. Substituting (3.8) in (3.7) results, after partitioning, in

$$\begin{aligned} 0 = {} & A_{11}P_1 + A_{12}P_2^{\mathrm{T}} + P_1A_{11}^{\mathrm{T}} + P_2A_{12}^{\mathrm{T}} + F_1F_1^{\mathrm{T}} \\ & - (P_1C_1^{\mathrm{T}} + P_2C_2^{\mathrm{T}})V^{-1}(C_1P_1 + C_2P_2^{\mathrm{T}}), \end{aligned} \tag{3.9a}$$

$$\begin{aligned} 0 = {} & \varepsilon A_{11}P_2 + A_{12}P_3 + P_1A_{21}^{\mathrm{T}} + P_2A_{22}^{\mathrm{T}} + F_1F_2^{\mathrm{T}} \\ & - (P_1C_1^{\mathrm{T}} + P_2C_2^{\mathrm{T}})V^{-1}(\varepsilon C_1P_2 + C_2P_3), \end{aligned} \tag{3.9b}$$

$$\begin{aligned} 0 = {} & \varepsilon A_{21}P_2 + A_{22}P_3 + \varepsilon P_2^{\mathrm{T}}A_{21}^{\mathrm{T}} + P_3A_{22}^{\mathrm{T}} + F_2F_2^{\mathrm{T}} \\ & - (\varepsilon P_2^{\mathrm{T}}C_1^{\mathrm{T}} + P_3C_2^{\mathrm{T}})V^{-1}(\varepsilon C_1P_2 + C_2P_3). \end{aligned} \tag{3.9c}$$

† The matrix X^{T} denotes the transpose of the matrix X.

Comparison of (3.9) with the partitioned form of the regulator Riccati equation (equation (5.4) of Chapter 3) shows the duality between the two sets of equations. Therefore, the properties of the solution of (3.9) for small ε are given by the following lemma, which is the dual of Theorem 4.2 of Chapter 3.

Lemma 3.1

If A_{22} is nonsingular and the triples (A_0, F_0, C_0) and (A_{22}, F_2, C_2) are stabilizable–detectable, where

$$A_0 = A_{11} - A_{12}A_{22}^{-1}A_{21}, \quad F_0 = F_1 - A_{12}A_{22}^{-1}F_2, \quad C_0 = C_1 - C_2A_{22}^{-1}A_{21},$$

then for sufficiently small ε (3.9) has a unique stabilizing solution which possesses a power series expansion at $\varepsilon = 0$. Furthermore,

$$P_1(0) = P_s, \quad P_2(0) = P_m, \quad P_3(0) = P_f,$$

where $P_s \geq 0$ is the stabilizing solution of the slow Riccati equation

$$\begin{aligned} 0 = {} & [A_0 - F_0S_0^TV_0^{-1}C_0]P_s + P_s[A_0 - F_0S_0^TV_0^{-1}C_0]^T \\ & + F_0[I - S_0^TV_0^{-1}S_0]F_0^T - P_sC_0^TV_0^{-1}C_0P_s, \\ S_0 = {} & -C_2A_{22}^{-1}F_2, \quad V_0 = V + S_0S_0^T, \end{aligned} \tag{3.10}$$

P_f is the stabilizing solution of the fast Riccati equation

$$0 = A_{22}P_f + P_fA_{22}^T + F_2F_2^T - P_fC_2^TV^{-1}C_2P_f, \tag{3.11}$$

and P_m is given by

$$\begin{aligned} P_m = {} & [P_s(C_1^TV^{-1}C_2P_f - A_{21}^T) - (A_{12}P_f + F_1F_2^T)] \\ & \times (A_{22} - P_fC_2^TV^{-1}C_2)^{-T}. \end{aligned} \tag{3.12}$$

All the conditions of Lemma 3.1 are implied by the stability assumptions of (3.4). Partitioning the filter gain Q as

$$Q(\varepsilon) = \begin{bmatrix} P_1(\varepsilon) & P_2(\varepsilon) \\ P_2^T(\varepsilon) & \dfrac{1}{\varepsilon}P_3(\varepsilon) \end{bmatrix} \begin{bmatrix} C_1^T \\ C_2^T \end{bmatrix} V^{-1} = \begin{bmatrix} Q_1(\varepsilon) \\ \dfrac{1}{\varepsilon}Q_2(\varepsilon) \end{bmatrix},$$

the Kalman–Bucy filter (3.5) can be rewritten as

$$\dot{\hat{x}}(t) = A_{11}\hat{x}(t) + A_{12}\hat{z}(t) + Q_1(\varepsilon)[y(t) - C_1\hat{x}(t) - C_2\hat{z}(t)], \tag{3.13}$$

$$\varepsilon\dot{\hat{z}}(t) = A_{21}\hat{x}(t) + A_{22}\hat{z}(t) + Q_2(\varepsilon)[y(t) - C_1\hat{x}(t) - C_2\hat{z}(t)]. \tag{3.14}$$

From Lemma 3.1 we can see that

$$Q_1(\varepsilon) = (P_1(\varepsilon)C_1^T + P_2(\varepsilon)C_2^T)V^{-1} = (P_sC_1^T + P_mC_2^T)V^{-1} + O(\varepsilon), \quad (3.15a)$$

$$Q_2(\varepsilon) = (\varepsilon P_2^T(\varepsilon)C_1^T + P_3(\varepsilon)C_2^T)V^{-1} = P_fC_2^TV^{-1} + O(\varepsilon). \quad (3.15b)$$

Thus the calculation of the filter gains Q_1 and Q_2 can be greatly simplified by employing Lemma 3.1. Solving the high-dimensional ill-conditioned Riccati equation (3.7) is replaced by solving two lower-order ε-independent Riccati equations (3.10) and (3.11). Furthermore, higher-order approximations can be obtained as in Section 3.5. Simplification of off-line Riccati equation computations is not the only simplification that can be achieved here. Further simplification can be achieved by decomposing the Kalman–Bucy filter into two lower-order filters: a slow filter to estimate the slow variables and a fast filter to estimate the fast variables. Such restructuring of the system gives a deeper insight into the problem, and leads to two-time-scale hierarchical structures.

The desired decomposition will be derived by formally defining slow and fast subsystems, as was done in Section 4.2. Slow and fast filters are designed to estimate the slow and fast variables respectively. Then slow and fast estimates are used to recover estimates of the original variables x and z.

A slow subsystem is defined by replacing z in (3.1) and (3.3) by its steady-state component, which is the solution of (3.2) with $\varepsilon = 0$. The slow subsystem is given by

$$\dot{x}_s(t) = A_0x_s(t) + F_0w(t), \quad (3.16a)$$

$$y(t) = C_0x_s(t) + S_0w(t) + v(t). \quad (3.16b)$$

The corresponding steady-state Kalman–Bucy filter is

$$\dot{\hat{x}}_s(t) = A_0\hat{x}_s(t) + Q_s[y(t) - C_0\hat{x}_s(t)],$$

where

$$Q_s = (P_sC_0^T + F_0S_0^T)V_0^{-1},$$

and $P_s \geq 0$ is the stabilizing solution of the slow Riccati equation (3.10).

A fast subsystem is defined by treating x as a constant variable. The bias caused by x in z and y can be removed by defining

$$z_f = z + A_{22}^{-1}A_{21}x$$

and

$$y_f = y - C_0x = C_2z_f + v.$$

The fast subsystem is given by

$$\varepsilon\dot{z}_f(t) = A_{22}z_f(t) + F_2w(t), \quad (3.17a)$$

$$y_f(t) = C_2z_f(t) + v(t). \quad (3.17b)$$

The corresponding steady-state Kalman–Bucy filter is

$$\varepsilon\dot{\hat{z}}_f(t) = A_{22}\hat{z}_f(t) + Q_f[y_f(t) - C_2\hat{z}_f(t)],$$

where

$$Q_f = P_f C_2^T V^{-1}$$

and $P_f \geq 0$ is the stabilizing solution of the fast Riccati equation (3.11). In implementing the fast filter, y_f is replaced by $y - C_0\hat{x}_s$. Thus wc have proposed thc following decomposition:

$$\dot{\hat{x}}_s(t) = A_0\hat{x}_s(t) + Q_s[y(t) - C_0\hat{x}_s(t)], \tag{3.18a}$$

$$\varepsilon\dot{\hat{z}}_f(t) = A_{22}\hat{z}_f(t) + Q_f[y(t) - C_0\hat{x}_s(t) - C_2\hat{z}_f(t)], \tag{3.18b}$$

$$\underline{\hat{x}}(t) = \hat{x}_s(t), \tag{3.18c}$$

$$\underline{\hat{z}}(t) = -A_{22}^{-1}A_{21}\hat{x}_s(t) + \hat{z}_f(t). \tag{3.18d}$$

A block-diagram representation of the decomposition (3.18) is shown in Fig. 4.4, and its near-optimality is established in Theorem 3.1.

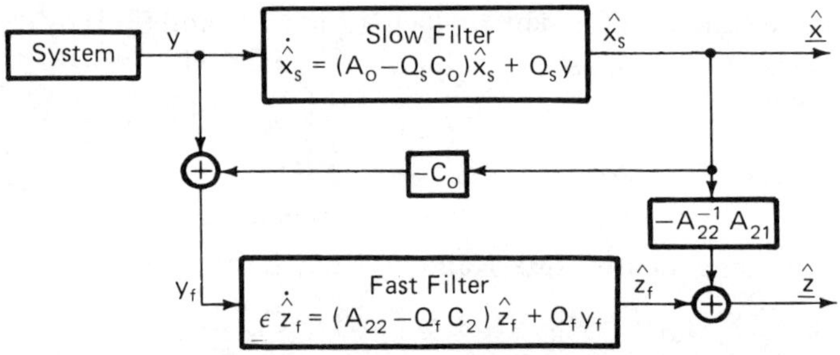

Fig. 4.4. Two-time-scale hierarchical state estimation.

Theorem 3.1

If A_0 and A_{22} are Hurwitz matrices then the decomposition (3.18) is near-optimal for sufficiently small ε. In particular, there exists an $\varepsilon^* > 0$ such that, for all $\varepsilon \in (0, \varepsilon^*)$, (3.19) holds at steady state (as $t \to \infty$):

$$\left.\begin{aligned} \hat{x}(t) - \underline{\hat{x}}(t) &= O(\varepsilon^{1/2}), \\ \hat{z}(t) - \underline{\hat{z}}(t) &= O(\varepsilon^{1/2}). \end{aligned}\right\} \tag{3.19}$$

Furthermore, if the optimal filter is initialized at certain initial conditions

$\hat{x}_0$ and $\hat{z}_0$, and the near-optimal filter (3.18) is initialized at $\hat{x}_s(0) = \hat{x}_0$ and $\hat{z}_f(0) = \hat{z}_0 + A_{22}^{-1}A_{21}\hat{x}_0$, then (3.19) holds for all $t \geq 0$.

Proof The idea of the proof is similar to that of Theorem 2.1. First, a decoupling transformation is used to transform the optimal filter into new coordinates where it can be verified that the decomposition (3.18) follows from the optimal filter by perturbation of right-hand-side coefficients. Then Theorem A.2 of Appendix A is employed to prove near-optimality. Although the decoupling transformation was discussed in Chapter 2 for singularly perturbed systems whose right-hand-side coefficients are independent of ε, extension to the case where the coefficients are analytic functions of ε at $\varepsilon = 0$ is straightforward (see Exercise 4.6). This extension is needed here because the Kalman–Bucy filter gains $Q_1(\varepsilon)$ and $Q_2(\varepsilon)$ are analytic at $\varepsilon = 0$. In the proof of Theorem 2.1, the matrices L and H of the decoupling transformation were chosen exactly as in Chapter 2. This time, the choice of L and H is more involved. So, let us derive the transformation step by step. A comparison of the fast filter (3.18b) with the second equation of the Kalman–Bucy filter, (3.14), shows that when both equations are viewed as systems driven by the innovation process $y - C_1\hat{x} - C_2\hat{z} \sim y - C_0\hat{x}_s - C_2\hat{z}_f$, the essential difference between them is the term $A_{21}\hat{x}$ on the right-hand side of (3.14). That term can be eliminated via the transformation

$$\hat{\eta}(t) = \hat{z}(t) + L\hat{x}(t), \tag{3.20}$$

where L satisfies

$$0 = A_{22}L - A_{21} - \varepsilon L(A_{11} - A_{12}L). \tag{3.21}$$

This equation is the same as equation (2.8) of Chapter 2. Hence, for sufficiently small ε, $L(\varepsilon)$ exists and $L(0) = A_{22}^{-1}A_{21}$. Using (3.20), the Kalman–Bucy filter (3.13), (3.14) is transformed into

$$\begin{aligned}\dot{\hat{x}}(t) &= (A_{11} - A_{12}L)\hat{x}(t) + A_{12}\hat{\eta}(t) \\ &\quad + Q_1[y(t) - (C_1 - C_2L)\hat{x}(t) - C_2\hat{\eta}(t)],\end{aligned} \tag{3.22}$$

$$\begin{aligned}\varepsilon\dot{\hat{\eta}}(t) &= (A_{22} + \varepsilon LA_{12})\hat{\eta}(t) \\ &\quad + (Q_2 + \varepsilon LQ_1)[y(t) - (C_1 - C_2L)\hat{x}(t) - C_2\hat{\eta}(t)].\end{aligned} \tag{3.23}$$

It is apparent that

$$C_1 - C_2L(\varepsilon) = C_1 - C_2A_{22}^{-1}A_{21} + O(\varepsilon) = C_0 + O(\varepsilon),$$

which confirms that the right-hand-side coefficients of the fast filter (3.18b) are $O(\varepsilon)$ perturbations of the right-hand-side coefficients of (3.23). Consider now the slow equations. Comparison of the slow filter (3.18a) with (3.22) shows that the term $(A_{12} - Q_1C_2)\hat{\eta}$ on the right-hand-side of (3.22) should be eliminated. This can be achieved by the transformation

$$\hat{\xi}(t) = \hat{x}(t) - \varepsilon H_1 \hat{\eta}(t), \tag{3.24}$$

where H_1 is chosen to satisfy

$$\begin{aligned} 0 = &- H_1(A_{22} - Q_2C_2 + \varepsilon LA_{12} - \varepsilon LQ_1C_2) + (A_{12} - Q_1C_2) \\ &+ \varepsilon(A_{11} - A_{12}L)H_1 - \varepsilon(Q_1 - H_1Q_2 + \varepsilon H_1LQ_1) \\ &(C_1 - C_2L)H_1. \end{aligned} \tag{3.25}$$

Setting $\varepsilon = 0$ in (3.25) shows that

$$\begin{aligned} H_1(0) &= [A_{12} - Q_1(0)C_2][A_{22} - Q_2(0)C_2]^{-1} \\ &= [A_{12} - (P_sC_1^T + P_mC_2^T)V^{-1}C_2][A_{22} - P_fC_2^TV^{-1}C_2]^{-1}, \end{aligned} \tag{3.26}$$

which is well-defined because $A_{22} - P_fC_2^TV^{-1}C_2$ is a Hurwitz matrix. Application of the implicit function theorem (Dieudonné, 1982) shows the existence of $H_1(\varepsilon)$ satisfying (3.25) for sufficiently small ε. Combining (3.20) and (3.24), we conclude that the transformation

$$\begin{bmatrix} \hat{\xi}(t) \\ \hat{\eta}(t) \end{bmatrix} = \begin{bmatrix} I_n - \varepsilon H_1 L & -\varepsilon H_1 \\ L & I_m \end{bmatrix} \begin{bmatrix} \hat{x}(t) \\ \hat{z}(t) \end{bmatrix} \tag{3.27}$$

transforms the Kalman–Bucy filter (3.13), (3.14) into the new form

$$\begin{aligned} \dot{\hat{\xi}}(t) = &(A_{11} - A_{12}L)\hat{\xi}(t) + (Q_1 - H_1Q_2 - \varepsilon H_1LQ_1) \\ &\times [y(t) - (C_1 - C_2L)\hat{\xi}(t)], \end{aligned} \tag{3.28a}$$

$$\begin{aligned} \varepsilon\dot{\hat{\eta}}(t) = &(A_{22} + \varepsilon LA_{12})\hat{\eta}(t) + (Q_2 + \varepsilon LQ_1)[y(t) - (C_1 - C_2L)\hat{\xi}(t) \\ &- (C_2 + \varepsilon C_1H_1 - \varepsilon C_2LH_1)\hat{\eta}(t)], \end{aligned} \tag{3.28b}$$

$$\hat{x}(t) = \hat{\xi}(t) + \varepsilon H_1\hat{\eta}(t), \tag{3.28c}$$

$$\hat{z}(t) = -L\hat{\xi}(t) + (I_m - \varepsilon LH_1)\hat{\eta}(t). \tag{3.28d}$$

It is important to notice that there is no approximation involved in deriving (3.28). It is merely a similarity transformation of the Kalman–Bucy filter equations, of which the estimates $\hat{x}(t)$ and $\hat{z}(t)$, provided by (3.28), are the optimal estimates.

Lemma 3.2

The right-hand-side coefficients of (3.18) are $O(\varepsilon)$ perturbations of the right-hand-side coefficients of (3.28).

The truth of this lemma can be easily seen by inspection of (3.18) and (3.28) and use of the expressions $A_{11} - A_{12}L(0) = A_0$, $Q_2(0) = Q_f$, $C_1 - C_2L(0) = C_0$ and $Q_1(0) - H_1(0)Q_2(0) = Q_s$, where the last one is shown in Appendix B. To apply Theorem A.2, we augment both (3.18) and (3.28) with the state equations (3.1), (3.2) to obtain systems of equations driven by white noise. For the optimal filter (3.28) we get

$$\begin{bmatrix} \dot{x} \\ \dot{\hat{\xi}} \\ \\ \varepsilon\dot{z} \\ \varepsilon\dot{\hat{\eta}} \end{bmatrix} = \begin{bmatrix} \mathscr{A}_{11}(\varepsilon) & \mathscr{A}_{12}(\varepsilon) \\ \\ \mathscr{A}_{21}(\varepsilon) & \mathscr{A}_{22}(\varepsilon) \end{bmatrix} \begin{bmatrix} x \\ \hat{\xi} \\ \\ z \\ \hat{\eta} \end{bmatrix} + \begin{bmatrix} \mathscr{B}_1(\varepsilon) \\ \\ \mathscr{B}_2(\varepsilon) \end{bmatrix} \begin{bmatrix} w \\ \\ v \end{bmatrix}. \tag{3.29}$$

The corresponding equation for the proposed filter (3.18) is

$$\begin{bmatrix} \dot{x} \\ \dot{\hat{x}}_s \\ \\ \varepsilon\dot{z} \\ \varepsilon\dot{\hat{z}}_f \end{bmatrix} = \begin{bmatrix} \underline{\boldsymbol{\mathscr{A}}}_{11} & \underline{\boldsymbol{\mathscr{A}}}_{12} \\ \\ \underline{\boldsymbol{\mathscr{A}}}_{21} & \underline{\boldsymbol{\mathscr{A}}}_{22} \end{bmatrix} \begin{bmatrix} x \\ \hat{x}_s \\ \\ z \\ \hat{z}_f \end{bmatrix} + \begin{bmatrix} \bar{\boldsymbol{\mathscr{B}}}_1 \\ \\ \boldsymbol{\mathscr{B}}_2 \end{bmatrix} \begin{bmatrix} w \\ \\ v \end{bmatrix}. \tag{3.30}$$

The matrices $\mathscr{A}_{ij}$, $\mathscr{B}_i$, $\boldsymbol{\mathscr{A}}_{ij}$ and $\boldsymbol{\mathscr{B}}_i$ are obtained in an obvious way. Moreover, by Lemma 3.2 we have

$$\mathscr{A}_{ij}(\varepsilon) - \underline{\boldsymbol{\mathscr{A}}}_{ij} = O(\varepsilon), \quad \mathscr{B}_i(\varepsilon) - \underline{\boldsymbol{\mathscr{B}}}_i = O(\varepsilon). \tag{3.31}$$

To apply Theorem A.2 we need to verify that $\mathscr{A}_{11}(0) - \mathscr{A}_{12}(0)\mathscr{A}_{22}^{-1}(0)\mathscr{A}_{21}(0)$ and $\mathscr{A}_{22}(0)$ are Hurwitz matrices. The matrix $\mathscr{A}_{22}(0)$ is given by the lower block-triangular matrix

$$\mathscr{A}_{22}(0) = \begin{bmatrix} A_{22} & 0 \\ Q_f C_2 & A_{22} - Q_f C_2 \end{bmatrix},$$

which is Hurwitz since both A_{22} and $A_{22} - Q_f C_2$ are Hurwitz matrices. The matrix $\mathscr{A}_{11}(0) - \mathscr{A}_{12}(0)\mathscr{A}_{22}^{-1}(0)\mathscr{A}_{21}(0)$ is given by

$$\mathscr{A}_{11}(0) - \mathscr{A}_{12}(0)\mathscr{A}_{22}^{-1}(0)\mathscr{A}_{21}(0) = \begin{bmatrix} A_0 & 0 \\ Q_s C_0 & A_0 - Q_s C_0 \end{bmatrix}.$$

Again this matrix is Hurwitz since both A_0 and $A_0 - Q_s C_0$ are so. Applying Theorem A.2, we conclude that (as $t \to \infty$)

$$\hat{\xi}(t) - \hat{x}_s(t) = O(\varepsilon),$$
$$\hat{\eta}(t) - \hat{z}_f(t) = O(\varepsilon^{1/2}).$$

Equation (3.19) follows by writing

$$\hat{x}(t) - \underline{\hat{x}}(t) = \hat{\xi}(t) - x_s(t) + \varepsilon H_1 \hat{\eta}(t),$$
$$\hat{z}(t) - \underline{\hat{z}}(t) = -A_{22}^{-1}A_{21}(\hat{\xi}(t) - \hat{x}_s(t)) - (L(\varepsilon) - L(0))\hat{\xi}(t) + (\hat{\eta}(t) - \hat{z}_f(t)) - \varepsilon L H_1 \hat{\eta}(t)$$

and applying the triangle inequality. If the slow and fast filters are initialized at $\hat{x}_s(0) = \hat{x}_0$ and $\hat{z}_f(0) = \hat{z}_0 + A_{22}^{-1}A_{21}\hat{x}_0$ then Theorem A.1 implies that (3.19) holds for all $t \geq 0$ since $\hat{x}_s(0) - \hat{\xi}(0) = O(\varepsilon)$ and $\hat{z}_f(0) - \hat{\eta}(0) = O(\varepsilon)$. □

One important application of Theorem 3.1 lies in the establishment of the validity of reduced-order filtering. Notice that in the decomposition (3.18) the estimate of the slow variable x depends only on the output of the slow filter (3.18a); it is independent of the fast filter (3.18b). Recall that the slow filter (3.18a) is derived on the basis of the reduced-order slow subsystem, which has been obtained by neglecting the fast dynamics of the singularly perturbed system (3.1)–(3.3). Thus, if our objective is to estimate the slow variable x only, then in designing the Kalman–Bucy filter the full model (3.1)–(3.3) can be replaced by the reduced-order slow model (3.16). Theorem 3.1 shows that such reduced-order filtering is near-optimal as far as the slow variable x is concerned.

4.4 The Steady-State LQG Controller

Chapter 3 introduced the composite control method where the control is sought as the sum of a slow control u_s and fast control u_f. The slow control is designed using the slow model and the fast control is designed using the fast model. In this section we extend the near-optimal composite control method of Section 3.4 to the linear–quadratic Gaussian (LQG) optimal regulator. For this problem, the system model takes the stochastic time-invariant singularly perturbed form

$$\dot{x}(t) = A_{11}x(t) + A_{12}z(t) + B_1u(t) + F_1w(t), \tag{4.1}$$

$$\varepsilon\dot{z}(t) = A_{21}x(t) + A_{22}z(t) + B_2u(t) + F_2w(t), \tag{4.2}$$

$$y(t) = C_1x(t) + C_2z(t) + v(t), \tag{4.3}$$

where x, z, y, w and v are as defined in Section 4.3 and $u \in R^r$ is the control input. The steady-state linear–quadratic Gaussian control problem is defined as the problem of finding $u(t)$ as a function of the observed output $\{y(\sigma); \sigma \leq t\}$ so as to minimize the performance index

$$J = \lim_{t_f - t_0 \to \infty} \frac{1}{t_f - t_0} \mathrm{E}\left\{\frac{1}{2}\int_{t_0}^{t_f} [y_c^{\mathrm{T}}(t)y_c(t) + u^{\mathrm{T}}(t)Ru(t)]\,\mathrm{d}t\right\}, \quad R > 0, \tag{4.4}$$

where $y_c \in R^p$ is the controlled output to be regulated to zero, which is given by

$$y_c(t) = M_1x(t) + M_2z(t). \tag{4.5}$$

This control problem is the stochastic version of the optimal regulator problem of Section 3.4, except that the feedback solution is sought under imperfect and partial observations instead of perfect and full state feedback. Under standard stabilizability–detectability conditions, which are guaranteed to hold for sufficiently small ε if the conditions of Theorem 4.1 of Chapter 3 and Lemma 3.1 are satisfied, the optimal control law is given by (Kwakernaak and Sivan, 1972)

$$u(t) = -[G_1(\varepsilon)\hat{x}(t) + G_2(\varepsilon)\hat{z}(t)], \tag{4.6}$$

where $\hat{x}(t)$ and $\hat{z}(t)$ are the steady-state optimal estimates provided by the Kalman–Bucy filter:

$$\dot{\hat{x}}(t) = A_{11}\hat{x}(t) + A_{12}\hat{z}(t) + B_1u(t) + Q_1(\varepsilon)[y(t) - C_1\hat{x}(t) - C_2\hat{z}(t)], \tag{4.7}$$

$$\varepsilon\dot{\hat{z}}(t) = A_{21}\hat{x}(t) + A_{22}\hat{z}(t) + B_2u(t) + Q_2(\varepsilon)[y(t) - C_1\hat{x}(t) - C_2\hat{z}(t)]. \tag{4.8}$$

The regulator gains $G_1(\varepsilon)$ and $G_2(\varepsilon)$ are given† by

$$G_1(\varepsilon) = R^{-1}(B_1^{\mathrm{T}}K_1(\varepsilon) + B_2^{\mathrm{T}}K_2(\varepsilon)),$$

$$G_2(\varepsilon) = R^{-1}(\varepsilon B_1^{\mathrm{T}}K_2(\varepsilon) + B_2^{\mathrm{T}}K_3(\varepsilon)), \tag{4.9}$$

† Note the notational difference of sign between the regulator gain matrices G_1 and G_2 of (4.9) in this chapter and (5.2), (5.3) of Chapter 3.

where K_1, K_2 and K_3 comprise the solution of the regulator Riccati equation (4.5) or (5.4) of Chapter 3. The filter gains Q_1 and Q_2 are given by (3.15).

A composite feedback control approach to the LQG control problem can be developed in a way similar to the composite control of the deterministic LQ control problem of Section 3.4. The composite control is sought in the form

$$u_c(t) = u_s(t) + u_f(t), \tag{4.10}$$

where u_s is a slow control component and u_f is a fast control component. In Chapter 3, the slow control u_s is a feedback function of the slow variable x_s, and the fast control u_f is a feedback function of the fast variable z_f. It is reasonable to seek the composite control (4.10) in the same form, but with x_s and x_f replaced by their estimates. Applying (4.10) to the system (4.1), (4.2) yields

$$\dot{x}(t) = A_{11}x(t) + A_{12}z(t) + B_1(u_s(t) + u_f(t)) + F_1w(t), \tag{4.11}$$

$$\varepsilon\dot{z}(t) = A_{21}x(t) + A_{22}z(t) + B_2(u_s(t) + u_f(t)) + F_2w(t). \tag{4.12}$$

To dcfine a slow subsystem we set $\varepsilon = 0$ in (4.12), solve the resulting algebraic equation in z and eliminate z from (4.11) and the output equation (4.3) to get

$$\dot{x}_s(t) = A_0x_s(t) + B_0(u_s(t) + u_f(t)) + F_0w(t), \tag{4.13a}$$

$$y(t) = C_0x_s(t) + D_0(u_s(t) + u_f(t)) + S_0w(t) + v(t), \tag{4.13b}$$

where $D_0 = -C_2A_{22}^{-1}B_2$. In other words, the slow subsystem is defined by replacing z by its steady-state component. A fast subsystem is defined by removing the slow bias from z and the output y. The slow term on the right-hand-side of (4.12) is $A_{21}x + B_2u_s$. Therefore, roughly speaking, the slow bias of z is given by

$$z_s(t) = -A_{22}^{-1}(A_{21}x(t) + B_2u_s(t)). \tag{4.14}$$

The fast component of z, denoted by z_f, and the fast component of y, denoted by y_f, are defined by

$$z_f(t) = z(t) + A_{22}^{-1}(A_{21}x(t) + B_2u_s(t)), \tag{4.15}$$

$$\begin{aligned} y_f(t) &= y(t) - C_1x(t) + C_2A_{22}^{-1}(A_{21}x(t) + B_2u_s(t)) \\ &= y(t) - C_0x(t) - D_0u_s(t) \\ &= C_2z_f(t) + v(t). \end{aligned} \tag{4.16}$$

Computing the derivative $\varepsilon\dot{z}_f$ with $z_s(t)$ treated as a constant, i.e. $\varepsilon\dot{z}_s(t)$

~ 0, we obtain the fast subsystem

$$\varepsilon \dot{z}_f(t) = A_{22} z_f(t) + B_2 u_f(t) + F_2 w(t), \tag{4.17a}$$

$$y_f(t) = C_2 z_f(t) + v(t). \tag{4.17b}$$

The next task is to associate with each subsystem an appropriately selected performance index. Using (4.15), the controlled output $y_c(t)$ in (4.5) can be expressed as

$$\begin{aligned} y_c(t) &= M_1 x(t) - M_2 A_{22}^{-1}(A_{21} x(t) + B_2 u_s(t)) + M_2 z_f(t) \\ &= M_0 x(t) + N_0 u_s(t) + M_2 z_f(t), \end{aligned}$$

which decomposes as the sum of a slow component $M_0 x + N_0 u_s$ and a fast component $M_2 z_f$. Thus for the slow subsystem (4.13) we consider the performance index

$$\begin{aligned} J_s = \lim_{t_f - t_0 \to \infty} \frac{1}{t_f - t_0} \mathrm{E}\Big\{ \int_{t_0}^{t_f} ([M_0 x_s(t) + N_0 u_s(t)]^{\mathrm{T}} [M_0 x_s(t) \\ + N_0 u_s(t)] + u_s^{\mathrm{T}}(t) R u_s(t))\, \mathrm{d}t \Big\}, \end{aligned} \tag{4.18}$$

and for the fast subsystem (4.17) we consider the performance index

$$J_f = \lim_{t_f - t_0 \to \infty} \frac{1}{t_f - t_0} \mathrm{E}\Big\{ \int_{t_0}^{t_f} [z_f^{\mathrm{T}}(t) M_2^{\mathrm{T}} M_2 z_f(t) + u_f^{\mathrm{T}}(t) R u_f(t)]\, \mathrm{d}t \Big\}. \tag{4.19}$$

The performance indices J_s and J_f are stochastic versions of those used in Chapter 3 (compare (4.18) and (4.19) with (4.14) and (4.20) of Chapter 3 respectively). Thus we have replaced the original *LQG* problem (4.1)–(4.4) by two *LQG* problems: the slow problem is defined by (4.13), (4.18) and the fast problem is defined by (4.17), (4.19). The slow *LQG* problem is a standard one except for the presence of the input $u_f(t)$. However, since by construction $u_f(t)$ is a function of $y(t)$, the well known separation principle of stochastic control (Kwakernaak and Sivan, 1972) yields the optimal solutions

$$u_s(t) = -G_s \hat{x}_s(t), \tag{4.20}$$

where, for any $u_f(t)$, $\hat{x}_s(t)$ is the optimal estimate provided by the slow Kalman–Bucy filter

$$\begin{aligned} \dot{\hat{x}}_s(t) &= A_0 \hat{x}_s(t) + B_0(u_s(t) + u_f(t)) \\ &\quad + Q_s[y(t) - D_0(u_s(t) + u_f(t)) - C_0 \hat{x}_s(t)]. \end{aligned} \tag{4.21}$$

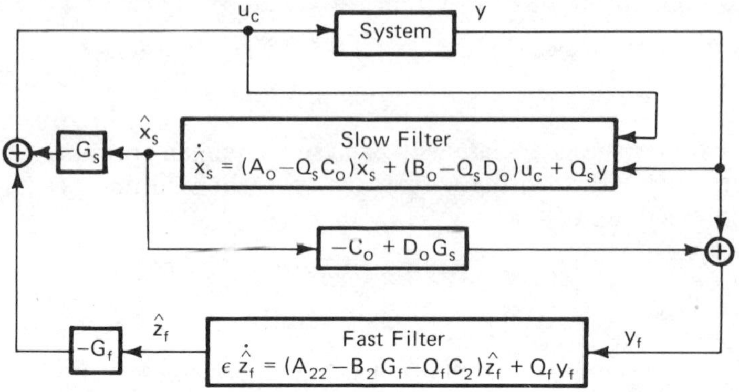

Fig. 4.5. Block diagram representation of the composite stochastic control.

The fast LQG problem is a standard one whose optimal solution is given by

$$u_f(t) = -G_f\hat{z}_f(t), \tag{4.22}$$

$$\varepsilon\dot{\hat{z}}_f(t) = A_{22}\hat{z}_f(t) + B_2u_f(t) + Q_f[y_f(t) - C_2\hat{z}_f(t)]. \tag{4.23}$$

The filter gains Q_s and Q_f are as defined in Section 4.3 and the regulator gains† G_s and G_f are as defined in Section 3.4. In implementing the fast filter (4.23), $y_f(t)$ is replaced by $y(t) - C_0\hat{x}_s(t) - D_0u_s(t)$. Thus the composite control takes the final form (4.24), whose block-diagram representation is shown in Fig. 4.5:

$$u_c(t) = -G_s\hat{x}_s(t) - G_f\hat{z}_f(t), \tag{4.24a}$$

$$\dot{\hat{x}}_s(t) = A_0\hat{x}_s(t) + B_0u_c(t) + Q_s[y(t) - D_0u_c(t) - C_0\hat{x}_s(t)], \tag{4.24b}$$

$$\varepsilon\dot{\hat{z}}_f(t) = (A_{22} - B_2G_f - Q_fC_2)\hat{z}_f(t) + Q_f[y(t) + D_0G_s\hat{x}_s(t) - C_0\hat{x}_s(t)]. \tag{4.24c}$$

The near-optimality of the composite control is established by Theorem 4.1.

Theorem 4.1

Suppose that the conditions of Theorem 4.1 of Chapter 3 and Lemma 3.1 are satisfied. Let $x(t)$ and $z(t)$ be the optimal trajectories and J be the optimal value of the performance index. Let $\underline{x}(t)$, $\underline{z}(t)$ and $\underline{J}$ be the corresponding quantities under the composite control (4.24). Then there exists an $\varepsilon^* > 0$

† It should be noted, as a matter of notation, that the regulator gains G_s and G_f replace $-G_0$ and $-G_2^{(0)}$ of Section 3.4.

such that for all $\varepsilon \in (0, \varepsilon^*)$, (4.25)–(4.28) hold:

$$\frac{\Delta J}{J} = \frac{\underline{J} - J}{J} = O(\varepsilon), \tag{4.25}$$

$$x(t) - \underline{x}(t) = O(\varepsilon) \qquad (\text{as } t \to \infty), \tag{4.26}$$

$$z(t) - \underline{z}(t) = O(\varepsilon^{1/2}) \quad (\text{as } t \to \infty). \tag{4.27}$$

Furthermore, if the optimal Kalman filter (4.7), (4.8) is initialized at certain initial conditions $\hat{x}_0$ and $\hat{z}_0$, and the slow and fast filters of the composite control are initialized at

$$\hat{x}_s(0) = \hat{x}_0, \tag{4.28a}$$

$$\hat{z}_f(0) = \hat{z}_0 + A_{22}^{-1}(A_{21} - B_2 G_s)\hat{x}_0, \tag{4.28b}$$

then (4.26) and (4.27) hold for all $t \geq 0$.

Proof The steps of the proof are similar to those of the proof of Theorem 3.1. We show that the coefficients of composite control (4.24) are $O(\varepsilon)$ perturbations of the coefficients of the optimal control (4.6)–(4.8) when the latter is expressed in appropriately selected coordinates. Identifying the appropriate coordinates is done exactly as in the proof of Theorem 3.1. Omitting details, the required transformation is

$$\begin{bmatrix} \hat{\xi}(t) \\ \hat{\eta}(t) \end{bmatrix} = \begin{bmatrix} I_n - \varepsilon H_3 L_2 & -\varepsilon H_3 \\ L_2 & I_m \end{bmatrix} \begin{bmatrix} \hat{x}(t) \\ \hat{z}(t) \end{bmatrix}, \tag{4.29}$$

where L_2 satisfies an equation similar to (3.21) with A_{ij} replaced by $A_{ij} - B_i G_j$, and H_3 satisfies an equation similar to (3.25) with L replaced by L_2. It can be verified that the use of (4.29) transforms the optimal control into the new form

$$u(t) = -(G_1 - G_2 L_2)\hat{\xi}(t) - (G_2 + \varepsilon G_1 H_3 - \varepsilon G_2 L_2 H_3)\hat{\eta}(t), \tag{4.30a}$$

$$\begin{aligned} \dot{\hat{\xi}}(t) = {} & [(A_{11} - B_1 G_1) - (A_{12} - B_1 G_2) L_2]\hat{\xi}(t) \\ & - [(B_1 - H_3 B_2) G_2 - \varepsilon H_3 L_2 B_1 G_2 + \varepsilon B_1 G_1 H_3 \\ & - \varepsilon B_1 G_2 L_2 H_3]\hat{\eta}(t) + (Q_1 - H_3 Q_2 - \varepsilon H_3 L_2 Q_1) \\ & \times [y(t) - (C_1 - C_2 L_2)\hat{\xi}(t)], \end{aligned} \tag{4.30b}$$

$$\begin{aligned} \varepsilon\dot{\hat{\eta}}(t) = {} & [(A_{22} - B_2 G_2) + \varepsilon L_2 (A_{12} - B_1 G_2)]\hat{\eta}(t) \\ & + (Q_2 + \varepsilon L_2 Q_1)[y(t) - (C_1 - C_2 L_2)\hat{\xi}(t) \\ & - (C_2 + \varepsilon C_1 H_3 - \varepsilon C_2 L_2 H_3)\hat{\eta}(t)]. \end{aligned} \tag{4.30c}$$

Lemma 4.1

The right-hand-side coefficients of the composite control (4.24) are $O(\varepsilon)$ perturbations of the right-hand-side coefficients of the optimal control (4.30).

This lemma can be seen by inspection of (4.24) and (4.30) and use of the following expressions, which follow either from Section 3.4 or from Section 4.3:

$$\begin{aligned} A_{11} - B_1G_1(0) - (A_{12} - B_1G_2(0))L_2(0) &= A_0 - B_0G_s, \\ G_1(0) - G_2(0)L_2(0) &= G_s, \\ Q_1(0) - H_3(0)Q_2(0) &= Q_s, \\ G_2(0) &= G_f, \\ Q_2(0) &= Q_f, \end{aligned}$$

as well as the expression

$$\begin{aligned} B_1 - H_3(0)B_2 &= B_1 - H_3(0)A_{22}A_{22}^{-1}B_2 \\ &= B_1 + [-H_3(0)Q_2(0)C_2 - A_{12} + Q_1(0)C_2]A_{22}^{-1}B_2 \\ &= B_0 + [Q_1(0) - H_3(0)Q_2(0)]C_2A_{22}^{-1}B_2 \\ &= B_0 - Q_sD_0 \end{aligned}$$

and its dual

$$C_1 - C_2L_2(0) = C_0 - D_0G_s.$$

Next, the closed-loop system under optimal control is represented as a system driven by white noise similar to (3.29) except for different $\mathcal{A}_{ij}$, $\mathcal{B}_i$ matrices. It can be verified that the use of the composite control results in $O(\varepsilon)$ perturbations in $\mathcal{A}_{ij}$ and $\mathcal{B}_i$. The stability requirements on $\mathcal{A}_{11}(0) - \mathcal{A}_{12}(0)\mathcal{A}_{22}^{-1}(0)\mathcal{A}_{21}(0)$ and $\mathcal{A}_{22}(0)$ follow by showing that

$$\begin{aligned} \mathcal{A}_{22}(0) &= \begin{bmatrix} A_{22} & -B_2G_f \\ Q_fC_2 & A_{22} - B_2G_f - Q_fC_2 \end{bmatrix} \\ &= \begin{bmatrix} I_m & 0 \\ I_m & I_m \end{bmatrix} \begin{bmatrix} A_{22} - B_2G_f & -B_2G_f \\ 0 & A_{22} - Q_fC_2 \end{bmatrix} \\ &\quad \times \begin{bmatrix} I_m & 0 \\ -I_m & I_m \end{bmatrix} \end{aligned}$$

and

$$\mathscr{A}_{11}(0) - \mathscr{A}_{12}(0)\mathscr{A}_{22}^{-1}(0)\mathscr{A}_{21}(0)$$

$$= \begin{bmatrix} I_n & 0 \\ I_n & I_n \end{bmatrix} \begin{bmatrix} A_0 - B_0 G_s & -B_0 G_s + \Delta \\ 0 & A_0 - Q_s C_0 \end{bmatrix} \begin{bmatrix} I_n & 0 \\ -I_n & I_n \end{bmatrix}, \qquad (4.31)$$

where $\Delta = B_0 G_f (A_{22} - B_2 G_f)^{-1} A_{22} (A_{22} - Q_f C_2)^{-1} Q_f C_0$.

The use of Theorem A.2 of Appendix A gives (4.26) and (4.27). Theorem A.3 leads to (4.25) after verifying that when J and $\underline{J}$ are expressed as quadratic forms in $(x, \hat{\xi}, z, \hat{\eta})$ and $(\underline{x}, \hat{x}_s, \underline{z}, \hat{z}_f)$ respectively, the matrices in $\underline{J}$ are $O(\varepsilon)$ perturbations of the matrices in J. Finally, Theorem A.1 shows that if $\hat{x}(0)$ and $\hat{z}(0)$ are chosen as in (4.28) then (4.26) and (4.27) hold for all $t \geq 0$ since $\hat{x}(0) - \hat{\xi}(0) = O(\varepsilon)$ and $\hat{z}_f(0) - \hat{\eta}(0) = O(\varepsilon)$. □

The composite control (4.24) and Theorem 4.1 extend, in an intuitively appealing way, to the LQG problem the composite control method of the deterministic LQ problem of Section 3.4. There are, however, differences between the LQ and the LQG results worthy of note. First, the use of the composite control in the LQ problem results in $\Delta J/J = O(\varepsilon^2)$ versus $\Delta J/J = O(\varepsilon)$ in the LQG problem. Second, if x and z are the states of the system under optimal control, then in the LQ problem $x - \underline{x} = O(\varepsilon)$ and $z - \underline{z} = O(\varepsilon)$, whereas in the LQG problem $x - \underline{x} = O(\varepsilon)$ and $z - \underline{z} = O(\varepsilon^{1/2})$, in a mean-square sense. Third, and this is an important difference, in the LQ problem it is shown (Theorem 4.4 of Chapter 3) that a reduced control law derived by solving only the slow optimal control problem, i.e. $u_r = -G_s x$, is near-optimal with $\Delta J/J = O(\varepsilon)$. This fact justifies using the slow reduced-order model as a basis for solving the LQ problem. In the LQG problem a similar reduced control will not be near-optimal. This point can be explained, loosely speaking, in the following way. In the LQ problem the optimal value of J takes the form $J = J_s + O(\varepsilon)$ (recall Theorem 4.3 of Chapter 3); so optimization of the slow control problem alone is enough to achieve near-optimality. In the LQG problem the optimal value of J takes the form $J = J_f/\varepsilon + O(1)$ (see the proof of Theorem A.3 of Appendix A and replace σ by J); so optimization of the fast control problem is necessary to the achievement of near-optimality. The use of the reduced control will be near-optimal only in the special case when $J_f = 0$, since in that case $J = J_s + O(\varepsilon)$ (see Exercise 4.7). Examples where $J_f = 0$ are the case $M_2 = 0$ (the fast variables are not penalized in the performance index) and the case $F_2 = 0$ (no white noise input in fast equation).

4.5 An Aircraft Autopilot Case Study

The near-optimality of the composite stochastic control of Theorem 4.1 will now be demonstrated on the autopilot control of the longitudinal motion of an F-8 aircraft introduced in Case Study 7.2 of Chapter 1. The numerical data are taken from Elliott (1977). The equations of motion are given by

$$\frac{d}{dt}\begin{bmatrix} v \\ \theta \\ \alpha \\ q \end{bmatrix} = \begin{bmatrix} X_v & \dfrac{-g}{V_0} & \dfrac{X_\alpha}{V_0} & 0 \\ 0 & 0 & 0 & 1 \\ Z_v V_0 & 0 & Z_\alpha & 1 \\ M_v V_0 & 0 & M_\alpha & M_q \end{bmatrix} \begin{bmatrix} v \\ \theta \\ \alpha \\ q \end{bmatrix} + \begin{bmatrix} \dfrac{X_\delta}{V_0} \\ 0 \\ Z_\delta \\ M_\delta \end{bmatrix} \delta + \begin{bmatrix} \dfrac{X_\alpha}{V_0} \\ 0 \\ Z_\alpha \\ M_\alpha \end{bmatrix} \alpha_N, \quad (5.1)$$

where all the variables and parameters, except α_N, are defined in Section 1.7; α_N is the nose angle of attack due to wind disturbance. It is usually modeled as exponentially correlated noise with power spectral density of the form

$$\frac{\sigma^2 L/\pi V_0^3}{1+\omega^2 (I/2V_0)^2}.$$

Although it is not realistic to assume that the bandwidth of this noise is much wider than the "short-period" frequency, for the sake of simplifying the presentation of the example we model α_N as white noise with intensity $\sigma^2 I/\pi V_0^3$. The numbers used in our calculations are representative for flight of the F-8 aircraft at 20 000 ft with $V_0 = 620$ ft/s (Mach number = 0.6) and $\alpha_0 = 0.078$ rad. They are

$$X_v = -0.015, \quad g = 32.2, \quad X_\alpha = -14.0, \quad X_\delta = -1.1,$$

$$Z_v = -0.00019, \quad Z_\alpha = -0.84, \quad Z_\delta = -0.11,$$

$$M_v = 0.00005, \quad M_\alpha = -4.8, \quad M_q = -0.49, \quad M_\delta = -8.7,$$

$$\frac{\sigma^2 L}{\pi V_0^3} = 7.51273 \times 10^{-4}.$$

The angles α, θ, δ and σ_N are in rad, q in rad/s and v is dimensionless ($v = u/V_0$ where u is the incremental velocity in ft/s). Together with the equations of motion we have to choose a set of measurements. Suppose we measure the pitch angle θ and the pitch rate q. Then the measurement equations are

$$y_1 = \theta + v_1', \tag{5.2}$$

$$y_2 = q + v_2'. \tag{5.3}$$

Each of the measurement noise processes v_i' is modeled as an exponentially correlated process with power spectral density $2\sigma_i^2\tau_i/(1+\omega^2\tau_i^2)$, where σ_i and τ_i are given by

pitch angle sensor: $\tau_1 = 0.08\,\text{s}, \quad \sigma_1 = 0.5°,$

pitch angle gyro: $\tau_2 = 0.16\,\text{s}, \quad \sigma_2 = 0.2°\,\text{s}^{-1}.$

Thus v_1' and v_2' can be reasonably modeled as white noise processes with intensities $2\sigma_1^2\tau_1 = 12.184697 \times 10^{-6}\,\text{rad}^2.\text{s}$ and $2\sigma_2^2\tau_2 = 3.8991 \times 10^{-6}\,\text{rad}^2/\text{s}$ respectively.

Finally, the performance index is chosen as

$$J = \lim_{t_f - t_0 \to \infty} \frac{1}{t_f - t_0} \mathrm{E}\left\{\int_{t_0}^{t_f} \left[\frac{v^2}{v_{max}^2} + \frac{\theta^2}{\theta_{max}^2} + \frac{\alpha^2}{\alpha_{max}^2} + \frac{q^2}{q_{max}^2} + \rho\frac{\delta^2}{\delta_{max}^2}\right] \mathrm{d}t\right\}. \tag{5.4}$$

We give results for the case $v_{max} = 0.1$, $\theta_{max} = \alpha_{max} = \delta_{max} = 0.1\,\text{rad}$, $q_{max} = 0.1\,\text{rad/s}$ and $\rho = 40$.

Now, the LQG problem is completely defined, and our first task is to bring it into the singularly perturbed form (4.1)–(4.5). This modeling step is done as in Section 1.7. The angle θ is replaced by

$$\psi = \theta + \frac{M_\alpha}{Z_\alpha M_q - M_\alpha}\alpha - \frac{Z_\alpha}{Z_\alpha M_q - M_\alpha}q,$$

the time scale is changed from t to $t' = t\omega_0$ and the singular perturbation parameter is chosen as $\varepsilon = \omega_0/\omega_1$, where ω_0 and ω_1 are the undamped frequencies of phugoid and short period modes and are given, respectively, by (7.25) and (7.31) of Chapter 1. The numerical values for these quantities are

$$\psi = \theta - 0.921022\alpha + 0.161179q,$$

$$\omega_0 = 0.076774\,\text{rad/s},$$

$$\omega_1 = 2.28289\,\text{rad/s},$$

$$\varepsilon = 0.0336.$$

To take care of the effect of changing the time scale on the white noise processes, and to match the problem statement (4.1)–(4.5), the white noise

$w(t')$ is defined as

$$w(t') = \left[\frac{\pi V_0^3}{\sigma^2 L}\right]^{1/2} \frac{\alpha_N(t'/\omega_0)}{\omega_0^{1/2}},$$

and the intensity matrix of the measurement noise, V, is taken as

$$V = \begin{bmatrix} 2\sigma_1^2\tau_1\omega_0 & 0 \\ 0 & 2\sigma_2^2\tau_2\omega_0 \end{bmatrix} = \begin{bmatrix} 0.935473 \times 10^{-6} & 0 \\ 0 & 0.299351 \times 10^{-6} \end{bmatrix}.$$

The matrices appearing in (4.1)–(4.5) are given by

$$A_{11} = \begin{bmatrix} -0.195378 & -0.676469 \\ 1.478265 & 0 \end{bmatrix}, \qquad A_{12} = \begin{bmatrix} -0.917160 & 0.109033 \\ 0 & 0 \end{bmatrix},$$

$$A_{21} = \begin{bmatrix} -0.051601 & 0 \\ 0.013579 & 0 \end{bmatrix}, \qquad A_{22} = \begin{bmatrix} -0.367954 & 0.438041 \\ -2.102596 & -0.214640 \end{bmatrix},$$

$$B_1 = \begin{bmatrix} -0.023109 \\ -16.945030 \end{bmatrix}, \qquad B_2 = \begin{bmatrix} -0.048184 \\ -3.810954 \end{bmatrix},$$

$$C_1 = \begin{bmatrix} 0 & 1 \\ 0 & 0 \end{bmatrix}, \qquad C_2 = \begin{bmatrix} 0.921022 & -0.161179 \\ 0 & 1 \end{bmatrix},$$

$$F_1 = \begin{bmatrix} -0.223371 \times 10^{-2} \\ 0 \end{bmatrix}, \qquad F_2 = \begin{bmatrix} -0.279448 \times 10^{-2} \\ -1.596845 \times 10^{-2} \end{bmatrix},$$

$$M_1 = \begin{bmatrix} 10 & 0 \\ 0 & 10 \\ 0 & 0 \\ 0 & 0 \end{bmatrix}, \qquad M_2 = \begin{bmatrix} 0 & 0 \\ 9.21022 & -1.61179 \\ 10 & 0 \\ 0 & 10 \end{bmatrix},$$

$R = 4000.$

Numerical results for both the optimal and composite controllers are presented in Tables 4.1–4.3. Table 4.1 compares the solutions of the full Riccati equation with the solutions of the slow and fast Riccati equations. Table 4.2 compares the closed-loop eigenvalues, while Table 4.3 compares the steady-state root-mean-square values of the state variables and the values of the performance index for closed-loop optimal and composite

TABLE 4.1
Comparison of Solutions of Riccati Equations

$$K_1 = \begin{bmatrix} 1.4903 & -0.09082 \\ -0.09082 & 0.35355 \end{bmatrix}, \quad K_s = \begin{bmatrix} 1.48854 & -0.08216 \\ -0.08216 & 0.33774 \end{bmatrix}$$

$$K_3 = \begin{bmatrix} 4.4481 & -0.31804 \\ -0.31804 & 0.94598 \end{bmatrix}, \quad K_f = \begin{bmatrix} 4.31257 & -0.32132 \\ -0.32132 & 0.96069 \end{bmatrix}$$

$$P_1 = \begin{bmatrix} 0.07869 & 0.18133 \\ 0.18133 & 1.63093 \end{bmatrix}, \quad P_s = \begin{bmatrix} 0.07592 & 0.17150 \\ 0.17150 & 1.57401 \end{bmatrix}$$

$$P_3 = \begin{bmatrix} 0.11835 & 0.15702 \\ 0.15702 & 0.85509 \end{bmatrix}, \quad P_f = \begin{bmatrix} 0.05733 & 0.16163 \\ 0.16163 & 0.85553 \end{bmatrix}$$

TABLE 4.2
Open-Loop and Closed-Loop Eigenvalues

	Slow	Fast		
Open-loop	−0.0069 ± j0.0765	−0.6656 ± j2.1821		
	Regulator poles		Filter poles	
	Slow	Fast	Slow	Fast
Closed-loop-optimal	−0.084, −0.137	−1.071 ± j2.13	−0.02 ± j0.02	−0.572, −66.561
Closed-loop-composite	−0.083, −0.14	−1.055 ± j2.084	−0.021 ± j0.019	−0.589, −66.536

TABLE 4.3
Performance Comparison

	Optimal	Composite	Error	Relative error
rms v	0.004363	0.004822	0.459×10^{-3}	0.1052
rms ψ	0.005766	0.005991	0.225×10^{-3}	0.039
rms α	0.034471	0.035133	0.662×10^{-3}	0.0192
rms q	0.064482	0.064989	0.507×10^{-3}	0.0079
J	0.370759	0.372158	1.399×10^{-3}	0.0038

control systems. In order to improve the conditioning of the Riccati equations in computations, the performance index is multiplied by 10^{-2} while the noise intensities are multiplied by 10^4. Such scaling does not affect the feedback control law; see Section 4.7 for further discussion of scaling. The Riccati equations solutions of Table 4.1 are given for the scaled problem, while the r.m.s. values and performance index of Table 4.3 are the actual ones. The eigenvalues of Table 4.2 are given in the original time scale t.

4.6 Corrected LQG Design and the Choice of the Decoupling Transformation

Should a closer approximation than that provided by the LQG design of Theorem 4.1 be desired, a corrected LQG design may be pursued in the manner of Section 3.5. In the proof of Theorem 4.1 we have shown that the composite control (4.24) is nothing but an $O(\varepsilon)$ perturbation of the coefficients of the optimal control (4.30), in which the Kalman–Bucy filter variables are restructured by the decoupling transformation (4.29). By Theorems A.1–A.3 of Appendix A, all that is needed to obtain closer approximations of the system trajectory and performance index is to retain higher-order terms in the expansions of the ε-dependent coefficients of the optimal control (4.30). We will illustrate this by presenting the case of an $O(\varepsilon^2)$ approximation of coefficients.

The optimal control (4.30) is given in terms of six matrices: the regulator gains $G_1(\varepsilon)$ and $G_2(\varepsilon)$, the filter gains $Q_1(\varepsilon)$ and $Q_2(\varepsilon)$, $L_2(\varepsilon)$, which satisfies (3.21) with A_{ij} replaced by $A_{ij} - B_iG_j(\varepsilon)$ and $H_3(\varepsilon)$, which satisfies (3.25) with L replaced by L_2. Each of these six matrices is analytic at $\varepsilon = 0$. Writing their Maclaurin series as $G_1(\varepsilon) = G_1^{(0)} + \varepsilon G_1^{(1)} + \dots$, with similar expressions for the other matrices, substituting in (4.30) and dropping $O(\varepsilon^2)$ coefficients, we arrive at the corrected design

$$\begin{aligned} u(t) = &-[G_s + \varepsilon(G_1^{(1)} - G_2^{(0)}L_2^{(1)} - G_2^{(1)}L_2^{(0)})]\hat{\xi}_1(t) \\ &- [G_f + \varepsilon(G_2^{(1)} + G_sH_3^{(0)})]\hat{\eta}_1(t), \end{aligned} \tag{6.1a}$$

$$\begin{aligned} \dot{\hat{\xi}}_1(t) = &[A_0 - B_0G_s - \varepsilon(A_{12}L_2^{(1)} - B_1G_2^{(0)}L_2^{(1)} + B_1G_1^{(1)} \\ &- B_1G_2^{(1)}L_2^{(0)})]\hat{\xi}_1(t) \\ &- [(B_0 - Q_sD_0)G_f - \varepsilon((Q_sD_0 - B_0)G_2^{(1)} + H_3^{(1)}B_2G_2^{(0)} \\ &+ H_3^{(0)}L_2^{(0)}B_1G_2^{(0)} - B_1G_sH_3^{(0)})]\hat{\eta}_1(t) \\ &+ [Q_s + \varepsilon(Q_1^{(1)} - H_3^{(0)}Q_2^{(1)} - H_3^{(1)}Q_2^{(0)} - H_3^{(0)}L_2^{(0)}Q_1^{(0)})] \\ &\times [y(t) - (C_0 - D_0G_s - \varepsilon C_2L_2^{(1)})\hat{\xi}_1(t)], \end{aligned} \tag{6.1b}$$

$$\varepsilon\dot{\hat{\eta}}_1(t) = [A_{22} - B_2G_{\mathrm{f}} - \varepsilon(B_2G_2^{(1)} - L_2^{(0)}A_{12} + L_2^{(0)}B_1G_{\mathrm{f}})]\hat{\eta}_1(t)$$
$$+ [Q_{\mathrm{f}} + \varepsilon(Q_2^{(1)} + L_2^{(0)}Q_1^{(0)})]$$
$$\times [y(t) - (C_0 - D_0G_{\mathrm{s}} - \varepsilon C_2L_2^{(1)})\hat{\xi}_1(t)$$
$$- (C_2 + \varepsilon(C_0 - D_0G_{\mathrm{s}})H_3^{(0)})\hat{\eta}_1(t)]. \tag{6.1c}$$

The computation of the matrix corrections $G_1^{(1)}$, $G_2^{(1)}$, $Q_1^{(1)}$, $Q_2^{(1)}$, $L_2^{(1)}$ and $H_3^{(1)}$ proceeds as follows. First, the regulator gain corrections are specified as in Section 3.5 with $K_2^{(0)} = K_m$ and a notational change of sign by

$$G_1^{(1)} = R^{-1}(B_1^{\mathrm{T}}K_1^{(1)} + B_2^{\mathrm{T}}K_2^{(1)\mathrm{T}}), \tag{6.2a}$$
$$G_2^{(1)} = R^{-1}(B_1^{\mathrm{T}}K_{\mathrm{m}} + B_2^{\mathrm{T}}K_3^{(1)}), \tag{6.2b}$$

where $K_1^{(1)}$ satisfies the Lyapunov equation

$$0 = K_1^{(1)}(A_0 - B_0G_{\mathrm{s}}) + (A_0 - B_0G_{\mathrm{s}})^{\mathrm{T}}K_1^{(1)}$$
$$- (A_{21} - B_2G_1^{(0)})^{\mathrm{T}}(A_{22} - B_2G_{\mathrm{f}})^{-\mathrm{T}}K_{\mathrm{m}}^{\mathrm{T}}(A_0 - B_0G_{\mathrm{s}})$$
$$- (A_0 - B_0G_{\mathrm{s}})^{\mathrm{T}}K_{\mathrm{m}}(A_{22} - B_2G_{\mathrm{f}})^{-1}(A_{21} - B_2G_1^{(0)}), \tag{6.3a}$$

$K_3^{(1)}$ satisfies the Lyapunov equation

$$0 = K_3^{(1)}(A_{22} - B_2G_{\mathrm{f}}) + (A_{22} - B_2G_{\mathrm{f}})^{\mathrm{T}}K_3^{(1)}$$
$$+ K_{\mathrm{m}}^{\mathrm{T}}(A_{12} - B_1G_{\mathrm{f}}) + (A_{12} - B_1G_{\mathrm{f}})^{\mathrm{T}}K_{\mathrm{m}}, \tag{6.3b}$$

and $K_2^{(1)}$, which is dependent on $K_1^{(1)}$ and $K_3^{(1)}$, is given by

$$K_2^{(1)} = -[K_1^{(1)}(A_{12} - B_1G_{\mathrm{f}}) + (A_{11} - B_1G_1^{(0)})^{\mathrm{T}}K_{\mathrm{m}}$$
$$+ (A_{21} - B_2G_1^{(0)})^{\mathrm{T}}K_3^{(1)}](A_{22} - B_2G_{\mathrm{f}})^{-1}. \tag{6.3c}$$

As the dual of the expressions (6.2) and (6.3), the filter gain corrections are specified by

$$Q_1^{(1)} = (P_1^{(1)}C_1^{\mathrm{T}} + P_2^{(1)}C_2^{\mathrm{T}})V^{-1}, \tag{6.4a}$$
$$Q_2^{(1)} = (P_{\mathrm{m}}^{\mathrm{T}}C_1^{\mathrm{T}} + P_M^{(1)}C_2^{\mathrm{T}})V^{-1}, \tag{6.4b}$$

where $P_1^{(1)}$, $P_3^{(1)}$ and $P_2^{(1)}$ respectively satisfy the linear equations

$$0 = (A_0 - Q_{\mathrm{s}}C_0)P_1^{(1)} + P_1^{(1)}(A_0 - Q_{\mathrm{s}}C_0)^{\mathrm{T}}$$
$$- (A_{12} - Q_1^{(0)}C_2)(A_{22} - Q_{\mathrm{f}}C_2)^{-1}P_{\mathrm{m}}^{\mathrm{T}}(A_0 - Q_{\mathrm{s}}C_0)^{\mathrm{T}}$$
$$- (A_0 - Q_{\mathrm{s}}C_0)P_{\mathrm{m}}(A_{22} - Q_{\mathrm{f}}C_2)^{-\mathrm{T}}(A_{12} - Q_1^{(0)}C_2)^{\mathrm{T}}, \tag{6.5a}$$

$$0 = (A_{22} - Q_{\mathrm{f}}C_2)P_3^{(1)} + P_3^{(1)}(A_{22} - Q_{\mathrm{f}}C_2)^{\mathrm{T}}$$
$$+ (A_{21} - Q_{\mathrm{f}}C_1)P_{\mathrm{m}} + P_{\mathrm{m}}^{\mathrm{T}}(A_{21} - Q_{\mathrm{f}}C_1)^{\mathrm{T}}, \tag{6.5b}$$

$$P_2^{(1)} = -[P_1^{(1)}(A_{21} - Q_f C_1)^T + (A_{11} - Q_1^{(0)} C_1)P_m + (A_{12} - Q_1^{(0)} C_2)P_3^{(1)}](A_{22} - Q_f C_2)^{-T}. \tag{6.5c}$$

Thus we are left with the calculation of $L_2^{(1)}$ and $H_3^{(1)}$. Substituting Maclaurin expansions for the respective variables in the equations of L_2 and H_3 and matching coefficients of like powers of ε, we obtain the expressions for $L_2^{(1)}$ and $H_3^{(1)}$, which are given by

$$L_2^{(1)} = (A_{22} - B_2 G_f)^{-1}[-B_2(G_1^{(1)} - G_2^{(1)} L_2^{(0)}) + L_2^{(0)}(A_0 - B_0 G_s)] \tag{6.6}$$

$$H_3^{(1)} = -[(Q_1^{(1)} - H_3^{(0)} Q_2^{(1)})C_2 + H_3^{(0)} L_2^{(0)}(A_{12} - Q_1^{(0)} C_2) - (A_{11} - A_{12} L_2^{(0)})H_3^{(0)} + Q_s(C_0 - D_0 G_s)H_3^{(0)}] \times (A_{22} - Q_f C_2)^{-1}. \tag{6.7}$$

The following theorem gives the order of approximation of the state trajectories and the performance index.

Theorem 6.1

Suppose the conditions of Theorem 4.1 hold. Let $x(t)$ and $z(t)$ be the optimal trajectories and let J be the optimal value of the performance index. Let $x_1(t)$, $z_1(t)$ and J_1 be the corresponding quantities under the corrected controller (6.1). Then there exists an $\varepsilon^* > 0$ such that for all $\varepsilon \in (0, \varepsilon^*)$, (6.8)–(6.10) hold:

$$\frac{\Delta J}{J} = \frac{J_1 - J}{J} = O(\varepsilon^2), \tag{6.8}$$

$$x(t) - x_1(t) = O(\varepsilon^2) \quad (\text{as } t \to \infty), \tag{6.9}$$

$$z(t) - z_1(t) = O(\varepsilon^{3/2}) \quad (\text{as } t \to \infty). \tag{6.10}$$

Proof Similar to the proof of Theorem 4.1. □

An initial condition statement similar to that of Theorem 4.1 is also possible.

Example 6.1

As a numerical illustration of Theorem 6.1, let us consider the second-order example

$$\dot{x} = x + 2z + u + w,$$

$$\varepsilon\dot{z} = z + 2u + w,$$

$$y = x + z + v,$$

$$J = \lim_{t_f - t_0 \to \infty} \frac{1}{t_f - t_0} \mathrm{E}\left\{\frac{1}{2}\int_{t_0}^{t_f} (4x^2 + 4xz + z^2 + u^2)\,\mathrm{d}t\right\},$$

where both w and v are white noise processes with intensity one. Table 4.4 compares values of the performance index J under the optimal, the composite (uncorrected) and corrected designs for five different values of ε. The reader may verify that the numerical results agree with the orders of approximation established by Theorems 4.1 and 6.1. As $\varepsilon \to 0$ both designs yield acceptable approximations. For a specified value of ε one may have to correct the design to get acceptable results. For example, at $\varepsilon = 0.2$ the uncorrected design results in 19.23% error which may not be acceptable. One step of correction brings this error down to 1.598%. The numerical results demonstrate also the fact that J is $O(1/\varepsilon)$.

TABLE 4.4
Performance Index Comparison

ε	Optimal	Composite	Corrected	Percentage error composite	Percentage error corrected
0.2	59.036772	70.391063	59.980312	19.23	1.598
0.1	94.468821	98.890743	94.663604	4.68	0.206
0.05	165.65066	167.638049	165.718247	1.2	0.04
0.025	308.182981	309.128464	308.212322	0.31	0.01
0.01	735.918503	736.285953	735.929465	0.05	0.001

Let us conclude this section by re-examining the role of the decoupling transformation used in deriving the composite control (4.24) or the corrected control (6.1). Arriving at (4.24) or (6.1) can be described as a two-step process. In the first step, the decoupling transformation

$$\begin{bmatrix} \hat{\xi}(t) \\ \hat{\eta}(t) \end{bmatrix} = \begin{bmatrix} I_n - \varepsilon HL & -\varepsilon H \\ L & I_m \end{bmatrix} \begin{bmatrix} \hat{x}(t) \\ \hat{z}(t) \end{bmatrix} \triangleq T \begin{bmatrix} \hat{x} \\ \hat{z} \end{bmatrix}, \qquad (6.11)$$

with the particular choice $L = L_2$ and $H = H_3$, was used to restructure the Kalman–Bucy filter. In the second step, the coefficients of the optimal control law are approximated; an $O(\varepsilon)$ approximation yields (4.24), while an $O(\varepsilon^2)$ approximation yields (6.1). The choice $L = L_2$ and $H = H_3$ was motivated by our desire that the $O(\varepsilon)$ approximation coincides with the composite control that has been intuitively derived. A question that could be posed here is whether there are choices of L and H that could restructure the Kalman–Bucy filter in a different way. In fact, there are several choices of L and H that could be given a physical interpretation. For example, L and H could be chosen to block-diagonalize any of the matrices A, the open-loop matrix $A - QC$, the homogeneous part of the Kalman–Bucy filter when viewed as a system driven by both the control input u and the observed output y, and $A - BG$, the homogeneous part of the Kalman–Bucy filter when viewed as a system driven by the innovation process. Each of these matrices corresponds to a standard singularly perturbed system model since $\det(A_{22}) \neq 0$, $\det(A_{22} - Q_2(0)C_2) \neq 0$ and $\det(A_{22} - B_2G_2(0)) \neq 0$. Furthermore, we could have mixed situations. For example, the reader may verify that the choice of $L = L_2$ and $H = H_3$ corresponds to the following situation. In choosing L_2 the Kalman–Bucy filter is viewed as a system driven by the innovation process and L_2 is chosen to put it in the actuator form. Then, in choosing H_3 the Kalman–Bucy filter is viewed as a system driven by u and y and H_3 is chosen to put it in the sensor form. Obviously, there are several meaningful choices of L and H. Therefore, instead of studying the effect of using a particular transformation, we consider a class of transformations of the form (6.11), where the matrices L and H are required only to be analytic functions of ε at $\varepsilon = 0$. Each member of the class corresponds to a choice of coordinate system for representing the optimal Kalman–Bucy filter (4.7), (4.8). Since in this class either L or H or both could be zero, the class includes the original coordinate system in which the Kalman–Bucy filter (4.7), (4.8) is represented.

To gain some insight into the effect of using the transformation (6.11) on the optimal Kalman–Bucy filter, we compute the power spectral density of $(\hat{\xi}, \hat{\eta})$, the state vector of the transformed filter. The power spectral density of $(\hat{\xi}, \hat{\eta})$ can be determined by studying the equation

$$\begin{pmatrix} \dot{\hat{\xi}} \\ \dot{\hat{\eta}} \end{pmatrix} = T(A - BG)T^{-1}\begin{pmatrix} \hat{\xi} \\ \hat{\eta} \end{pmatrix} + TQ\nu. \tag{6.12}$$

Since the innovation process ν is white noise, the power spectral density of $(\hat{\xi}, \hat{\eta})$ is determined by the transfer-function matrix

$$W(s) = T(sI_{n+m} - A + BG)^{-1}Q \triangleq \begin{pmatrix} W_1(s) \\ W_2(s) \end{pmatrix}, \quad W_1 \in C^{n\times l}, \quad W_2 \in C^{m\times l}. \tag{6.13}$$

To compute the resolvent matrix $(sI - A + BG)^{-1}$, the transformation (6.11), with appropriately selected $L = L_2$ and $H = H_2$, is used to block-diagonalize $A - BG$, resolvent matrices of the diagonal blocks are computed, and then the inverse transformation is applied to recover $(sI - A + BG)^{-1}$. Multiplying $(sI - A + BG)^{-1}$ from the right by Q and from the left by T, where T is any member in the class, we get

$$W_1(s) = (I_n - \varepsilon HL + \varepsilon HL_2)\Gamma_1(s)(Q_1 - \varepsilon H_2 L_2 Q_1 - H_2 Q_2) + \varepsilon(H_2 - H - \varepsilon HLH_2 + \varepsilon HL_2H_2)\Gamma_2(\varepsilon s)(Q_2 + \varepsilon L_2 Q_1), \quad (6.14)$$

$$W_2(s) = (L - L_2)\Gamma_1(s)(Q_1 - \varepsilon H_2 L_2 Q_1 - H_2 Q_2) + (I_m + \varepsilon LH_2 - \varepsilon L_2 H_2)\Gamma_2(\varepsilon s)(Q_2 + \varepsilon L_2 Q_1), \quad (6.15)$$

where

$$\Gamma_1(s) = [sI_n - (A_{11} - B_1 G_1) + (A_{12} - B_1 G_2)L_2]^{-1},$$
$$\Gamma_2(\varepsilon s) = [\varepsilon s I_m - (A_{22} - B_2 G_2) - \varepsilon L_2(A_{12} - B_1 G_2)]^{-1}.$$

Examination of expressions (6.14) and (6.15) reveals that all transformations of the form (6.11) share two properties. First, the fast variable $\hat{\eta}$ has a slow (low-frequency) bias. Second, the fast (high-frequency) component in $W_1(s)$ is multiplied by ε; hence, for sufficiently small ε, $\hat{\xi}$ is predominantly slow. Inspection of (6.15) shows also that the choice $L = L_2$ (which was adopted in Section 4.4) eliminates the slow bias of $\hat{\eta}$. The final word on the choice of the appropriate coordinates is problem-dependent, and implementation factors, which are not considered here, will affect the choice. It is clear, however, that, irrespective of which transformation is used, near-optimum controllers can be obtained by truncating coefficients of Maclaurin expansions since L and H are analytic at $\varepsilon = 0$.

4.7 Scaled White-Noise Inputs

Frequently in applications we are faced with problems where the white noise input to a singularly perturbed system is scaled by a fractional power of the singular perturbation parameter, e.g. $\varepsilon^{1/2}$. As an example of a mechanism through which such scaling factors arise, let us consider a zero-mean stochastic process with correlation function $R(\tau) = \sigma^2 e^{-|\tau|/\theta}$, where σ is of order one and θ is small. One can model this process as the state of a first-order differential system driven by white noise:

$$\theta \dot{z} = -z + (2\theta)^{1/2} w, \quad (7.1)$$

where w has intensity σ. Taking $\theta = \varepsilon$, we see that we have a singularly perturbed equation similar to the one we studied earlier except for the $\varepsilon^{1/2}$ factor multiplying the white noise input. In this section, we study the effect of scaling the white noise input on the results obtained throughout the chapter. We start with the simple case studied in Section 4.2. Consider the singularly perturbed system

$$\dot{x}(t) = A_{11}x(t) + A_{12}z(t) + \mu B_1 w(t), \tag{7.2}$$

$$\varepsilon\dot{z}(t) = A_{21}x(t) + A_{22}z(t) + \mu B_2 w(t), \tag{7.3}$$

where all the variables are as defined in Section 4.2 and μ is a positive parameter that may depend on ε. It is only required that μ be a bounded function of ε in the neighborhood of $\varepsilon = 0$. Typical forms of μ are $\mu = \varepsilon^\alpha$ for $\alpha \in [0, 1]$. In view of the analysis of Section 4.2, it should not be a surprise to see that μ has no effect on the slow–fast decomposition (2.13), except of course, for scaling w by μ. In particular, defining x_s and z_f as

$$\dot{x}_s(t) = A_0 x_s(t) + \mu B_0 w(t), \quad x_s(0) = x_0, \tag{7.4a}$$

$$\varepsilon\dot{z}_f(t) = A_{22} z_f(t) + \mu B_2 w(t), \quad z_f(0) = z_0 + A_{22}^{-1} A_{21} x_0, \tag{7.4b}$$

we have the following theorem.

Theorem 7.1

Suppose that A_0 and A_{22} are Hurwitz matrices and μ is as defined above. Then there exists an $\varepsilon^* > 0$ such that for all $\varepsilon \in (0, \varepsilon^*)$, (7.5) is satisfied for all $t \geq 0$:

$$\left.\begin{aligned} x(t) - x_s(t) &= O(\varepsilon^{1/2}), \\ z(t) + A_{22}^{-1}A_{21}x_s(t) - z_f(t) &= O(\varepsilon^{1/2}). \end{aligned}\right\} \tag{7.5}$$

Moreover, at the steady state (as $t \to \infty$) we have

$$\left.\begin{aligned} x(t) - x_s(t) &= \mu O(\varepsilon^{1/2}), \\ z(t) + A_{22}^{-1}A_{21}x_s(t) - z_f(t) &= \mu O(\varepsilon^{1/2}). \end{aligned}\right\} \tag{7.6}$$

The proof of this theorem is the same as that of Theorem 2.1, and makes use of Theorems A.1 and A.2 of Appendix A. Notice that (7.6) gives a sharper bound on the approximation error whenever $\mu = \varepsilon^\alpha$ with $\alpha > 0$.

Moving now to the estimation and control problems of Sections 4.3 and 4.4, we notice that scaling the white noise inputs may have a crucial role, since it will affect the filter Riccati equation and the filter gain. There is one significant special case, however, where the results of Sections 4.3 and 4.4 remain invariant with respect to white noise scaling. This is the case when all the white noise inputs are scaled by the same factor μ. Let us examine the LQG controller. The system equations are taken as

$$\dot{x}(t) = A_{11}x(t) + A_{12}z(t) + B_1u(t) + \mu F_1 w(t), \tag{7.7}$$

$$\varepsilon\dot{z}(t) = A_{21}x(t) + A_{22}z(t) + B_2u(t) + \mu F_2 w(t), \tag{7.8}$$

$$y(t) = C_1x(t) + C_2z(t) + \mu v(t), \tag{7.9}$$

where all the variables are as defined in Section 4.4 and μ is defined above. The performance index is defined by (4.4) and (4.5). The factor μ has no effect on the solution of the regulator part of the problem. It affects the filter part. However, since both the state excitation noise and the observation noise are scaled by the same factor μ, the effect of μ will be to scale the solution of the filter Riccati equation but it will not affect the filter gain. In particular, the solution of the filter Riccati equation takes the form

$$0 = AP + PA^{\mathrm{T}} + \mu^2 FF^{\mathrm{T}} - \frac{1}{\mu^2} PC^{\mathrm{T}}V^{-1}CP, \tag{7.10}$$

$$P = \mu^2 \begin{bmatrix} P_1 & P_2 \\ P_2^{\mathrm{T}} & \dfrac{1}{\varepsilon}P_3 \end{bmatrix}, \tag{7.11}$$

where P_1, P_2 and P_3 satisfy (3.9). The filter gains are given by (3.15). In other words, μ does not affect the optimal control but affects the optimal value of J and the variances of the states variables. As a consequence, the composite control (4.24) is near-optimal, as stated in the following theorem.

Theorem 7.2

Suppose that the conditions of Theorem 4.1 hold and μ is as defined above. Let $x(t)$, $z(t)$ and J be the optimal trajectories and cost of the LQG problem (7.7)–(7.9) and (4.4), (4.5). Let $\underline{x}(t)$, $\underline{z}(t)$ and $\underline{J}$ be the corresponding quantities under the composite control (4.24). Then there exists $\varepsilon^* > 0$ such that for all $\varepsilon \in (0, \varepsilon^*)$, (7.12)–(7.14) hold:

$$\frac{\Delta J}{J} = \frac{\underline{J} - J}{J} = O(\varepsilon), \tag{7.12}$$

$$x(t) - \underline{x}(t) = \mu O(\varepsilon) \quad (\text{as } t \to \infty), \tag{7.13}$$

$$z(t) - \underline{z}(t) = \mu O(\varepsilon) \quad (\text{as } t \to \infty). \tag{7.14}$$

Again, the proof of this Theorem is similar to that of Theorem 4.1 and uses Theorem A.2 and A.3 of Appendix A.

The more general cases when w and v are scaled by different factors or when the white-noise inputs to the slow and fast equations are scaled by different factors, e.g. $\mu_1 F_1$ and $\mu_2 F_2$, are more difficult and less understood. The difficulty arises from the fact that the solution of the filter Riccati equation and the filter gain will depend on ratios between various scaling parameters. There has been some effort to study such problems which will be surveyed in the notes and references section.

4.8 Exercises

Exercise 4.1

Consider the second-order singularly perturbed system

$$\dot{x}(t) = -x(t) + z(t),$$

$$\varepsilon \dot{z}(t) = -z(t) + v(t),$$

where $v(t)$ is a zero-mean stationary Gaussian exponentially correlated process with power spectral density $2\sigma^2\theta/(1 + \omega^2\theta^2)$. Three cases are of interest here:

(i) $\theta \ll \varepsilon$,
(ii) $\theta = O(\varepsilon)$,
(iii) $\theta = O(1)$.

(a) In which of these cases can $v(t)$ be modeled as white noise? If $v(t)$ cannot be modeled as white noise, model it as the solution of a differential equation driven by white noise.

(b) For each case, write the augmented state equation in the singularly perturbed form and obtain slow–fast decompositions.

Exercise 4.2

The noise voltage sources in the RC-circuit of Fig. 4.6 represent the thermal or Johnson noise associated with the resistors R_1 and R_2, due to the random motion of charges, which is independent of their mean or average motion. For an Ohmic resistance R at thermal equilibrium, the thermal noise has a constant spectrum $4kTR$, where k is Boltzmann's constant and T is temperature in Kelvin. Let $R_1 = R_2 = R$, $C_1 = \varepsilon C$, $C_2 = C$.

(a) Taking $x = v_2$ and $z = v_1$, show that the state equations are in the singularly perturbed form.

(b) Obtain a slow–fast decomposition using (2.13).

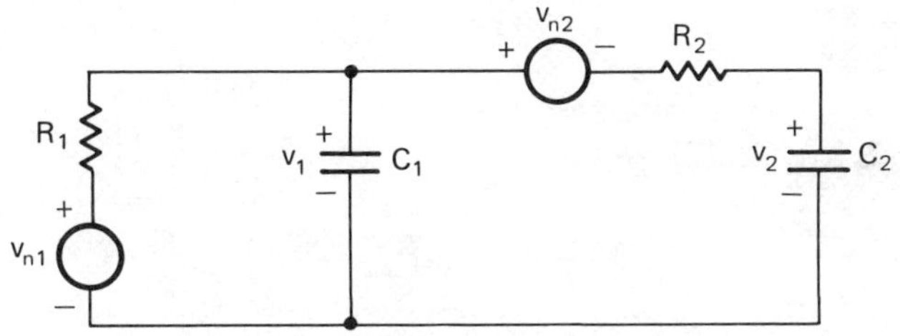

Fig. 4.6. RC-circuit with noise voltage sources of Exercise 4.2.

(c) Suppose now that $\varepsilon = 10^{-5}$, which is a reasonably small number, and $R = 1000\ \Omega$. Taking into consideration that $k = 1.38 \times 10^{-23}$ J/K and T (at room temperature) $= 300$ K, the constant $(4kTR)^{1/2} \sim 4.07 \times 10^{-9}$ is much smaller than ε. How would this fact affect the validity of the approximations obtained in part (b)?

Exercise 4.3

Let

$$\dot{x}(t) = Ax(t) + Bw(t), \quad x(0) = x_0,$$

where $w(t)$ is a zero-mean stationary white noise process with intensity matrix V, and x_0 is a zero-mean random vector independent of w with variance Q_0.

(a) Show that $Q(t)$, the variance of $x(t)$, is given by

$$Q(t) = \mathrm{E}\{x(t)x^{\mathrm{T}}(t)\} = \mathrm{e}^{At}\, Q_0\, \mathrm{e}^{A^{\mathrm{T}}t} + \int_0^t \mathrm{e}^{A(t-\tau)} BVB^{\mathrm{T}}\, \mathrm{e}^{A^{\mathrm{T}}(t-\tau)}\, \mathrm{d}\tau.$$

(b) Show that $Q(t)$ satisfies the Lyapunov equation

$$\dot{Q}(t) = AQ(t) + Q(t)A^{\mathrm{T}} + BVB^{\mathrm{T}}, \quad Q(0) = Q_0.$$

(c) Taking

$$A = \begin{bmatrix} A_{11} & A_{12} \\ \dfrac{A_{21}}{\varepsilon} & \dfrac{A_{22}}{\varepsilon} \end{bmatrix}, \quad B = \begin{bmatrix} B_1 \\ \dfrac{B_2}{\varepsilon} \end{bmatrix},$$

where $\det A_{22} \neq 0$, show that $Q(t, \varepsilon)$ takes the form

$$Q(t, \varepsilon) = \begin{bmatrix} Q_1(t, \varepsilon) & Q_2(t, \varepsilon) \\ Q_2^{\mathrm{T}}(t, \varepsilon) & \dfrac{1}{\varepsilon} Q_3(t, \varepsilon) \end{bmatrix}$$

Hint: Use the decoupling transformation of Section 2.4.

Exercise 4.4

Consider (2.16), (2.17) and let

$$\sum_{i=0}^{\infty} \frac{\varepsilon^i}{i!} L^{(i)} \text{ and } \sum_{i=0}^{\infty} \frac{\varepsilon^i}{i!} H^{(i)}$$

be the Maclaurin series of L and H respectively.

(a) Show that

$$L^{(0)} = A_{22}^{-1} A_{21}, \quad L^{(1)} = A_{22}^{-1} L^{(0)} A_0,$$
$$H^{(0)} = A_{12} A_{22}^{-1}, \quad H^{(1)} = A_0 H^{(0)} A_{22}^{-1} - H^{(0)} L^{(0)} H^{(0)}.$$

(b) Let $x(t)$, $z(t)$ be the solution of (2.1), (2.2), and let $\bar{x}(t)$, $\bar{z}(t)$ be given by

$$\bar{x}(t) = \bar{\xi}(t) + \varepsilon H^{(0)} \bar{\eta}(t)$$
$$\bar{z}(t) = -(L^{(0)} + \varepsilon L^{(1)})\bar{\xi}(t) + [I_m - \varepsilon L^{(0)} H^{(0)}]\bar{\eta}(t)$$

where $\bar{\xi}(t)$, $\bar{\eta}(t)$ respectively satisfy

$$\dot{\bar{\xi}}(t) = [A_{11} - A_{12}(L^{(0)} + \varepsilon L^{(1)})]\bar{\xi}(t)$$
$$+ [B_1 - (H^{(0)} + \varepsilon H^{(1)})B_2 - \varepsilon H^{(0)} L^{(0)} B_1] w(t),$$
$$\bar{\xi}(0) = [I_n - \varepsilon H^{(0)} L^{(0)}] x_0 - \varepsilon H^{(0)} z_0,$$
$$\varepsilon \dot{\bar{\eta}}(t) = [A_{22} + \varepsilon L^{(0)} A_{12}]\bar{\eta}(t) + [B_2 + \varepsilon L^{(0)} B_1] w(t),$$
$$\bar{\eta}(0) = (L^{(0)} + \varepsilon L^{(1)}) x_0 + z_0.$$

Show that

$$x(t) - \bar{x}(t) = O(\varepsilon^{3/2}),$$
$$z(t) - \bar{z}(t) = O(\varepsilon^{3/2}).$$

Exercise 4.5

(a) Define and solve the slow and fast steady-state filtering problems for

$$\dot{x}(t) = -x(t) + z(t),$$
$$\varepsilon \dot{z}(t) = -z(t) + w(t),$$
$$y(t) = x + z + v(t),$$

where $w(t)$ and $v(t)$ have intensity one.

(b) Solve the original steady-state filtering problem, parameterized in ε.

(c) Compute the steady-state root-mean-square estimation errors in both cases for $\varepsilon = 0.5$, 0.1 and 0.05 and compare them.

Exercise 4.6

Consider the singularly perturbed system

$$\dot{x}(t) = A_{11}(\varepsilon)x(t) + A_{12}(\varepsilon)z(t),$$
$$\varepsilon\dot{z}(t) = A_{21}(\varepsilon)x(t) + A_{22}(\varepsilon)z(t),$$

where $A_{ij}(\varepsilon)$ are analytic at $\varepsilon = 0$ and $\det A_{22}(0) \neq 0$. Consider the similarity transformation (4.8) of Chapter 2, where L and H satisfy equations similar to (2.8) and (4.3) respectively, of Chapter 2, but with A_{ij} replaced by $A_{ij}(\varepsilon)$.

(a) Using the implicit function theorem, prove that there is an $\varepsilon^* > 0$ such that $\forall\varepsilon \in (0, \varepsilon^*)$ there exist $L(\varepsilon)$ and $H(\varepsilon)$ satisfying their respective equations with $L(0) = A_{22}^{-1}(0)A_{21}(0)$ and $H(0) = A_{12}(0)A_{22}^{-1}(0)$.

(b) Show that the similarity transformation decouples the above system into slow and fast parts.

Exercise 4.7

With reference to the variance analysis of Appendix A, consider the system (A 1) and the quadratic cost (A 18). Let σ_s and σ_f be defined by

$$\sigma_s = \mu^2 \operatorname{tr}[C_0^T C_0 Q_0], \quad \sigma_f = \mu^2 \operatorname{tr}[C_2^T(0)C_2(0)Q_f],$$

where Q_s and Q_f respectively satisfy

$$0 = A_0 Q_0 + Q_0 A_0^T + B_0 V B_0^T,$$
$$0 = A_{22}(0)Q_f + Q_f A_{22}^T(0) + B_2(0)VB_2^T(0),$$

and where

$$A_0 = A_{11}(0) - A_{12}(0)A_{22}^{-1}(0)A_{21}(0), \quad B_0 = B_1(0) - A_{12}(0)A_{22}^{-1}(0)B_2(0)$$
$$C_0 = C_1(0) - C_2(0)A_{22}^{-1}(0)A_{21}(0).$$

(a) Verify that $Q_0 = \bar{Q}_{11}(0)$ and $Q_f = \bar{Q}_{33}(0)$ ($\bar{Q}_{ij}(0)$ is defined in Appendix A).

It is shown in Appendix A that $\sigma = \sigma_f/\varepsilon + O(1)$ and

$$\sigma_f = 0 \Leftrightarrow C_2(0)\, e^{A_{22}(0)t}\, B_2(0) \equiv 0.$$

(b) Show that

$$\sigma_f = 0 \Rightarrow \begin{aligned} &\text{(i)} & C_2(0)\bar{Q}_{33}(0) &= 0, \\ &\text{(ii)} & C_2(0)\bar{Q}_{13}(0) &= -C_2(0)A_{22}^{-1}(0)A_{21}(0), \\ &\text{(iii)} & C_2(0)\bar{Q}_{33}^{(1)}C_2^{\mathrm{T}}(0) &= C_2(0)A_{22}^{-1}(0)A_{21}(0)Q_0 \\ &&& \quad \times A_{21}^{\mathrm{T}}(0)A_{22}^{-\mathrm{T}}(0)C_2^{\mathrm{T}}(0). \end{aligned}$$

(c) Show that $\sigma_f = 0 \Rightarrow \sigma = \sigma_s + O(\varepsilon)$.

Exercise 4.8

Consider the LQG control problem

$$\dot{x} = z + w_1,$$

$$\varepsilon\dot{z} = -x - z + u + w_2,$$

$$y_1 = x + v_1,$$

$$y_2 = z + v_2,$$

$$J = \lim_{t_f - t_0 \to \infty} \frac{1}{t_f - t_0} \mathrm{E} \int_{t_0}^{t_f} [x^2(t) + z^2(t) + u^2(t)]\,\mathrm{d}t$$

where $V = I_2$.

(a) Determine the composite control design.

(b) Determine the corrected design of Section 4.6.

(c) Determine the optimal design parameterized in ε.

(d) Find J for each of the three designs for the following values of ε: 0.5, 0.25, 0.125, 0.0625 and 0.03125.

4.9 Notes and References

Research in filtering and control problems for singularly perturbed models with white-noise inputs has revealed difficulties not present in deterministic problems. This is due to the fact that, as we have seen in Section 4.2, the input white-noise process "fluctuates" faster than the fast dynamic variables, which as $\varepsilon \to 0$ themselves tend to white-noise processes. Because of these difficulties, other formulations of stochastic problems for singularly perturbed systems have been investigated. Blankenship and Sachs (1979) studied a linear system of the form (2.1), (2.2) where the white-noise input is replaced by wideband noise. Their work deals with two asymptotic phenomena; one due to small time constants and the other due to small correlation times.

Section 4.3 is a steady-state treatment of a singularly perturbed Kalman–Bucy filtering problem of Haddad (1976). The singularly perturbed LQG control problem of Section 4.4 is a steady-state version of Haddad and Kototović (1977); see also Teneketzis and Sandell (1977). The treatment of the filtering and control problems of Sections 4.3 and 4.4 is different from the original papers. It is based on a more recent approach by Khalil and Gajic (1984), which makes use of the decoupling transformation of Section 2.4 and performs a variance analysis of singularly perturbed systems driven by white noise by studying singularly perturbed Lyapunov equations. This variance analysis is included in Appendix A. The corrected LQG design of Section 4.6, which appears in the literature for the first time, is based again on Khalil and Gajic (1984), where general procedures for obtaining higher-order approximations are discussed.

The issue of scaling the white noise input disturbances, which is addressed in Section 4.7 for the special case of uniform scaling, has been motivated by two considerations. First, it arises in physical models through mechanisms similar to the one discussed in Section 4.7. Second, it has been called upon as a way to reformulate LQG control problems so as to avoid divergent ($O(1/\varepsilon)$) performance indices. Various alternative formulations of the linear stochastic regulator problem, where several scaling parameters are used to scale white-noise contributions and the fast part of the performance index, have been reported in the literature. Interested readers may consult the surveys by Kokotović (1984) and Saksena *et al.* (1984). A nonlinear version of the problem is treated in Bensoussan (1981).

The treatment of this chapter is limited to standard linear singularly perturbed models driven by white noise. It does not cover stochastic averaging and small-noise problems; sources for which are the surveys by Blankenship (1979) and Schuss (1980) and the book by Kushner (1984).

5 LINEAR TIME-VARYING SYSTEMS

5.1 Introduction

For linear time-invariant models the only source of multiple time-scale properties, as we have witnessed in Chapter 2, is the separation of the system eigenvalues into two (or more) groups of eigenvalues of different orders of magnitude. In the case of linear time-varying systems, another source of multiple time-scale properties is the variation of model parameters. The simplest example of this, studied in Section 5.2, is the linear system

$$\varepsilon \dot{z}(t) = A(t)z(t), \tag{1.1}$$

where the model coefficient matrix $A(t)$ is a function of time t over some time interval of interest. Section 5.2 establishes conditions, including bounds on ε, for the asymptotic stability of the system (1.1).

Later sections parallel the time-invariant development of Chapter 2 through dealing with the more general linear time-varying singularly perturbed system model

$$\dot{x}(t) = A_{11}(t)x(t) + A_{12}(t)z(t), \tag{1.2}$$

$$\varepsilon \dot{z}(t) = A_{21}(t)x(t) + A_{22}(t)z(t). \tag{1.3}$$

Section 5.3 provides a transformation for decomposing the system (1.2), (1.3) into two subsystems in separate time scales, while Section 5.4 establishes conditions for system asymptotic stability. An important class of systems with time-varying parameters is the class of adaptive control systems. The robustness issue in Section 5.5 is whether the considered adaptive system remains stable in the presence of unmodeled high-frequency dynamics.

Proceeding to linear time-varying singularly perturbed systems with inputs, Section 5.6 develops state approximations similar to those of Section 2.5 for time-invariant systems. Controllability properties of such systems are explored in Section 5.7. In anticipation of the trajectory optimization problems of Chapter 6, this controllability study is extended to the analysis of the asymptotic behavior of a control which transfers the system state from a given initial point to a desired final point in the state space. Finally, observability properties are established in Section 5.8.

5.2 Slowly Varying Systems

In this section, we study time-varying systems described by (1.1). When $\|\dot{A}(t)\| = O(1)$, the parameters are slowly varying relative to the dynamics of the system. One would intuitively expect that for ε sufficiently small the dynamic behavior of the system could be predicted by studying a family of time-invariant systems with parameters frozen at a particular time instant t. In particular, when $\operatorname{Re}\lambda(A(t)) < -c < 0$ for all t, intuition suggests that as $\varepsilon \to 0$ the above system should be asymptotically stable.

Consider the system

$$\varepsilon\dot{z}(t) = A(t)z(t), \quad t \in I, \quad z \in R^m, \tag{2.1}$$

where the time interval I could be a finite interval $[t_0, t_f]$ or an infinite interval $[t_0, \infty)$. We make the following assumption.

Assumption 2.1

$$\operatorname{Re}\lambda(A(t)) \leq -c_1 < 0 \quad \forall t \in I.$$

This assumption guarantees that the family of frozen-time systems associated with (2.1) is asymptotically stable. Hence the exponential matrix exp-$[A(t)\theta]$ satisfies

$$\|\exp[A(t)\theta]\| \leq K(A)e^{-\alpha\theta} \quad \forall \theta \geq 0, \quad \forall t \in I, \tag{2.2}$$

where $\alpha > 0$ is independent of t, but $K(A) > 0$ depends on t. To extend the exponentially decaying bound (2.2) from a frozen-time system to the actual time-varying system (2.1), the inequality (2.2) should hold uniformly in t. Towards that goal we assume the following.

Assumption 2.2

$$\|A(t)\| \leq c_2 \quad \forall t \in I.$$

Lemma 2.1

Under Assumptions 2.1 and 2.2, there exist positive constants α_1 and K_1 such that

$$\|\exp[A(t)\theta]\| \le K_1 e^{-\alpha_1\theta} \quad \forall\theta > 0, \quad \forall t \in I. \tag{2.3}$$

Proof The matrices $A(t)$ satisfying Assumptions 2.1 and 2.2 form a compact set, which we denote by S. Consider

$$e^{(A+B)\theta} = e^{A\theta} + \int_0^\theta e^{A(\theta-\sigma)} B e^{(A+B)\sigma}\, d\sigma.$$

Using (2.2), we get

$$\|e^{(A+B)\theta}\| \le K(A)e^{-\alpha\theta} + \int_0^\theta K(A)e^{-\alpha(\theta-\sigma)}\|B\|\,\|e^{(A+B)\sigma}\|\, d\sigma,$$

$$e^{\alpha\theta}\|e^{(A+B)\theta}\| \le K(A) + K(A)\|B\| \int_0^\theta e^{\alpha\sigma}\|e^{(A+B)\sigma}\|\, d\sigma.$$

Application of Gronwall's Lemma (Vidyasagar, 1978) yields

$$\|e^{(A+B)\theta}\| \le K(A)e^{-(\alpha - K(A)\|B\|)\theta} \quad \forall\theta \ge 0. \tag{2.4}$$

By (2.4), there exists a positive constant $\alpha_1 < \alpha$ and a neighborhood $N(A)$ of A such that if $C \in N(A)$ then

$$\|e^{C\theta}\| \le K(A)e^{-\alpha_1\theta} \quad \forall\theta \ge 0.$$

Since S is compact it is covered by a finite number of these neighborhoods, from which (2.3) follows. □

With the uniform inequality (2.3) established, we are now ready to establish the asymptotic stability of the time-varying system (2.1) when ε is small and $\dot{A}(t) = O(1)$. The last assumption, which corresponds to assuming that the variation in model parameters is slow, takes the following explicit form.

Assumption 2.3

$$\|\dot{A}(t)\| \le \beta \quad \forall t \in I.$$

Lemma 2.2, below, shows that the transition matrix of (2.1), denoted by $\phi(t, s)$, can be uniformly approximated by the exponential matrix $\exp[A(s)(t-s)/\varepsilon]$, corresponding to a system in which the parameters of

$A(t)$ are frozen at $t = s$. The lemma gives an exponentially decaying bound on the error

$$\Psi(t, s) = \phi(t, s) - \exp[A(s)(t - s)/\varepsilon]. \tag{2.5}$$

Lemma 2.2

Under Assumptions 2.1–2.3 there exist positive constants ε^*, α_2 and K_2 such that, for all $\varepsilon \in (0, \varepsilon^*)$,

$$\|\Psi(t, s)\| \leq \varepsilon K_2 e^{-\alpha_2(t-s)/\varepsilon}, \quad t \geq s, \quad t, s \in I. \tag{2.6}$$

Proof The definition of $\Psi(t, s)$ implies

$$\begin{aligned} &\Psi(s, s) = 0, \\ &\frac{\partial}{\partial t}\Psi(t, s) = \frac{1}{\varepsilon}A(t)\Psi(t, s) + \frac{1}{\varepsilon}(A(t) - A(s))\exp\left[A(s)\left(\frac{t - s}{\varepsilon}\right)\right]. \end{aligned} \tag{2.7}$$

Applying the variations of constants formula to (2.7), we obtain

$$\begin{aligned} \Psi(t, s) = &\frac{1}{\varepsilon}\int_s^t \Psi(t, \tau)(A(\tau) - A(s))\exp\left[A(s)\left(\frac{\tau - s}{\varepsilon}\right)\right]\mathrm{d}\tau \\ &+ \frac{1}{\varepsilon}\int_s^t \exp\left[A(\tau)\left(\frac{t - \tau}{\varepsilon}\right)\right](A(\tau) - A(s))\exp\left[A(s)\left(\frac{\tau - s}{\varepsilon}\right)\right]\mathrm{d}\tau. \end{aligned}$$

Multiplying through by $e^{\alpha_2(t-s)/\varepsilon}$, where $0 < \alpha_2 < \alpha_1$, and letting $\eta(t, s) = e^{\alpha_2(t-s)/\varepsilon}\Psi(t, s)$, we obtain

$$\begin{aligned} \eta(t, s) = &\frac{1}{\varepsilon}\int_s^t \eta(t, \tau)e^{\alpha_2(\tau-s)/\varepsilon}(A(\tau) - A(s))\exp\left[A(s)\left(\frac{\tau - s}{\varepsilon}\right)\right]\mathrm{d}\tau \\ &+ \frac{1}{\varepsilon}\int_s^t e^{\alpha_2(t-s)/\varepsilon}\exp\left[A(\tau)\left(\frac{t - \tau}{\varepsilon}\right)\right](A(\tau) - A(s)) \\ &\times \exp\left[A(s)\left(\frac{\tau - s}{\varepsilon}\right)\right]\mathrm{d}\tau. \end{aligned} \tag{2.8}$$

Viewing $\eta(t, s)$ as a functional on $[s, t]$ with norm $\|\eta\| = \sup_{s \leq \tau \leq t}\|\eta(t, \tau)\|$, (2.8) is rewritten in the form

$$\eta = F(\eta) + B = T(\eta).$$

We shall prove that $T(\cdot)$ is a contraction mapping over a set of the form

$\{\eta : \|\eta\| \le \varepsilon\rho\}$, where ρ will be chosen later. By Assumptions 2.1–2.3, we have

$$\|B\| \le \frac{1}{\varepsilon}\int_s^t \mathrm{e}^{\alpha_2(t-s)/\varepsilon} K_1 \mathrm{e}^{-\alpha_1(t-\tau)/\varepsilon}\beta(\tau - s)K_1\mathrm{e}^{-\alpha_1(\tau-s)/\varepsilon}\,\mathrm{d}\tau$$

$$= \frac{\varepsilon K_1^2\beta}{2}\left(\frac{t-s}{\varepsilon}\right)^2 \mathrm{e}^{-(\alpha_1-\alpha_2)(t-s)/\varepsilon} \le \frac{2K_1^2\beta}{(\alpha_1-\alpha_2)^2 e^2}\,\varepsilon, \qquad (2.9)$$

where we have used the fact that $x^2\mathrm{e}^{-ax} \le 4/a^2\mathrm{e}^2 \quad \forall x \ge 0$. For $\|\eta\| \le \varepsilon\rho$ we have

$$\|F(\eta)\| \le \frac{1}{\varepsilon}\int_s^t \varepsilon\rho\mathrm{e}^{\alpha_2(\tau-s)/\varepsilon}\beta(\tau - s)K_1\mathrm{e}^{-\alpha_1(\tau-s)/\varepsilon}\,\mathrm{d}\tau$$

$$\le \varepsilon^2\rho\beta K_1\int_0^\infty w\mathrm{e}^{-(\alpha_1-\alpha_2)w}\,\mathrm{d}w = \frac{\rho\beta K_1}{(\alpha_1-\alpha_2)^2}\,\varepsilon^2. \qquad (2.10)$$

Combining (2.9) and (2.10), we get

$$\|T(\eta)\| \le \frac{\rho\beta K_1}{(\alpha_1-\alpha_2)^2}\,\varepsilon^2 + \frac{2K_1^2\beta}{(\alpha_1-\alpha_2)^2\mathrm{e}^2}\,\varepsilon.$$

Choose

$$\rho = \frac{2K_1^2\beta}{(\alpha_1-\alpha_2)^2\mathrm{e}^2(1-\gamma)}, \qquad \varepsilon^* = \frac{(\alpha_1-\alpha_2)^2\gamma}{\beta K_1},$$

where $0 < \gamma < 1$. Then, for all $\varepsilon < \varepsilon^*$, $\|T(\eta)\| < \varepsilon\rho$. Moreover,

$$\|T(\eta_2) - T(\eta_1)\| \le \frac{1}{\varepsilon}\int_s^t \mathrm{e}^{\alpha_2(\tau-s)/\varepsilon}\beta(\tau-s)K_1\mathrm{e}^{-\alpha_1(\tau-s)/\varepsilon}\,\mathrm{d}\tau\,\|\eta_2 - \eta_1\|$$

$$\le \varepsilon\beta K_1\int_0^\infty w\mathrm{e}^{-(\alpha_1-\alpha_2)w}\,\mathrm{d}w\,\|\eta_2-\eta_1\|$$

$$= \frac{\varepsilon\beta K_1}{(\alpha_1-\alpha_2)^2}\|\eta_2-\eta_1\|$$

$$< \gamma\|\eta_2-\eta_1\| \quad \forall\varepsilon < \varepsilon^*.$$

Thus $T(\cdot)$ is a contraction mapping, and the unique solution of (2.8) satisfies

$$\|\eta(t,s)\| \le \varepsilon\rho \Rightarrow \|\Psi(t,s)\| \le \varepsilon\rho\mathrm{e}^{-\alpha_2(t-s)/\varepsilon}. \qquad \square$$

Lemma 2.2 confirms the intuitively expected result that the system (2.1) is uniformly asymptotically stable for sufficiently small ε. The proof of the lemma gives an estimate $\varepsilon^* = (\alpha_1 - \alpha_2)^2\gamma/\beta K_1$ such that for all $\varepsilon < \varepsilon^*$ uniform asymptotic stability holds. The arbitrary positive constants $\alpha_2 < \alpha_1$ and $\gamma < 1$ can be adjusted to improve ε^* while maintaining a useful exponentially decaying bound on $\|\phi(t, s)\|$ (see Exercise 5.2). It is obvious, however, that within the analysis of Lemma 2.2 the best bound on ε guaranteeing uniform asymptotic stability can be obtained by letting $\alpha_2 \to 0$ and $\gamma \to 1$ to get $\varepsilon^* = \alpha_1^2/\beta K_1$.

An alternative bound on ε guaranteeing uniform asymptotic stability can be obtained by using a quadratic Lyapunov function $v(t, z) = z^{\mathrm{T}}P(t)z$. Let $P(t)$ be the solution of the algebraic Lyapunov equation (2.11), parameterized in t:

$$P(t)A(t) + A^{\mathrm{T}}(t)P(t) = -I_m. \tag{2.11}$$

It can be shown (Exercise 5.3) that Assumptions 2.1–2.3 on $I = [t_0, \infty)$ guarantee that for every $t \geq t_0$, (2.11) has a unique positive-definite solution and that $P(t)$ is bounded from above. Let

$$V(t, z) = z^{\mathrm{T}}(t)P(t)z(t), \tag{2.12}$$

$$\begin{aligned}\dot{V} &= -\frac{1}{\varepsilon}z^{\mathrm{T}}(PA + A^{\mathrm{T}}P)z + z^{\mathrm{T}}\dot{P}z \\ &\leq -\frac{1}{\varepsilon}(1 - \varepsilon\|\dot{P}\|_2)\|z\|_2^2.\end{aligned}$$

Therefore $\dot{V}$ is negative-definite for all $\varepsilon < (\sup_{t \geq t_0}\|\dot{P}\|_2)^{-1}$. The calculation of these bounds is illustrated by the following example.

Example 2.1

Consider $\varepsilon\dot{z} = A(t)z$, where

$$A(t) = \begin{bmatrix} -1 + a\cos^2 t & 1 - a\sin t\cos t \\ -1 - a\sin t\cos t & -1 + a\sin^2 t \end{bmatrix}.$$

The characteristic equation of $A(t)$ is

$$\lambda^2 + (2 - a)\lambda + (2 - a) = 0.$$

For all $a < 2$ the eigenvalues of $A(t)$ satisfy $\operatorname{Re}\lambda < 0$. Moreover, $A(t)$ and $\dot{A}(t)$ are bounded for all t. Thus $A(t)$ satisfies Assumptions 2.1–2.3. This, however, does not guarantee asymptotic stability if ε is not small. For example, when $\varepsilon = 1$, $\phi(t, 0)$ is given by

$$\phi(t,0) = \begin{bmatrix} e^{(a-1)t}\cos t & e^{-t}\sin t \\ -e^{(a-1)t}\sin t & e^{-t}\cos t \end{bmatrix},$$

which shows that for $1 < a < 2$ the system is unstable even though Assumptions 2.1–2.3 are satisfied. As $\varepsilon \to 0$ Lemma 2.2 guarantees exponential stability for all $\varepsilon < \varepsilon^*$, where ε^* is estimated by Lemma 2.2 as $\varepsilon_1^* = \alpha_1^2/\beta K_1$ and by the Lyapunov function (2.11) as $\varepsilon_2^* = (\sup\|\dot{P}(t)\|_2)^{-1}$. To calculate ε_1^* we evaluate the exponential matrix

$$\exp[A(t)\theta] = e^{-(1-a/2)\theta}\begin{bmatrix} \Psi_{11}(t)\cos(\omega\theta - \delta(t)) & \Psi_{12}(t)\sin\omega\theta \\ \Psi_{21}(t)\sin\omega\theta & \Psi_{22}(t)\cos(\omega\theta + \delta(t)) \end{bmatrix},$$

where

$$\begin{aligned} \omega^2 &= 1 - \tfrac{1}{4}a^2, \\ \tan\delta(t) &= \frac{a\cos 2t}{2\omega}, \\ \Psi_{11}(t) &= \Psi_{22}(t) = \sec\delta(t), \\ \Psi_{12}(t) &= (1 - \tfrac{1}{2}a\sin 2t)/\omega, \\ \Psi_{21}(t) &= -(1 + \tfrac{1}{2}a\sin 2t)/\omega. \end{aligned}$$

Taking $\|A\| = \|A\|_\infty = \max_i \sum_j |a_{ij}|$, it can be shown that

$$\|\exp[A(t)\theta]\| \le e^{-(1-a/2)\theta}\frac{4+a}{(4-a^2)^{1/2}}.$$

The matrix $\dot{A}(t)$ is given by

$$\dot{A}(t) = \begin{bmatrix} -a\sin 2t & -a\cos 2t \\ -a\cos 2t & a\sin 2t \end{bmatrix},$$

and $\|\dot{A}(t)\| \le a\sqrt{2}$. Therefore

$$\varepsilon_1^* = \frac{(2-a)^2(4-a^2)^{1/2}}{4\sqrt{2}(4+a)a}.$$

On the other hand, $P(t)$ satisfying the Lyapunov equation (2.11) is given by

$$P(t) = \frac{1}{2(2-a)^2} \times \begin{bmatrix} (2-a)(1-a\sin^2 t) + (2 + a\sin 2t) & -\frac{1}{2}a(2-a)\sin 2t + a\cos 2t \\ -\frac{1}{2}a(2-a)\sin 2t + a\cos 2t & [(2-a)(1-a\cos^2 t) + (2 - a\sin 2t)] \end{bmatrix}.$$

Then

$$\dot{P}(t) = \frac{a[4+(2-a)^2]^{1/2}}{2(2-a)^2} \begin{bmatrix} \cos(2t+\phi) & -\sin(2t+\phi) \\ -\sin(2t+\phi) & -\cos(2t+\phi) \end{bmatrix},$$

where $\tan\phi = \frac{1}{2}(2-a)$, and

$$\|\dot{P}(t)\|_2 = \frac{a[4+(2-a)^2]^{1/2}}{2(2-a)^2}.$$

Therefore

$$\varepsilon_2^* = \frac{2(2-a)^2}{a[4+(2-a)^2]^{1/2}}.$$

The expressions for ε_1^* and ε_2^* vanish as $a \to 2$, which is expected since the eigenvalues of $A(t)$ tend to the imaginary axis as $a \to 2$. Comparison of ε_1^* and ε_2^* shows that $\varepsilon_2^* > \varepsilon_1^*$ for all $1 < a < 2$. For example, at $a = 1.1$, $\varepsilon_1^* = 0.042$ and $\varepsilon_2^* = 0.671$. The superiority of ε_2^* over ε_1^* is typical.

We conclude this section by showing that regular perturbations of the right-hand-side of (2.1) do not destroy the uniform exponential bound established by Lemma 2.2. Consider the system

$$\varepsilon\dot{z}(t) = [A(t) + \varepsilon^\gamma \Gamma(t, \varepsilon)]z(t), \tag{2.13}$$

where $\|\Gamma(t, \varepsilon)\| \leq c_3 \; \forall t \in I$, and $\gamma > 0$. Let $\phi_1(t, s)$ be the transition matrix of (2.13).

Lemma 2.3

Under Assumptions 2.1–2.3, there exist positive constants ε^*, α_3 and K_3 such that for all $\varepsilon \in (0, \varepsilon^*)$

$$\left\| \phi_1(t, s) - \phi(t, s) \right\| \leq \varepsilon^\gamma K_3 e^{-\alpha_3(t-s)/\varepsilon}, \quad t \geq s, \quad t, s \in I. \tag{2.14}$$

where $\phi(t, s)$ is the transition matrix of (2.1), which, by Lemma 2.2, satisfies the bound $\|\phi(t, s)\| \leq K e^{-\alpha(t-s)/\varepsilon}$.

Proof

$$\phi_1(t, s) = \phi(t, s) + \int_s^t \phi(t, \tau)\varepsilon^{\gamma-1}\Gamma(\tau, \varepsilon)\phi_1(\tau, s)\, d\tau, \tag{2.15}$$

From (2.15) we have

$$\begin{aligned}\phi_1(t, s) - \phi(t, s) = &\int_s^t \phi(t, \tau)\varepsilon^{\gamma-1}\Gamma(\tau, \varepsilon)\phi(\tau, s)\, d\tau \\ &+ \int_s^t \phi(t, \tau)\varepsilon^{\gamma-1}\Gamma(\tau, \varepsilon)(\phi_1(\tau, s) - \phi(\tau, s))\, d\tau.\end{aligned}$$

Taking norms,

$$\begin{aligned}\|\phi_1(t, s) - \phi(t, s)\| \leq\ &K^2 c_3 \varepsilon^{\gamma-1} e^{-\alpha(t-s)/\varepsilon}(t - s) \\ &+ K c_3 \varepsilon^{\gamma-1} \int_s^t e^{-\alpha(t-\tau)/\varepsilon}\|\phi_1(\tau, s) - \phi(\tau, s)\|\, d\tau.\end{aligned}$$

Multiplying throughout by $e^{\alpha_3(t-s)/\varepsilon}$, where $0 < \alpha_3 < \alpha$, and setting $\eta(t, s) = e^{\alpha_3(t-s)/\varepsilon}\|\phi_1(t, s) - \phi(t, s)\|$ yields

$$\eta(t, s) \leq \varepsilon^\gamma C_4 + \varepsilon^{\gamma-1} C_5 \int_s^t e^{-(\alpha-\alpha_3)(t-\tau)/\varepsilon}\eta(\tau, s)\, d\tau.$$

By Gronwall's Lemma (Vidyasagar 1978), we have

$$\begin{aligned}\eta(t, s) &\leq \varepsilon^\gamma C_4 \exp\left(\varepsilon^{\gamma-1} C_5 \int_s^t e^{-(\alpha-\alpha_3)(t-\tau)/\varepsilon}\, d\tau\right) \\ &\leq \varepsilon^\gamma C_4 \exp\left(\frac{\varepsilon^\gamma C_5}{\alpha - \alpha_3}\right) \leq \varepsilon^\gamma K_3.\end{aligned}$$

□

5.3 Decoupling Transformation

In this section, we consider the singularly perturbed system

$$\left.\begin{aligned}\dot{x}(t) &= A_{11}(t)x(t) + A_{12}(t)z(t), \quad x \in R^n, \\ \varepsilon\dot{z}(t) &= A_{21}(t)x(t) + A_{22}(t)z(t), \quad x \in R^m,\end{aligned}\right\} \tag{3.1}$$

where $A_{22}(t)$ satisfies Assumptions (2.1)–(2.3) and A_{ij} are smooth functions of t; smoothness assumptions will be spelled out later. It is shown that, for

ε sufficiently small, there exists a Lyapunov transformation that transforms (3.1) into a block-diagonal form, decoupling the slow and fast dynamics in complete analogy with the similarity transformation (4.8) of Chapter 2. The decoupling transformation is derived in two steps exactly as in Chapter 2. First, $\eta(t) = z(t) + L(t)x(t)$ is substituted, where L is chosen to satisfy

$$\varepsilon \dot{L} = A_{22}L - A_{21} - \varepsilon L(A_{11} - A_{12}L), \tag{3.2}$$

so that the lower left corner block is eliminated. Second, $\xi(t) = x(t) - \varepsilon H(t)\eta(t)$ is substituted, where H is chosen to satisfy

$$-\varepsilon \dot{H} = HA_{22} - A_{12} + \varepsilon HLA_{12} - \varepsilon(A_{11} - A_{12}L)H, \tag{3.3}$$

so that the upper right corner block is eliminated. If there exist bounded continuously differentiable matrices $L(t)$ and $H(t)$, whose derivatives are bounded on I, which satisfy (3.2) and (3.3), then the transformation

$$\begin{bmatrix} \xi(t) \\ \eta(t) \end{bmatrix} = \begin{bmatrix} I_n - \varepsilon H(t)L(t) & -\varepsilon H(t) \\ L(t) & I_m \end{bmatrix} \begin{bmatrix} x(t) \\ z(t) \end{bmatrix} \triangleq T^{-1}(t) \left[x(t)z(t) \right] \tag{3.4}$$

is a Lyapunov transformation that transforms (3.1) into

$$\left.\begin{aligned} \dot{\xi}(t) &= [A_{11}(t) - A_{12}(t)L(t)]\xi(t), \\ \varepsilon\dot{\eta}(t) &= [A_{22}(t) + \varepsilon L(t)A_{12}(t)]\eta(t). \end{aligned}\right\} \tag{3.5}$$

It is left to the reader to verify that $\det T^{-1}(t) = 1$ and

$$T(t) = \begin{bmatrix} I_n & \varepsilon H(t) \\ -L(t) & I_m - \varepsilon L(t)H(t) \end{bmatrix}.$$

Our task now is to show that, for sufficiently small ε, there exist $L(t)$ and $H(t)$ with the abovementioned properties. It helps to notice that there are no boundary conditions imposed on (3.2) and (3.3). We are going to make use of this to simplify the solution of these equations. In particular, boundary conditions will be chosen to eliminate the boundary layers of $L(t)$ and $H(t)$ so that as $\varepsilon \to 0$ they converge, uniformly on I, to the reduced solutions obtained by setting $\varepsilon = 0$ in (3.2) and (3.3). These reduced solutions are $L_0(t) = A_{22}^{-1}(t)A_{21}(t)$ and $H_0(t) = A_{12}(t)A_{22}^{-1}(t)$. Let us start by seeking the solution of (3.2) in the series form

$$L(t) = \sum_{j=0}^{N-1} \varepsilon^j L_j(t) + \varepsilon^N R_L(t). \tag{3.6}$$

Substituting (3.6) into (3.2) and matching coefficients of like powers in ε, we obtain equations determining $L_0, \ldots, L_{N-1}$ in the form

$$0 = A_{22}L_0 - A_{21} \Rightarrow L_0 = A_{22}^{-1}A_{21}, \tag{3.7}$$

$$\dot{L}_0 = A_{22}L_1 - L_0(A_{11} - A_{12}L_0) \Rightarrow L_1 = A_{22}^{-1}[\dot{L}_0 + L_0(A_{11} - A_{12}L_0)], \tag{3.8}$$

and so on. The remainder term R_L satisfies an equation of the form

$$\begin{aligned} \varepsilon\dot{R}_L &= A_{22}R_L - \dot{L}_{N-1} + f_1(L_0, \ldots, L_{N-1}) \\ &\quad + \varepsilon f_2(R_L, L_0, \ldots, L_{N-1}, \varepsilon). \end{aligned} \tag{3.9}$$

By imposing appropriate differentiability and boundedness conditions on the matrices A_{ij} and their derivatives, it can be established that (3.9) has a bounded solution R_L. This then shows the existence of L satisfying (3.2) and gives $\sum_{j=0}^{N-1} \varepsilon^j L_j$ as a uniform $O(\varepsilon^N)$ approximation of L. Equation (3.3) can be treated in the same manner. Its solution is sought in the form

$$H(t) = \sum_{j=0}^{M-1} \varepsilon^j H_j(t) + \varepsilon^M R_H(t), \quad M \leq N. \tag{3.10}$$

$H_0, \ldots, H_{M-1}$ are determined from the equations

$$0 = H_0A_{22} - A_{12} \Rightarrow H_0 = A_{12}A_{22}^{-1}, \tag{3.11}$$

$$\begin{aligned} -\dot{H}_0 &= H_1A_{22} + H_0L_0A_{12} - (A_{11} - A_{12}L_0)H_0 \Rightarrow H_1 \\ &= [-\dot{H}_0 - H_0L_0A_{12} + (A_{11} - A_{12}L_0)H_0]A_{22}^{-1}, \end{aligned} \tag{3.12}$$

etc., and R_H satisfies an equation of the form

$$\begin{aligned} -\varepsilon\dot{R}_H &= R_HA_{22} + \dot{H}_{M-1} + g_1(L_0, \ldots, L_{N-1}, H_0, \ldots, H_{M-1}) \\ &\quad + \varepsilon g_2(R_H, R_L, L_0, \ldots, L_{N-1}, H_0, \ldots, H_{M-1}, \varepsilon), \end{aligned} \tag{3.13}$$

which under appropriate conditions has a bounded solution.

The above discussions show that obtaining higher-order approximations for the solutions of (3.2) and (3.3) is a mechanical repetition of the case of first-order approximations. Therefore the next theorem finalizes the result for first-order approximations. The reader is encouraged to work out the case of second-order approximations (Exercise 5.4).

Theorem 3.1

Assume that $A_{22}(t)$ satisfies Assumptions 2.1–2.3, that $A_{ij}(t)$ are continuously differentiable and bounded on I and that $\dot{A}_{12}$, $\dot{A}_{21}$ and $\dot{A}_{22}$ are bounded on I; then there exists an $\varepsilon^* > 0$ such that for all $\varepsilon \in (0, \varepsilon^*)$ there exist continuously differentiable matrices L and H, bounded on I, satisfying (3.2) and (3.3). Moreover, $L(t) = L_0(t) + O(\varepsilon)$ and $H(t) = H_0(t) + O(\varepsilon)$.

Proof Continuing our discussion, we take $N = M = 1$. Then $f_i(\cdot)$ and $g_i(\cdot)$ of (3.9) and (3.13) are given by

$$f_1 = -L_0(A_{11} - A_{12}L_0),$$

$$f_2 = -R_L A_{11} + R_L A_{12} L_0 + L_0 A_{12} R_L + \varepsilon R_L A_{12} R_L,$$

$$g_1 = H_0 L_0 A_{12} - (A_{11} - A_{12} L_0) H_0,$$

$$g_2 = R_H L_0 A_{12} + H_0 R_L A_{12} - A_{11} R_H + A_{12} R_L H_0 + A_{12} L_0 R_H + \varepsilon R_H R_L A_{12} + \varepsilon A_{12} R_L R_H.$$

We start by studying (3.9). Consider the integral equation $R_L(t) = SR_L(t)$, where

$$SR_L(t) = \frac{1}{\varepsilon} \int_{t_0}^{t} \phi(t, \tau)[f_1(\tau) - \dot{L}_0(\tau) + \varepsilon f_2(R_L)(\tau)]\, d\tau, \tag{3.14}$$

where $\phi(t, \tau)$ is the transition matrix of $\varepsilon \dot{z} = A_{22}(t) z$. Let $\mathscr{L} = \{R_L : \|R_L\| < \rho\}$, where $\|R_L\| = \sup_I \|R_L(t)\|$ and ρ will be chosen later. Using the matrix identity

$$R_{L1} A_{12} R_{L1} - R_{L2} A_{12} R_{L2} = (R_{L1} - R_{L2}) A_{12} R_{L1} + R_{L2} A_{12} (R_{L1} - R_{L2})$$

and the boundedness of A_{11}, A_{12} and L_0 on I, it follows that for R_{L1}, $R_{L2} \in \mathscr{L}$

$$\|f_2(R_{L1}) - f_2(R_{L2})\| \le K_1 \|R_{L1} - R_{L2}\| \quad \text{for } \varepsilon < \varepsilon_1^*, \text{ for some } \varepsilon_1^* > 0.$$

Using Lemma 2.2, there exists $\varepsilon_2^* > 0$ such that for $\varepsilon < \varepsilon_2^*$

$$\|SR_{L1}(t) - SR_{L2}(t)\| \le \int_{t_0}^{t} K e^{-\sigma(t-\tau)/\varepsilon}\, d\tau\, K_1 \|R_{L1} - R_{L2}\|$$

$$\le \frac{\varepsilon K K_1}{\sigma} \|R_{L1} - R_{L2}\|.$$

Hence

$$\|SR_{L1} - SR_{L2}\| < \tfrac{1}{2}\|R_{L1} - R_{L2}\| \quad \text{for } \varepsilon < \varepsilon_3^* = \frac{\sigma}{2KK_1}. \tag{3.15}$$

Also, from the boundedness of A_{11}, A_{12}, L_0 and $\dot{L}_0$ we have

$$\|f_1(t) - \dot{L}_0(t)\| \leq K_2,$$

and for every $R_L \in \mathscr{L}$

$$\|f_2(R_L)\| \leq K_3\rho + \varepsilon K_4\rho^2.$$

Employing Lemma 2.2 once again, we conclude that for $\varepsilon < \varepsilon_2^*$

$$\|SR_L(t)\| \leq \frac{1}{\varepsilon}\int_{t_0}^{t} K e^{-\sigma(t-\tau)/\varepsilon}[\varepsilon K_3\rho + \varepsilon^2 K_4\rho^2 + K_2]\,d\tau$$

$$\leq \frac{K}{\sigma}(\varepsilon K_3\rho + \varepsilon^2 K_4\rho^2 + K_2).$$

Choose $\rho = 2KK_2/\sigma$ and $\varepsilon_4^* = \min(K_3/2K_4\rho, \sigma/3KK_3)$, then for all $\varepsilon < \varepsilon_4^*$

$$\|SR_L\| < \rho. \tag{3.16}$$

It follows from (3.15) and (3.16) that, for $\varepsilon < \varepsilon_5^* = \min_{1 \leq i \leq 4} \varepsilon_i^*$, SR_L is a contraction mapping on $\mathscr{L}$. Therefore by the contraction principle the integral equation $R_L(t) = SR_L(t)$ has a unique solution in $\|R_L\| \leq \rho$. It is easily seen by differentiation of (3.14) that $R_L(t)$ is a solution of (3.9). Obviously, $L = L_0 + \varepsilon R_L$ is continuously differentiable and bounded on I.

We turn now to (3.13). Consider the integral equation $R_H(t) = \hat{T}R_H(t)$, where

$$\hat{T}R_H(t) = \frac{1}{\varepsilon}\int_{t}^{t_f} [\dot{H}_0(\tau) + g_1(\tau) + \varepsilon g_2(R_H)(\tau)]\phi(\tau, t)\,d\tau. \tag{3.17}$$

Let $\mathscr{H} = \{R_H : \|R_H\| \leq \rho_1\}$, where $\|R_H\| = \sup_I \|R_H(t)\|$. It can be easily verified that

$$\|\dot{H}_0(t) + g_1 + \varepsilon g_2(R_H)(t)\| \leq \varepsilon K_5\|R_H\| + K_6, \quad \text{for } \varepsilon < \varepsilon_6^*.$$

Hence, for any $R_H \in \mathscr{H}$ the use of Lemma 2.2 yields

$$\|\hat{T}R_H(t)\| \leq \int_t^{t_f} \frac{1}{\varepsilon}(\varepsilon K_5\rho_1 + K_6)K e^{-\sigma(\tau-t)/\varepsilon}\, d\tau$$

$$\leq \frac{K}{\sigma}(\varepsilon K_5 \rho_1 + K_6), \quad \text{for } \varepsilon < \varepsilon_7^*.$$

Choose $\rho_1 = 2KK_6/\sigma$ and $\varepsilon_8^* = \sigma/2KK_5$, then

$$\|\hat{T}R_H(t)\| < \rho_1, \quad \text{for } \varepsilon < \varepsilon_8^*.$$

Moreover,

$$\|\hat{T}R_{H1}(t) - \hat{T}R_{H2}(t)\| \leq K_7 \int_t^{t_f} K e^{-\sigma(\tau-t)/\varepsilon}\, d\tau \|R_{H1} - R_{H2}\|, \quad \text{for } \varepsilon < \varepsilon_7^*$$

$$\leq \frac{\varepsilon K K_7}{\sigma}\|R_{H1} - R_{H2}\|$$

$$< \tfrac{1}{2}\|R_{H1} - R_{H2}\|, \quad \text{for } \varepsilon < \varepsilon_9^* = \frac{\sigma}{2KK_7}$$

Thus, for $\varepsilon < \varepsilon^* = \min_{5 \leq i \leq 9} \varepsilon_i^*$, $\hat{T}R_H$ is a contraction mapping on $\mathscr{H}$, and the integral equation $R_H(t) = \hat{T}R_H(t)$ has a unique solution in $\|R_H\| \leq \rho_1$. Notice that in showing that $\hat{T}R_H$ is a contraction mapping all the constants are independent of t_f, so $t_f \to \infty$ is allowed. Finally, by differentiation of (3.17) it is seen that $R_H(t)$ is a solution of (3.13). □

The decoupling transformation derived here provides us with a useful tool that simplifies the analysis of singularly perturbed systems. In the rest of this chapter, this transformation is employed to study stability, asymptotic approximations, controllability and observability. In developing these results, the transformation (3.4) is used with L and H satisfying (3.2) and (3.3) respectively. Recalling the discussion of Section 2.4, we notice again that even when L and H are functions of t the transformation (3.4) is nonsingular, independent of L and H. While in proving theorems it is more convenient to use L and H satisfying (3.2) and (3.3) so that the transformed system has zero off-diagonal blocks, in solving singular perturbation problems it is sometimes more convenient to use (3.4) with L and H that are approximations of the solutions of (3.2) and (3.3). For example, the transformation

$$\begin{bmatrix} \xi \\ \eta \end{bmatrix} = \begin{bmatrix} I_n - \varepsilon H_0 L_0 & -\varepsilon H_0 \\ L_0 & I_m \end{bmatrix} \begin{bmatrix} x \\ z \end{bmatrix} \tag{3.18}$$

results in a system whose diagonal blocks are $O(\varepsilon)$ perturbations of (3.5), while the off-diagonal blocks are multiplied by ε. Such a transformation is an exact one and holds for all ε. This idea will be illustrated in the adaptive control example treated in Section 5.5.

In all cases studied in this chapter $A_{22}(t)$ will be required to satisfy the stability Assumption 2.1, which is a reasonable requirement since boundary layers must be asymptotically stable for the time-scale decomposition to be useful. The stability Assumption 2.1, however, can be replaced by a more general conditional stability assumption. That is, the eigenvalues of $A_{22}(t)$ are bounded away from the imaginary axis with a fixed number of them in the right half-plane. In Chapter 6 a situation will arise where $A_{22}(t)$ is the Hamiltonian matrix associated with optimal control problems, which is conditionally stable. We give here one possible generalization of Theorem 3.1 that will be used in Chapter 6, where we restrict our attention to compact time intervals. First, we make the following assumption.

Assumption 3.1

There exists a nonsingular continuously differentiable matrix $W(t)$ such that

$$W^{-1}(t)A_{22}(t)W(t) = D(t) = \begin{bmatrix} D_1(t) & 0 \\ 0 & D_2(t) \end{bmatrix} \quad \forall t \in [t_0, t_f],$$

where $\operatorname{Re}\lambda(D_1(t)) \le -c_1$ and $\operatorname{Re}\lambda(D_2(t)) \ge c_2$.

Corollary 3.1

Suppose that Assumption 3.1 is satisfied and all the assumptions of Theorem 3.1 hold on $[t_0, t_f]$ except that $A_{22}(t)$ does not satisfy Assumption 2.1. Then, there exists an $\varepsilon^* > 0$ such that for all $\varepsilon \in (0, \varepsilon^*)$, there exist continuous differentiable matrices $L(t)$ and $H(t)$ satisfying (3.2) and (3.3). Moreover, $L(t) = L_0(t) + O(\varepsilon)$ and $H(t) = H_0(t) + O(\varepsilon) \quad \forall t \in [t_0, t_f]$.

Proof Substitute $L = W\hat{L}$ and $H = \hat{H}W^{-1}$ in (3.2) and (3.3) respectively.

$$\varepsilon\dot{\hat{L}} = D\hat{L} - W^{-1}A_{21} - \varepsilon W^{-1}\dot{W}\hat{L} - \varepsilon\hat{L}(A_{11} - A_{12}W\hat{L}),$$

$$-\varepsilon\dot{\hat{H}} = \hat{H}D - A_{12}W - \varepsilon\hat{H}W^{-1}\dot{W} + \varepsilon\hat{H}\hat{L}A_{12}W - \varepsilon(A_{11} - A_{12}W\hat{L})\hat{H}.$$

From this point on, the proof is a step-by-step repetition of that of Theorem

3.1, with the exception that the integral (3.14) is replaced by an integral of the form

$$SR_{\hat{L}}(t) = \frac{1}{\varepsilon}\int_{t_0}^{t}\begin{bmatrix}\phi_1(t,\tau) & 0\\ 0 & 0\end{bmatrix}[\cdot]\,d\tau - \frac{1}{\varepsilon}\int_{t}^{t_f}\begin{bmatrix}0 & 0\\ 0 & \phi_2(t,\tau)\end{bmatrix}[\cdot]\,d\tau,$$

where $\phi_1(\cdot,\cdot)$ is the transition matrix of $\varepsilon\dot{z}_1 = D_1(t)z_1$ and $\phi_2(\cdot,\cdot)$ is the transition matrix of $\varepsilon\dot{z}_2 = D_2(t)z_2$. Notice that, by Lemma 2.2, $\|\phi_2(t,\tau)\| \le K e^{-\alpha(\tau-t)/\varepsilon}$ for $\tau \ge t$. The integral (3.17) is modified similarly.

In fact, the transformation $W(t)$ called for in Assumption 3.1 exists whenever the eigenvalues of a continuously differentiable matrix $A_{22}(t)$ are bounded away from the imaginary axis. For more details see Coppel (1967).

5.4 Uniform Asymptotic Stability

Uniform asymptotic stability of the singularly perturbed system

$$\left.\begin{aligned}\dot{x}(t) &= A_{11}(t)x(t) + A_{12}(t)z(t),\\ \varepsilon\dot{z}(t) &= A_{21}(t)x(t) + A_{22}(t)z(t),\end{aligned}\right\} \tag{4.1}$$

when ε is sufficiently small, is implied by the uniform asymptotic stability of the slow (reduced) system

$$\dot{x}_s(t) = A_0(t)x_s(t), \quad A_0 = A_{11} - A_{12}A_{22}^{-1}A_{21} = A_{11} - A_{12}L_0 \tag{4.2}$$

and asymptotic stability, uniformly in t, of the time-frozen fast (boundary layer) system

$$\varepsilon\dot{z}_f(t) = A_{22}(t)z_f(t), \tag{4.3}$$

where t is treated as a parameter. Namely, if the transition matrix $\phi_s(t,s)$ of the slow system (4.2) satisfies the inequality $\|\phi_s(t,s)\| \le K_s e^{-\sigma_s(t-s)}$, $t \ge s$, and the eigenvalues of $A_{22}(t)$ are in left half-plane uniformly in t, i.e. $\mathrm{Re}\,\lambda(A_{22}(t)) \le -c_1 < 0$, then one would intuitively expect that for well behaved coefficients $A_{ij}(t)$, the singularly perturbed system (4.1) will be uniformly asymptotically stable as $\varepsilon \to 0$. Theorem 4.1 states this result.

Theorem 4.1

Let all the assumptions of Theorem 3.1 hold on $[t_0, \infty)$ and suppose that the slow system (4.2) is uniformly asymptotically stable. Then, there exists an $\varepsilon^* > 0$ such that for all $\varepsilon \in (0, \varepsilon^*)$ the singularly perturbed system (4.1) is uniformly asymptotically stable. In particular, $\phi(t,s)$, the transition matrix of (4.1), satisfies

$$\|\phi(t, s)\| \leq K e^{-\alpha(t-s)} \quad \forall t \geq s \geq t_0,$$

where $K > 0$ and $\alpha > 0$ are independent of ε.

Proof Transform the system into decoupled form (3.5). Let $\phi_1(\cdot, \cdot)$ be the transition matrix of $\dot{\xi} = (A_{11} - A_{12}L)\xi$ and $\phi_2(\cdot, \cdot)$ be the transition matrix of $\varepsilon\dot{\eta} = (A_{22} + \varepsilon L A_{12})\eta$. Then

$$\phi(t, s) = T \begin{bmatrix} \phi_1(t, s) & 0 \\ 0 & \phi_2(t, s) \end{bmatrix} T^{-1}, \quad t \geq s,$$

$$\|\phi(t, s)\| \leq K_3 \max[\|\phi_1(t, s)\|, \|\phi_2(t, s)\|].$$

The system $\dot{\xi} = (A_{11} - A_{12}L)\xi$ is a regular perturbation of the uniformly asymptotically stable slow system (4.2), so $\|\phi_1(t, s)\| \leq K_1 e^{-\sigma_1(t-s)}$. By Lemmas 2.2 and 2.3, $\|\phi_2(t, s)\| \leq K_2 e^{-\sigma_2(t-s)/\varepsilon}$. Thus, for sufficiently small ε,

$$\|\phi(t, s)\| \leq K_3 \max[K_1, K_2] e^{-\min[\sigma_1, \sigma_2](t-s)}. \qquad \square$$

Theorem 4.1 is a fundamental result which holds asymptotically as $\varepsilon \to 0$. In addition to this conceptual result, we calculate bounds ε^* such that the uniform asymptotic stability of the singularly perturbed system is guaranteed for all $\varepsilon < \varepsilon^*$. We can also correct the slow and/or fast models by retaining higher-order terms of the expansions of L or H. In the rest of this section, two ε^* bounds are given and illustrated by an example, while the model correction process is illustrated by an adaptive control example in the next section.

Bounds ε^* will be calculated using Lyapunov methods. The first bound will be calculated using a converse Lyapunov method with conceptual Lyapunov functions. The second bound uses a direct Lyapunov method with quadratic Lyapunov functions.

Lemma 4.1

Suppose the assumptions of Theorem 4.1 hold, and let $\|\phi_s(t, s)\|_2 \leq K_s e^{-\sigma_s(t-s)}$, $\|\phi_f(t, s)\|_2 \leq K_f e^{-\sigma_f(t-s)/\varepsilon}$, $\|A_{12}(t)\|_2 \leq M_1$, $\|L_0(t)A_{12}(t)\|_2 \leq M_2$ and $\|\dot{L}_0(t) + L_0(t)A_0(t)\| \leq M_3$, where $\phi_s(\cdot, \cdot)$ and $\phi_f(\cdot, \cdot)$ are the transition matrices of (4.2) and (4.3) respectively. Then the singularly perturbed system (4.1) is uniformly asymptotically stable for all $\varepsilon < \varepsilon_1^*$, where

$$\varepsilon_1^* = \frac{\sigma_s \sigma_f}{\sigma_s M_2 K_f + K_s K_f M_1 M_3}.$$

Proof Consider $\nu(t, x, z) = \alpha_1 V(t, x) + \alpha_2 W(t, x, z)$ as a conceptual Lyapunov function candidate, where $\alpha_1 > 0$, $\alpha_2 > 0$,

$$V(t, x(t)) = \int_t^\infty \|\phi_s(\tau, t)x(t)\| \, d\tau$$

and

$$W(t, x(t), z(t)) = \int_t^\infty \|\phi_f(\tau, t)(z(t) + L_0(t)x(t))\| \, d\tau.$$

where $\|\cdot\| = \|\cdot\|_2$

It can be shown (Exercise 5.5) that there are positive constants γ_1, γ_2, γ_3 and γ_4 such that

$$\gamma_1\|x\| \leq V(t, x) \leq \gamma_2\|x\| \tag{4.4}$$

and

$$\varepsilon\gamma_3\|z + L_0x\| \leq W(t, x, z) \leq \varepsilon\gamma_4\|z + L_0x\|. \tag{4.5}$$

Thus $\nu(t, x, z)$ is positive-definite and decrescent. Now,

$$\begin{aligned}
\dot{V}(t, x(t)) &= \int_t^\infty \frac{\partial}{\partial t}\|\phi_s(\tau, t)x(t)\| \, d\tau - \|x(t)\| \\
&= \int_t^\infty \frac{1}{\|\phi_s(\tau, t)x(t)\|} [x^T(t)\phi_s^T(\tau, t)\phi_s(\tau, t)(-A_0(t)x(t) \\
&\quad + A_{11}(t)x(t) + A_{12}(t)z(t))] \, d\tau - \|x(t)\| \\
&\leq \int_t^\infty \|\phi_s(\tau, t)\| \|A_{12}\| \|L_0x + z\| \, d\tau - \|x\| \\
&\leq \int_t^\infty K_s e^{-\sigma_s(\tau - t)} \, d\tau \, M_1 \|L_0x + z\| - \|x\| \\
&= \frac{K_s M_1}{\sigma_s} \|L_0x + z\| - \|x\|.
\end{aligned} \tag{4.6}$$

Similarly,

$$\begin{aligned}
\dot{W}(t, x(t), z(t)) = \int_t^\infty & \frac{1}{\|\phi_f(\tau, t)(z(t) + L_0(t)x(t))\|} \\
& \times \Big\{(z(t) + L_0(t)x(t))^T \phi_f^T(\tau, t)\phi_f(\tau, t) \\
& \times \Big[-\frac{1}{\varepsilon}A_{22}(t)(z(t) + L_0(t)x(t)) + \frac{1}{\varepsilon}A_{21}(t)x(t)
\end{aligned}$$

$$
\begin{aligned}
&+\frac{1}{\varepsilon}A_{22}(t)z(t)+\dot{L}_0(t)x(t)\\
&+L_0(t)A_{11}(t)x(t)+L_0(t)A_{12}(t)z(t)\Big]\Big\}\,\mathrm{d}\tau\\
&-\|z(t)+L_0(t)x(t)\|\\
\le{}&\int_t^\infty K_\mathrm{f}\mathrm{e}^{-\sigma_\mathrm{f}(\tau-t)/\varepsilon}[M_3\|x\|+M_2\|z+L_0x\|]\,\mathrm{d}\tau\\
&-\|z+L_0x\|\\
\le{}&\frac{\varepsilon K_\mathrm{f}M_3}{\sigma_\mathrm{f}}\|x\|-\left(1-\frac{\varepsilon K_\mathrm{f}M_2}{\sigma_\mathrm{f}}\right)\|z+L_0x\|.
\end{aligned}
\tag{4.7}
$$

Using (4.6) and (4.7), we obtain

$$
\dot{\nu}\le-(\alpha_1\quad\alpha_2)\begin{bmatrix}1 & -\dfrac{K_\mathrm{s}M_1}{\sigma_\mathrm{s}}\\[2ex] -\dfrac{\varepsilon K_\mathrm{f}M_3}{\sigma_\mathrm{f}} & 1-\dfrac{\varepsilon K_\mathrm{f}M_2}{\sigma_\mathrm{f}}\end{bmatrix}\begin{bmatrix}\|x\|\\[2ex]\|z+L_0x\|\end{bmatrix}.
$$

$\dot{\nu}$ will be negative-definite if α_1 and α_2 can be chosen such that

$$
\frac{\alpha_1}{\alpha_2}>\frac{\varepsilon K_\mathrm{f}M_3}{\sigma_\mathrm{f}},
$$

$$
\frac{K_\mathrm{s}M_1}{\sigma_\mathrm{s}}\frac{\alpha_1}{\alpha_2}<1-\frac{\varepsilon K_\mathrm{f}M_2}{\sigma_\mathrm{f}},
$$

which is possible if

$$
\frac{\varepsilon K_\mathrm{f}M_3}{\sigma_\mathrm{f}}<\frac{\sigma_\mathrm{s}}{K_\mathrm{s}M_1}\left(1-\frac{\varepsilon K_\mathrm{f}M_2}{\sigma_\mathrm{f}}\right)
$$

or

$$
\varepsilon<\frac{\sigma_\mathrm{s}/K_\mathrm{s}M_1}{\dfrac{K_\mathrm{f}M_3}{\sigma_\mathrm{f}}+\dfrac{\sigma_\mathrm{s}K_\mathrm{f}M_2}{\sigma_\mathrm{f}K_\mathrm{s}M_1}}=\varepsilon_1^*. \qquad \square
$$

Lemma 4.2

Suppose the assumptions of Theorem 4.1 hold. Let $P_\mathrm{s}(t)\ge c_1 I_n>0$ satisfy the Lyapunov differential equation

$$
-\dot{P}_\mathrm{s}(t)=P_\mathrm{s}(t)A_0(t)+A_0^\mathrm{T}(t)P_\mathrm{s}(t)+I_n \tag{4.8}
$$

and let $P_f(t) \geq c_2 I_m > 0$ be the solution of the algebraic Lyapunov equation

$$0 = P_f(t)A_{22}(t) + A_{22}^T(t)P_f(t) + I_m. \tag{4.9}$$

Then the singularly perturbed system (4.1) is uniformly asymptotically stable for all $\varepsilon < \varepsilon_2^* = 1/(\gamma + \beta_1\beta_2)$, where $[\|\dot{P}_f(t)\|_2 + 2\|P_f(t)L_0(t)A_{12}(t)\|_2] \leq \gamma$, $2\|P_s(t)A_{12}(t)\|_2 \leq \beta_1$ and $2\|P_f(t)(\dot{L}_0(t) + L_0(t)A_0(t))\|_2 \leq \beta_2$.

The proof of Lemma 4.2 appears later as a special case of a general result for nonlinear systems in Section 7.5 (see Example 5.1 of Chapter 7).

Example 4.1

Consider the singularly perturbed system (4.1) with

$$A_{11}(t) = -4 + \cos t, \quad A_{12}(t) = [1 \quad 0],$$

$$A_{21}(t) = \begin{bmatrix} 1 \\ 0 \end{bmatrix}, \quad A_{22}(t) = \begin{bmatrix} -1 + 1.1\cos^2 t & 1 - 1.1\sin t\cos t \\ -1 - 1.1\sin t\cos t & -1 + 1.1\sin^2 t \end{bmatrix}.$$

The matrix $A_{22}(t)$ is a special case of Example 2.1 with $a = 1.1$.

$$A_{22}^{-1}(t) = \frac{1}{0.9}\begin{bmatrix} -1 + 1.1\sin^2 t & -1 + 0.55\sin 2t \\ 1 + 0.55\sin 2t & -1 + 1.1\cos^2 t \end{bmatrix}$$

and

$$L_0(t) = A_{22}^{-1}(t)A_{21} = \frac{1}{0.9}\begin{bmatrix} -1 + 1.1\sin^2 t \\ 1 + 0.55\sin 2t \end{bmatrix}.$$

The slow system is given by

$$\dot{x}_s = A_0(t)x_s = [-3.5 + \cos t + 0.6111\cos 2t]x_s.$$

We calculate a bound on ε using Lemma 4.1:

$$\phi_s(t, s) = \exp\left[\int_s^t A_0(\tau)\,d\tau\right],$$

$$\int_s^t A_0(\tau)\,d\tau = -3.5(t - s) - (\sin t - \sin s) - 0.3056(\sin 2t - \sin 2s)$$

$$\leq -1.8889(t - s).$$

Hence $\phi_s(t, s) \leq K_s e^{-\sigma_s(t-s)}$ with $K_s = 1$, $\sigma_s = 1.8889$.

Using expressions from Example 2.1, it can be shown that in $\|\cdot\|_2$ $\exp[A_{22}(t)\theta]$ satisfies inequality (2.3) with $\alpha_1 = 0.45$ and $K_1 = 3.0533$, and $\dot{A}_{22}(t)$ satisfies Assumption (2.3) with $\beta = 1.1$. Thus, from Lemma 2.2 (Exercise 5.2), $\phi_f(t,s)$ satisfies the inequality $\|\phi_f(t, s)\| \leq K_f e^{-\sigma_f(t-s)/\varepsilon}$ for all $\varepsilon < \varepsilon_0^* = (\alpha_1 - \alpha_2)^2\gamma/\beta K_1$, where

$$K_f = K_1\left(1 + \frac{2\gamma}{(1-\gamma)e^2}\right), \quad \sigma_f = \alpha_2, \quad \alpha_2 < \alpha_1, \quad \gamma < 1.$$

The choice of α_2 and γ affects the ε-bound in two ways. First, it affects K_f and σ_f, which are used to determine ε_1^*. Second, it affects ε_0^*, which affects the final bound which is $\min(\varepsilon_0^*, \varepsilon_1^*)$. Because of this, several choices of α_2 and γ may be tried in order to improve the bound. Let us take $\alpha_2 = 0.5\alpha_1$ and take $\gamma = 0.787$ such that $K_f = 2K_1$. The corresponding value of ε_0^* is 0.0119. Calculating the matrix norms indicated in Lemma 4.1 in $\|\cdot\|_2$, we obtain $M_1 = 1$, $M_2 = 2.0495$ and $M_3 = 11.6974$; hence $\varepsilon_1^* = 0.0045 < \varepsilon_0^*$. Thus the singularly perturbed system is uniformly asymptotically stable for all $\varepsilon < 0.0045$. Since ε_0^* is much larger than ε_1^* we may rechoose γ and/or α_2 allowing a smaller value of ε_0^* in order to improve ε_1^*. Let us take α_2 as before, but take $\gamma = 0.48105$ such that $K_f = 1.25K_1$. The corresponding values of ε_0^* and ε_1^* are 0.00724 and 0.00715 respectively. Hence the system is uniformly asymptotically stable for all $\varepsilon < 0.00715$. Now, we calculate a bound on ε using Lemma 4.2.

$$P_s(t) = \int_t^\infty \phi_s^2(\tau, t)\,d\tau \leq \int_t^\infty K_s^2 e^{-2\sigma_s(\tau - t)}\,d\tau = \frac{K_s^2}{2\sigma_s}.$$

From Example 2.1 we have expressions for $P_f(t)$ and $\dot{P}_f(t)$, and calculating their norms we obtain $\|P_f\|_2 = 2.2292$ and $\|\dot{P}_f\|_2 = 1.4892$. The constants γ, β_1 and β_2 can now be estimated as $\gamma = 10.6267$, $\beta_1 = 0.5294$ and $\beta_2 = 52.152$, leading to $\varepsilon_2^* = 0.0267$, which is more than thrice the bound obtained by Lemma 4.1.

5.5 Stability of a Linear Adaptive System

Problems with time-varying parameters naturally arise in schemes with continuous adjustment of parameters that are used for identification of, or adaptation to, unknown plants. Among those particularly popular are the so-called "model reference" schemes in which the same reference input $r(t)$ is applied to the system and to a reference model, and the error between their output signals $e(t) = y(t) - y_m(t)$ is regulated to zero by parameter adjustment. For identification the parameter adjustment is performed on the model, while in adaptive control the adjustable parameters are controller gains in the feedback loop.

A fundamental result of the adaptive control theory is that the identification or adaptation process is uniformly asymptotically stable (u.a.s.) and $e(t) \to 0$ as $t \to \infty$, if (a) the reference input $r(t)$ is persistently exciting, and (b) the order of the model is not lower than that of the plant. The second assumption, however, can seldom be satisfied in applications where the order of the plant is unknown. A more realistic assumption is that only the order of the dominant (slow) part of the plant is known. Then the model–plant mismatch can be attributed to the unmodeled parts of the plant, the so called high-frequency parasitics, and the effect of this mismatch can be analyzed by singular perturbation methods. An example presented here will illustrate this possibility on a simple linear time-varying adaptive system. The result can be extended to most identification schemes, because they are also linear. Although extensions to other adaptive control systems, which are almost always nonlinear, are technically involved, the instability mechanisms revealed in this simple system are an inherent property of most direct adaptive control schemes.

The adaptive system shown in Fig. 5.1 is linear because the only adjustable parameter is the feedforward gain k, and, ideally, the only unknown parameter is the plant gain g. The goal of adaptation is to adjust the overall gain kg to be equal to that of the model, g_m. The ideal plant has the pole $s = -1$ identical with that of the model, but the real plant has an additional parasitic pole at $s = -\varepsilon^{-1}$. Without this parasitic pole, that is when $\varepsilon = 0$, the ideal adaptation is exponentially stable with any nonzero input $r(t)$, such as $r(t) = R \sin \omega t$ or even $r = \text{const} \neq 0$. Then $e(t) \to 0$, $k(t)g \to g_m$ as $t \to \infty$. The robustness issue is whether this stability property will be preserved in the presence of the parasitic pole, at least when this pole is large,

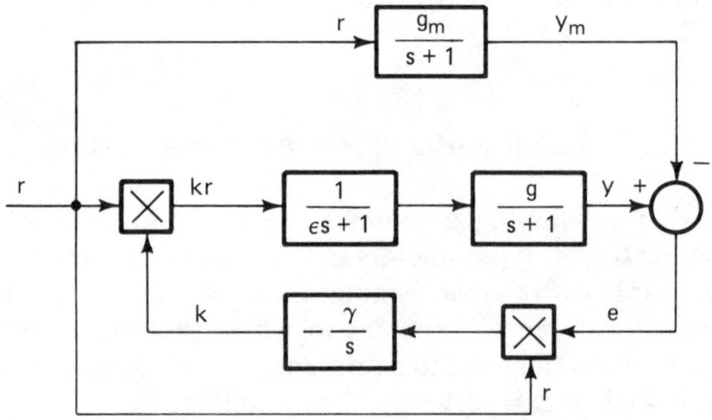

Fig. 5.1. A linear adaptive system.

that is when $\varepsilon > 0$ is small. If the stability is lost at some ε^* and beyond, then it would also be useful to give an estimate of ε^*, and show how it depends on system and signal properties. These issues are now addressed using the results of the preceding sections in this chapter.

The equations of the adaptive system in Fig. 5.1 with $g = 1$ are

$$\text{model:} \qquad \dot{y}_m = -y_m + g_m r, \tag{5.1}$$

$$\text{plant:} \qquad \begin{cases} \dot{y} = -y + z, & (5.2) \\ \varepsilon\dot{z} = -z + kr, & (5.3) \end{cases}$$

$$\text{adaptation:} \qquad \dot{k} = -\gamma(y - y_m)r, \tag{5.4}$$

where $\gamma \geq 0$ is the adaptation gain, which determines the speed of adaptation: high for fast adaptation, and low for slow adaptation. It is clear that the equations (5.2)–(5.4) constitute a third-order linear time-varying system, with the signal $y_m(t)$ as its forcing term. Owing to linearity the stability analysis is performed for the unforced part of the system, rewritten in the form of (4.1) with the matrices A_{11}, A_{12}, A_{21} and A_{22} as follows:

$$\dot{x} = \begin{bmatrix} -1 & 0 \\ -\gamma r & 0 \end{bmatrix} x + \begin{bmatrix} 1 \\ 0 \end{bmatrix} z, \quad x = \begin{bmatrix} y \\ k \end{bmatrix} \in R^2, \tag{5.5}$$

$$\varepsilon\dot{z} = [0 \quad r]x + [-1]z, \quad z \in R^1. \tag{5.6}$$

Since $A_{22} = -1$ is both nonsingular and Hurwitz, the corresponding assumptions in the preceding sections are satisfied and we can immediately apply the decoupling transformation (3.18). We first evaluate $L_0 = A_{22}^{-1}A_{21}$ and $H_0 = A_{12}A_{22}^{-1}$:

$$L_0 = [0 \quad -r], \quad H_0 = \begin{bmatrix} -1 \\ 0 \end{bmatrix}, \tag{5.7}$$

and substituting for L_0 and for H_0 in (3.18), we obtain

$$\eta = z + L_0 x = z - rk, \quad \xi = x - \varepsilon H_0 \eta = \begin{bmatrix} y + \varepsilon\eta \\ k \end{bmatrix}. \tag{5.8}$$

To simplify notation, instead of $\xi_2 = k$ we use k, and in $\xi_1 = y + \varepsilon\eta$ we drop the subscript, $\xi = y + \varepsilon\eta$. In terms of ξ, k and η the system (5.5), (5.6) becomes

$$\begin{bmatrix} \dot{\xi} \\ \dot{k} \end{bmatrix} = \begin{bmatrix} -1 + \varepsilon\gamma r^2 & r - \varepsilon\dot{r} \\ -\gamma r & 0 \end{bmatrix} \begin{bmatrix} \xi \\ k \end{bmatrix} + \begin{bmatrix} \varepsilon - \varepsilon^2\gamma r^2 \\ \varepsilon\gamma r \end{bmatrix} \eta, \tag{5.9}$$

$$\varepsilon\dot{\eta} = \varepsilon\gamma r^2 \xi - \varepsilon\dot{r}k + (-1 - \varepsilon^2\gamma r^2)\eta. \tag{5.10}$$

Its advantage over (5.5), (5.6) is that (5.9), (5.10) is decoupled in the sense that η appears in (5.9) multiplied by ε and so do ξ and k in (5.10). Had we chosen an $O(\varepsilon^2)$ approximation of L and H, rather than L_0 and H_0, this decoupling property would be enhanced to $O(\varepsilon^2)$. However, the $O(\varepsilon)$ decoupling of (5.9), (5.10) is sufficient to reveal the instabilities that can occur in fast and slow adaptation due to the presence of ε.

By Theorem 4.1 the u.a.s. of (5.9), (5.10) for ε small follows from the u.a.s. of the slow and fast systems. The fast system is $\varepsilon \dot{z}_f = -z_f$, which is obviously u.a.s. The slow system obtained from (5.9) by letting $\varepsilon = 0$,

$$\begin{bmatrix} \dot{y}_s \\ \dot{k}_s \end{bmatrix} = \begin{bmatrix} -1 & r \\ -\gamma r & 0 \end{bmatrix} \begin{bmatrix} y_s \\ k_s \end{bmatrix}, \tag{5.11}$$

can be analyzed using the Lyapunov function

$$v_s = \tfrac{1}{2} y_s^2 + \frac{1}{2\gamma} k_s^2, \tag{5.12}$$

whose derivative along its solutions is $\dot{v}_s = -y_s^2$. This proves that (5.11) is uniformly stable and, moreover, $y_s(t) \to 0$ as $t \to \infty$. To prove that also $k_s(t) \to 0$, we note from the first row of (5.11) that if $\dot{v}_s = -y_s^2 = 0$ then also $rk_s = 0$. Therefore, if after $y_s(t)$ has converged to zero, the signal $r(t)$ remains different from zero ("persistently exciting"), the fact that $\dot{v}_s \le 0$ can only be satisfied by $k_s = 0$ establishes that (5.11) is u.a.s. if $r(t)$ is not identically zero. Our conclusion from Theorem 4.1 applied to the adaptive system (5.5), (5.6) is as follows.

Corollary 5.1

If $r(t)$ and $\dot{r}(t)$ are bounded for all $t \ge 0$ and $r(t)$ is different from zero for almost all $t \ge 0$, then there exists an ε^* such that the unforced system (5.5), (5.6) is exponentially stable for all $\varepsilon \in (0, \varepsilon^*]$, and hence all the variables in the adaptive system (5.2)–(5.4) are bounded.

We see an impact of parasitics already. When $\varepsilon > 0$ in (5.2)–(5.4), the tracking error $e(t) = y(t) - y_m(t)$ will, in general, remain nonzero as $t \to \infty$, because the responses of a first-order model and a second-order system cannot exactly match. A further analysis would show that for small ε the tracking error will be small, if also $|\dot{r}|$ is $O(1)$ or less. A more significant impact of the parasitic pole is that it can lead to instabilities in both fast and slow adaptation. These phenomena will now be demonstrated by an analysis of (5.9) with $\eta = 0$.

Proposition 5.1

The system (5.9) with $\eta = 0$ will be unstable if

$$\gamma R_a^2 = \lim_{T\to\infty} \frac{\gamma}{T}\int_0^T r^2(t)\,\mathrm{d}t > \varepsilon^{-1}. \tag{5.13}$$

The proof follows from the classical result that the determinant $\Delta(t)$ of the state transition matrix of the system $\dot{x} = A(t)x$ satisfies the equation $\dot{\Delta} = (\operatorname{trace} A(t))\Delta$. The trace of the matrix in (5.9) is $-1 + \varepsilon\gamma r^2$, and when (5.13) is satisfied the corresponding $\Delta(t)$ tends to infinity. Another conclusion from (5.9) for $\dot{r} = \text{const}$, $\dot{r} = 0$, is simply derived from its characteristic equation

$$\lambda^2 + (1 - \varepsilon\gamma r^2)\lambda + \gamma r^2 = 0, \tag{5.14}$$

in which case the condition

$$\gamma r^2 < \varepsilon^{-1} \tag{5.15}$$

is both necessary and sufficient for the exponential stability of (5.9). Treating γR_a^2 as the average, and $\gamma r^2(t)$ as the instantaneous adaptation gain, we interpret (5.13) and (5.15) as the bounds on the adaptation speed imposed by the parasitic pole $-\varepsilon^{-1}$. Hence the mechanism of the fast adaptation instability is triggered by high gain γr^2.

A remarkable property of this adaptive system is that it can be unstable at a very low gain γr^2, if $\dot{r}$ is sufficiently large. A detailed analysis shows that for $r = \sin \omega t$ and with γ sufficiently small, the stability bound for slow adaptation is

$$\omega^2 < \varepsilon^{-1}, \tag{5.16}$$

that is, the input frequency must be lower than the square root of the parasitic pole.

The results of this analysis are illustrated by a simulation of the responses of the system (5.5), (5.6) with $r = \sin \omega t$ for three different sets of parameters. In all cases the parasitic pole is -10, that is, $\varepsilon = 0.1$.

The response in Fig. 5.2(a) for $\gamma = 1$ and $\omega = 1$ is exponentially stable, that is, the bound (5.13) is not violated and the Corollary 5.1 holds. However, when γ is increased an important instability occurs in Fig. 5.2(b), where $\gamma = 24$, $\omega = 0.5$ violate the bound (5.13). If, instead, γ is kept low, but ω is increased, a creeping instability is noticed along the axis $y = 0$; see Fig. 5.2(c), where $\gamma = 1$, $\omega = 4$ violate the bound (5.16). In conclusion, let us point out that it was the transformation of the original system (5.5), (5.6) into its decoupled form (5.9), (5.10) which led to the discovery of the

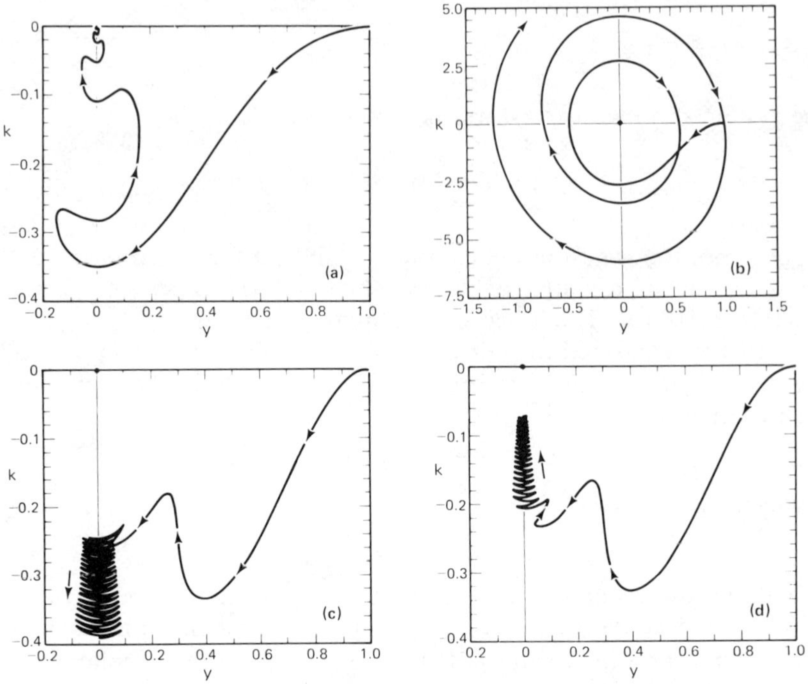

Fig. 5.2. Response of adaptive system.

two types of instability. Even a cursory inspection of the slow system matrix in (5.9) points to the two causes of instability: the terms $\varepsilon\gamma r^2$ and $\varepsilon\dot{r}$, which, in fact, appear in the stability bounds (5.13), (5.15) and (5.16). A similar qualitative analysis suggested that the instabilities can be counteracted by adding a term $-\sigma k$ in the adaptive law (5.4). In Fig. 5.2(d) the response of Fig. 5.2(c) is stabilized by choosing $\sigma = 0.05$. In Exercise 5.8 the reader is asked to repeat our analysis with this modification.

5.6 State Approximations

Paralleling the time-invariant treatment of Section 2.5, let us consider the singularly perturbed system

$$\left.\begin{aligned} \dot{x}(t) &= A_{11}(t)x(t) + A_{12}(t)z(t) + B_1u(t), \\ \varepsilon\dot{z}(t) &= A_{21}(t)x(t) + A_{22}(t)z(t) + B_2u(t), \end{aligned}\right\} \tag{6.1}$$

where the input $u(t)$ is a continuous bounded function of $t \in I$ and the initial state is given by $x(t_0) = x^0$ and $z(t_0) = z^0$. The solution of (6.1) for sufficiently small ε can be approximated by employing the decoupling transformation (3.4). Application of (3.4) transforms (6.1) into

$$\dot{\xi}(t) = [A_{11}(t) - A_{12}(t)L(t)]\xi(t) + [B_1(t) - \varepsilon H(t)L(t)B_1(t) - H(t)B_2(t)]u(t), \tag{6.2}$$

$$\varepsilon\dot{\eta}(t) = [A_{22}(t) + \varepsilon L(t)A_{12}(t)]\eta(t) + [B_2(t) + \varepsilon L(t)B_1(t)]u(t), \tag{6.3}$$

$$\xi(t_0) = [I_n - \varepsilon H(t_0)L(t_0)]x^0 - \varepsilon H(t_0)z^0, \tag{6.4}$$

$$\eta(t_0) = L(t_0)x^0 + z^0. \tag{6.5}$$

The solution of (6.2)–(6.5), when transformed back using the inverse of (3.4), yields the exact solution of (6.1). Asymptotic approximations to the solution of (6.1) with $O(\varepsilon^N)$ error can be obtained by retaining $O(\varepsilon^N)$ approximations of the right-hand-side coefficients of (6.2)–(6.5). In other words, the decoupling transformation of time-varying singularly perturbed systems, similar to its time-invariant counterpart of Section 2.5, reduces the approximation problem to a problem with regular perturbations of right-hand-side coefficients. Recalling from Section 5.3 that MacLaurin expansions for L and H can be easily constructed, we have now an efficient method for deriving asymptotic approximations for the solution of (6.1). The derivation of $O(\varepsilon)$ approximations is particularly simple, since in that case L and H can be approximated by $L_0 = A_{22}^{-1}A_{21}$ and $H_0 = A_{12}A_{22}^{-1}$, and all $O(\varepsilon)$ terms on the right-hand sides of (6.2)–(6.5) can be dropped. This process yields

$$\dot{x}_s(t) = A_0(t)x_s(t) + B_0(t)u(t), \quad x_s(t_0) = x^0, \tag{6.6}$$

$$\frac{dz_f}{d\tau} = A_{22}(t_0)z_f + B_2(t_0 + \varepsilon\tau)u(t_0 + \varepsilon\tau), \quad z_f(0) = A_{22}^{-1}(t_0)A_{21}(t_0)x^0 + z^0, \tag{6.7}$$

where (6.7) is written in the fast time scale $\tau = (t - t_0)/\varepsilon$. The reader should notice that (6.6) and (6.7) are the slow and fast systems derived by formally neglecting ε in (6.1).

Theorem 6.1

Suppose for all $t \in [t_0, t_f]$ that the assumptions of Theorem 3.1 hold and $B_i(t)$, $i = 1, 2$, and $v(t)$ are continuous. Then there exists an $\varepsilon^* > 0$

such that for all $\varepsilon \in (0, \varepsilon^*]$ the following expressions hold uniformly on $[t_0, t_f]$:

$$x(t) = x_s(t) + O(\varepsilon) \tag{6.8}$$

$$z(t) = -A_{22}^{-1}(t)A_{21}(t)x_s(t) + z_f\left(\frac{t - t_0}{\varepsilon}\right) + O(\varepsilon). \tag{6.9}$$

Moreover, if the conditions of Theorem 4.1 hold and $B_i(t)$ and $v(t)$ are continuous and bounded, then the above expressions hold uniformly on $[t_0, \infty)$.

Proof From Lemmas 2.2 and 2.3 the transition matrix of the homogeneous part of (6.3) satisfies the inequality $\|\phi(t, s)\| \leq Ke^{-\alpha(t-s)/\varepsilon}$. Using this, it is obvious that the contribution of the input term $\varepsilon LB_1 u$ is $O(\varepsilon)$.

Using again Lemmas 2.2 and 2.3, we see that the homogeneous part of (6.3) can be approximated by $A_{22}(t_0)\eta(t)$, which yields

$$\eta(t) = z_f\left(\frac{t - t_0}{\varepsilon}\right) + O(\varepsilon).$$

Comparison of (6.2), (6.4) with (6.6) and use of regular perturbation results (Hale 1980) yields

$$\xi(t) = x_s(t) + O(\varepsilon).$$

Using the inverse of the decoupling transformation (3.4), we get

$$x(t) = \xi(t) + \varepsilon H(t)\eta(t) = x_s(t) + O(\varepsilon)$$

$$\begin{aligned} z(t) &= -L(t)\xi(t) + (I_n - \varepsilon L(t)H(t))\eta(t) \\ &= -A_{22}^{-1}(t)A_{21}(t)x_s(t) + z_f\left(\frac{t - t_0}{\varepsilon}\right) + O(\varepsilon), \end{aligned}$$

which complete the proof for finite-time intervals. The case of infinite time follows immediately upon utilizing Theorem 4.1. □

The derivation of $O(\varepsilon^2)$ approximations is given in Exercise 5.9.

5.7 Controllability

In Section 2.6 we have seen that controllability of the slow and fast systems implies controllability of the full singularly perturbed time-invariant system when ε is sufficiently small. In this section we show that the same result holds for time-varying systems. We actually go beyond that and study the asymptotic behaviour of the steering control that moves the state of the system from a given initial point to a given final point in the state space.

Returning to the singularly perturbed time-varying system (6.1), defined on a finite-time interval $[t_0, t_f]$, let us associate with this system the slow system

$$\dot{x}_s(t) = A_0(t)x_s(t) + B_0(t)u_s(t) \tag{7.1}$$

and the fast system

$$\frac{dz_f}{d\tau} = A_{22}(t^*)z_f(\tau) + B_2(t^*)u_f(\tau), \quad \tau = \frac{t-t^*}{\varepsilon}. \tag{7.2}$$

The fast system is a time-invariant system at a frozen time $t^* \in [t_0, t_f]$. The slow system (7.1) is controllable on $[t_1, t_2] \subset [t_0, t_f]$ if for any given x_s^0 and x_s^f there exists a finite energy control $u_s(t)$ that moves the state of the system from x_s^0 at $t = t_1$ to x_s^f at $t = t_2$. This is equivalent to the following assumption.

Assumption 7.1

$$W_s(t_1, t_2) = \int_{t_1}^{t_2} \phi_s(t_2, t)B_0(t)B_0^T(t)\phi_s^T(t_2, t)\,dt > 0,$$

where $\phi_s(\cdot\,,\cdot)$ is the transition matrix of (7.1) and $W_s(t_1, t_2)$ is the controllability Grammian on $[t_1, t_2]$. The control that achieves this task with minimum energy is given by

$$u_s(t) = B_0^T(t)\phi_s^T(t_2, t)W_s^{-1}(t_1, t_2)[x_s^f - \phi_s(t_2, t_1)x_s^0]. \tag{7.3}$$

Similarly, the fast system is controllable if the following assumption holds.

Assumption 7.2

$$\text{Rank}\,[B_2(t^*), A_{22}(t^*)B_2(t^*), \ldots, A_{22}^{m-1}(t^*)B_2(t^*)] = m,$$

and the minimum energy control that moves the state of the system from

z_f^0 at $\tau = \tau_1$ to z_f^f at $\tau = \tau_2$ is given by

$$u_f(\tau) = B_2^T(t^*) \exp [A_{22}^T(t^*)(\tau_2 - \tau)]\, W_f^{-1}(\tau_1, \tau_2; t^*) \times [z_f^f - \exp [A_{22}(t^*)(\tau_2 - \tau_1)]\, z_f^0], \tag{7.4}$$

where

$$W_f(\tau_1, \tau_2; t^*) = \int_{\tau_1}^{\tau_2} \exp [A_{22}(t^*)(\tau_2 - \tau)] B_2(t^*) B_2^T(t^*) \times (\exp [A_{22}(t^*)(\tau_2 - \tau)])^T \, d\tau > 0. \tag{7.5}$$

Notice that $[\tau_1, \tau_2]$ could extend to infinity, which is permissible since we are working with $\operatorname{Re} \lambda(A_{22}(t^*)) < 0$ (a precise statement of conditions is given in Theorem 7.1 below).

To study the controllability of the full singularly perturbed system (6.1), we use the decoupling transformation (3.4) to transform it into

$$\dot{\xi}(t) = [A_{11}(t) - A_{12}(t)L(t)]\xi(t) + B_s(t)u(t), \tag{7.6}$$

$$\varepsilon\dot{\eta}(t) = [A_{22}(t) + \varepsilon L(t)A_{12}(t)]\eta(t) + B_f(t)u(t), \tag{7.7}$$

where $B_s = B_1 - \varepsilon HLB_1 - HB_2$ and $B_f = B_2 + \varepsilon LB_1$. Let $\phi_1(\cdot,\cdot)$ be the transition matrix of (7.6) and let $\phi_2(\cdot,\cdot)$ be the transition matrix of (7.7). Using Lemma 2.3, a modified version of Lemma 2.2 which is given in Exercise 5.15, Theorem 3.1 and the continuous dependence of $\phi_1(\cdot,\cdot)$ on parameters, we have

$$\phi_1(t, s) = \phi_s(t, s) + O(\varepsilon) \tag{7.8}$$

$$\phi_2(t, s) = \exp \left[A_{22}(t)\left(\frac{t-s}{\varepsilon}\right)\right] + \varepsilon\Psi_1(t, s), \quad \|\Psi_1(t, s)\| \le K e^{-\alpha(t-s)/\varepsilon}. \tag{7.9}$$

Furthermore, assuming that $B_1(\cdot)$ and $B_2(\cdot)$ are continuously differentiable, we have

$$B_s(t) = B_0(t) + O(\varepsilon), \tag{7.10}$$

$$B_f(t) = B_2(t) + O(\varepsilon). \tag{7.11}$$

The controllability Grammian of (7.6), (7.7) is given by

$$W(t_1, t_2) = \int_{t_1}^{t_2} \begin{bmatrix} \phi_1(t_2, t) & 0 \\ 0 & \phi_2(t_2, t) \end{bmatrix} \begin{bmatrix} B_s(t) \\ \dfrac{B_f(t)}{\varepsilon} \end{bmatrix} \begin{bmatrix} B_s(t) \\ \dfrac{B_f(t)}{\varepsilon} \end{bmatrix}^T \begin{bmatrix} \phi_1(t_2, t) & 0 \\ 0 & \phi_2(t_2, t) \end{bmatrix}^T dt \triangleq \begin{pmatrix} W_{11} & W_{12} \\ W_{12}^T & \dfrac{W_{22}}{\varepsilon} \end{pmatrix}. \tag{7.12}$$

We study the components of the matrix W:

$$W_{11}(t_1, t_2) = \int_{t_1}^{t_2} [\phi_s(t_2, t) + O(\varepsilon)][B_0(t) + O(\varepsilon)][B_0(t) + O(\varepsilon)]^{\mathrm{T}}$$

$$\times [\phi_s(t_2, t) + O(\varepsilon)]^{\mathrm{T}}\, \mathrm{d}t = W_s(t_1, t_2) + O(\varepsilon),$$

$$W_{12}(t_1, t_2) = \frac{1}{\varepsilon}\int_{t_1}^{t_2} [\phi_s(t_2, t) + O(\varepsilon)][B_0(t) + O(\varepsilon)][B_2(t) + O(\varepsilon)]^{\mathrm{T}}$$

$$\times \left[\exp\left\{A_{22}(t_2)\left(\frac{t_2 - t}{\varepsilon}\right)\right\} + \varepsilon\Psi_1(t_2, t)\right]^{\mathrm{T}} \mathrm{d}t$$

$$= \frac{1}{\varepsilon}\int_{t_1}^{t_2} \phi_s(t_2, t)B_0(t)B_2^{\mathrm{T}}(t)\left[\exp\left\{A_{22}(t_2)\left(\frac{t_2 - t}{\varepsilon}\right)\right\}\right]^{\mathrm{T}} \mathrm{d}t + O(\varepsilon)$$

$$= \int_{-(t_2-t_1)/\varepsilon}^{0} \phi_s(t_2, t_2 + \varepsilon\sigma)B_0(t_2 + \varepsilon\sigma)B_2^{\mathrm{T}}(t_2 + \varepsilon\sigma)$$

$$\times [\exp\{-A_{22}(t_2\ \sigma\}]^{\mathrm{T}}\, \mathrm{d}\sigma + O(\varepsilon)$$

$$= \int_{-\infty}^{0} \phi_s(t_2, t_2)B_0(t_2)B_2^{\mathrm{T}}(t_2)[\exp(-A_{22}(t_2)\sigma)]^{\mathrm{T}}\, \mathrm{d}\sigma + O(\varepsilon)$$

$$= -B_0(t_2)B_2^{\mathrm{T}}(t_2)A_{22}^{-\mathrm{T}}(t_2) + O(\varepsilon),$$

$$W_{22}(t_1, t_2) = \frac{1}{\varepsilon}\int_{t_1}^{t_2} \left[\exp\left\{A_{22}(t_2)\left(\frac{t_2 - t}{\varepsilon}\right)\right\} + \varepsilon\Psi_1(t_2, t)\right]$$

$$\times [B_2(t) + O(\varepsilon)][B_2(t) + O(\varepsilon)]^{\mathrm{T}}$$

$$\times \left[\exp\left\{A_{22}(t_2)\left(\frac{t_2 - t}{\varepsilon}\right)\right\} + \varepsilon\Psi_1(t_2, t)\right]^{\mathrm{T}} \mathrm{d}t$$

$$= \frac{1}{\varepsilon}\int_{t_1}^{t_2} \exp\left\{A_{22}(t_2)\left(\frac{t_2 - t}{\varepsilon}\right)\right\} B_2(t)B_2^{\mathrm{T}}(t)$$

$$\times \left[\exp\left\{A_{22}(t_2)\left(\frac{t_2 - t}{\varepsilon}\right)\right\}\right]^{\mathrm{T}} \mathrm{d}t + O(\varepsilon)$$

$$= \int_{-\infty}^{0} \exp(-A_{22}(t_2)\sigma)\, B_2(t_2)B_2^{\mathrm{T}}(t_2)$$

$$\times [\exp(-A_{22}(t_2)\sigma)]^{\mathrm{T}}\, \mathrm{d}\sigma + O(\varepsilon)$$

$$= W_f(-\infty, 0; t_2) + O(\varepsilon).$$

Therefore

$$W = \begin{bmatrix} W_s + O(\varepsilon) & -B_0 B_2^T A_{22}^{-T} + O(\varepsilon) \\ -A_{22}^{-1} B_2 B_0^T + O(\varepsilon) & \frac{1}{\varepsilon}(W_f + O(\varepsilon)) \end{bmatrix},$$

$$\det W = \frac{1}{\varepsilon^m} \det [W_f + O(\varepsilon)] \det [W_s + O(\varepsilon)],$$

which is positive for sufficiently small ε. Thus we have proved the following theorem.

Theorem 7.1

Suppose that the assumptions of Theorem 3.1 hold on $[t_0, t_f]$, that Assumptions 7.1 and 7.2 (at $t^* = t_2$) are satisfied, and that $B_i(t)$ are continuously differentiable. Then there exists an $\varepsilon^* > 0$ such that for all $\varepsilon \in (0, \varepsilon^*)$ the singularly perturbed system (7.1) is controllable on thc interval $[t_1, t_2]$.

It is worth emphasizing that the controllability of the fast system is required only at the terminal time t_2.

Let us continue with our analysis to find the minimum energy control that steers the state of the system from (x^0, z^0) at $t = t_1$ to (x^f, z^f) at $t = t_2$. This control is given by

$$u(t) = \begin{bmatrix} B_s^T(t) & \dfrac{B_f^T(t)}{\varepsilon} \end{bmatrix} \begin{bmatrix} \phi_1(t_2, t) & 0 \\ 0 & \phi_2(t_2, t) \end{bmatrix}^T \times W^{-1}(t_1, t_2) \begin{bmatrix} \xi^f - \phi_1(t_2, t_1)\xi^0 \\ \eta^f - \phi_2(t_2, t_1)\eta^0 \end{bmatrix}, \tag{7.13}$$

where

$$\xi^0 = [I - \varepsilon H(t_1)L(t_1)]x^0 - \varepsilon H(t_1)z^0,$$

$$\xi^f = [I - \varepsilon H(t_2)L(t_2)]x^f - \varepsilon H(t_2)z^f,$$

$$\eta^0 = L(t_1)x^0 + z^0,$$

$$\eta^f = L(t_2)x^f + z^f.$$

Seeking $V = W^{-1}$ in the form

$$V = \begin{pmatrix} V_{11} & \varepsilon V_{12} \\ \varepsilon V_{12}^T & \varepsilon V_{22} \end{pmatrix},$$

it can be easily verified (Exercise 5.12) that

$$V_{11}(t_1, t_2) = W_s^{-1}(t_1, t_2) + O(\varepsilon), \tag{7.14a}$$

$$V_{12}(t_1, t_2) = W_s^{-1}(t_1, t_2)B_0(t_2)B_2^{\mathrm{T}}(t_2)A_{22}^{-\mathrm{T}}(t_2) \times W_f^{-1}(-\infty, 0; t_2) + O(\varepsilon), \tag{7.14b}$$

$$V_{22}(t_1, t_2) = W_f^{-1}(-\infty, 0; t_2) + O(\varepsilon). \tag{7.14c}$$

Substituting back in (7.13) yields

$$u(t) = B_0^{\mathrm{T}}(t)\phi_s^{\mathrm{T}}(t_2, t)W_s^{-1}(t_1, t_2)[x^{\mathrm{f}} - \phi_s(t_2, t_1)x^0] + B_2^{\mathrm{T}}(t_2) \times \left[\exp\left\{A_{22}(t_2)\left(\frac{t_2 - t}{\varepsilon}\right)\right\}\right]^{\mathrm{T}} W_f^{-1}(-\infty, 0; t_2)[z^{\mathrm{f}} + A_{22}^{-1}(t_2)A_{21}(t_2)x^{\mathrm{f}} + A_{22}^{-1}(t_2)B_2(t_2)B_0^{\mathrm{T}}(t_2)W_s^{-1}(t_1, t_2)(x^{\mathrm{f}} - \phi_s(t_2, t_1)x^0)] + O(\varepsilon). \tag{7.15}$$

Examining (7.15) shows that the dominant part of $u(t)$ comprises two terms. The first term contains $\phi_s(t_2, t)$ as a multiplier while the second term contains $\exp\{A_{22}(t_2)(t_2 - t)/\varepsilon\}$ as a multiplier. The second term creates a boundary layer at $t = t_2$ which is significant only within an $O(\varepsilon)$ interval of t_2. Denoting the first term by u_s for slow control, and the second term by u_f for fast control, the expression for u, dropping $O(\varepsilon)$, can be rewritten as

$$u(t) = u_s(t) + u_f\left(\frac{t_2 - t}{\varepsilon}\right), \tag{7.16}$$

where

$$u_s(t) = B_0^{\mathrm{T}}(t)\phi_s^{\mathrm{T}}(t_2, t)W_s^{-1}(t_1, t_2)[x^{\mathrm{f}} - \phi_s(t_2, t_1)x^0], \tag{7.17}$$

$$u_f(t) = B_2^{\mathrm{T}}(t_2)\left[\exp\left\{A_{22}(t_2)\left(\frac{t_2 - t}{\varepsilon}\right)\right\}\right]^{\mathrm{T}} W_f^{-1}(-\infty, 0; t_2) \times [z^{\mathrm{f}} + A_{22}^{-1}(t_2)A_{21}(t_2)x^{\mathrm{f}} + A_{22}^{-1}(t_2)B_2(t_2)u_s(t_2)]. \tag{7.18}$$

It is interesting now to see that the slow and fast controls can be intuitively derived from the slow and fast systems in the following manner. Postulating the control as the sum of the slow and fast components as in (7.16) where u_f decays exponentially as $(t_2 - t)/\varepsilon \to \infty$, we see that the effect of u_f on slow dynamics is $O(\varepsilon)$. Neglecting fast dynamics and dropping u_f, we arrive at the slow model (7.1). The solution of the slow control problem, given by (7.3), with $x_s^0 = x^0$ and $x_s^{\mathrm{f}} = x^{\mathrm{f}}$ yields the slow control u_s as in (7.17).

To see the effect of the postulated control on the fast dynamics we express the second equation of (6.1) in the fast time scale $\tau = (t - t_2)/\varepsilon$ and neglect terms $O(\varepsilon)$ to get

$$\frac{d\tilde{z}}{d\tau} = A_{21}(t_2)x(t_2) + A_{22}(t_2)\tilde{z}(\tau) + B_2(t_2)u_s(t_2) + B_2(t_2)\tilde{u}_f(\tau).$$

Removing the slow bias by defining z_f as

$$z_f(\tau) = \tilde{z}(\tau) + A_{22}^{-1}(t_2)[A_{21}(t_2)x(t_2) + B_2(t_2)u_s(t_2)],$$

we arrive at the fast model (7.2) with $t^* = t_2$. Notice now that, without the presence of the fast control, the terminal value of z is determined by its slow bias. The role of the fast control is to force z to terminate at z^f. Therefore the fast control problem is solved to steer z_f from an initial value $z_f^0 = 0$ to a final value $z_f^f = z^f + A_{22}^{-1}(t_2)[A_{21}(t_2)x^f + B_2(t_2)u_s(t_2)]$. Moreover, since the fast control problem is solved in the fast time scale $\tau = (t - t_2)/\varepsilon$, the terminal time is $\tau_2 = 0$, while the initial time $\to -\infty$ as $\varepsilon \to 0$ so that we can take $\tau_1 = -\infty$. With these values the solution of the fast control problem, given by (7.4), yields u_f as in (7.18). Figure 5.3 gives a pictorial representation of the trajectory. The trajectory consists of three parts, which, using guidance terminology, we called climb, cruise and descent arcs. The cruise arc is determined by the solution of the slow control problem to move the slow variable from x^0 to x^f. The climb arc is an initial boundary layer due to the deviation of z^0 from the slow value of z determined by the cruise arc. Since the fast system is open-loop asymptotically stable, the initial boundary layer decays exponentially in the fast time scale $(t - t_1)/\varepsilon$. If x^f is the only imposed terminal condition we can allow the trajectory to continue with the cruise arc until it hits the value

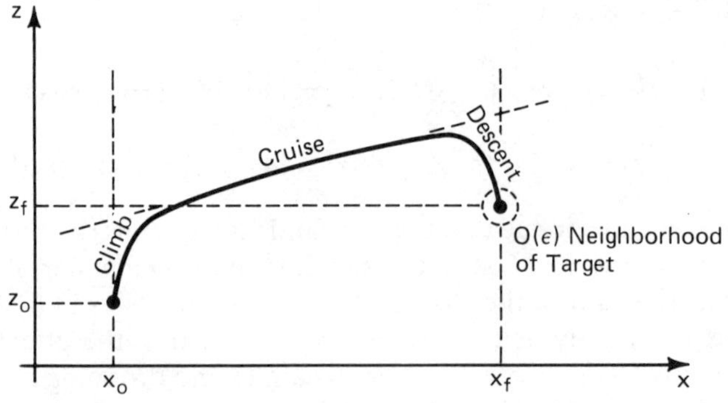

Fig. 5.3. State trajectory for minimum energy control.

$x = x^f$. However, when $z(t_f) = z^f$ is imposed as well, a fast control in the fast time scale $(t - t_2)/\varepsilon$ is applied forcing the trajectory to follow the descent arc towards the target point. Of course, owing to $O(\varepsilon)$ perturbations in deriving the control, there is no guarantee that the trajectory will hit the target exactly. It will be, however, within an $O(\varepsilon)$ neighborhood of the target which is represented by a dotted circle in Figure 5.3.

The calculation of the steering control is illustrated by a simple time-invariant example.

Example 7.1

Consider again the normalized DC-motor model of Example 3.1 of Chapter 3:

$$\dot{x} = z,$$

$$\varepsilon\dot{z} = -x - z + u,$$

where x, z and u are the speed, current and controlling voltage respectively. It is desired to steer the state from $x(0) = z(0) = 0$ to $x(1) = 1$, $z(1) = 0$. The steering control problem corresponds to achieving a desired speed at a given time while regulating the current to zero at that time.

The slow control problem is defined by

$$\dot{x}_s = -x_s + u_s, \quad x_s(0) = 0, \quad x_s(1) = 1,$$

and the slow control is given by

$$u_s(t) = 2.313\mathrm{e}^{-(1-t)}.$$

The fast control problem is defined by

$$\frac{\mathrm{d}z_f}{\mathrm{d}\tau} = -z_f + u_f, \quad z_f(-\infty) = 0, \quad z_f(0) = -(-1 + u_s(1)) = -1.313,$$

and the fast control is given by

$$y_f(\tau) = -2.626\mathrm{e}^{\tau}.$$

The total steering control is given by

$$u(t) = u_s(t) + u_f\left(\frac{t-1}{\varepsilon}\right) = 2.313\,\mathrm{e}^{-(1-t)} - 2.626\,\mathrm{e}^{-(1-t)/\varepsilon}. \tag{7.19}$$

The application of the control (7.19) to the system brings $(x(t), z(t))$ to within $O(\varepsilon)$ from the target point $(1, 0)$, for sufficiently small ε. To get a better feeling for the deviation from the target point we calculate $(x(1), z(1))$ for $\varepsilon = 0.1$, 0.05, and 0.01 (Table 5.1).

The results confirm that the final point is within an $O(\varepsilon)$ neighborhood

TABLE 5.1

	$\varepsilon = 0.1$	$\varepsilon = 0.05$	$\varepsilon = 0.01$
$x(1)$	0.731	0.903	0.98
$z(1)$	0.144	0.017	0.003

of the target. If 10% error is tolerable then $\varepsilon = 0.05$ is small enough for the control to be successful. If 2% error is tolerable then $\varepsilon = 0.01$ is small enough. The results also show that, in this example, for $\varepsilon = 0.1$ the error might not be tolerable. In this case the control will have to be corrected to account for $O(\varepsilon)$ terms that have been neglected throughout the derivations. To account for $O(\varepsilon)$ terms the slow and fast models will have to be corrected by including higher-order terms of L, and the effect of boundary conditions will be corrected by including higher-order terms of H. Also, the effect of the fast control on the slow system will have to be taken into consideration. Since correction by including higher-order terms of L and H is more or less straightforward, let us illustrate the correction by taking into consideration the effect of the fast control on the slow system. This correction, alone, might be enough to achieve the required accuracy and it is particularly interesting since it does not change the structure of the control.

In calculating the slow component of the postulated control $u_s + u_f$ we neglected the effect of u_f as $O(\varepsilon)$. In the current example, when $\varepsilon = 0.1$, the application of $u_f = -2.626\,e^{\tau}$ to the slow system results in a deviation of $x_s(1)$ that is -0.269. The error in $x_s(1)$ may be compensated for by shifting the boundary condition x^f to 1.269. Recalculating the slow control for boundary conditions $x_s(0) = 0$, $x_s(1) = 1.239$, we obtain

$$u_s(t) = 2.866\,e^{-(1-t)}.$$

The fast control problem is resolved with boundary conditions $x_f(-\infty) = 0$ and $x_f(0) = -(-1.239 + u_s(1)) = -1.627$ to obtain

$$u_f(\tau) = -3.254\,e^{\tau}$$

The total steering control is given by

$$u(t) = u_s(t) + u_f\left(\frac{t-1}{\varepsilon}\right) = 2.866\,e^{-(1-t)} - 3.254\,e^{-(1-t)/\varepsilon}. \tag{7.20}$$

Application of (7.20) to the system ($\varepsilon = 0.1$) results in

$$\begin{pmatrix} x(1) \\ z(1) \end{pmatrix} = \begin{pmatrix} 0.906 \\ 0.178 \end{pmatrix},$$

which improves $x(1)$ a lot without affecting $z(1)$ very much.

So far we have required that $A_{22}(t)$ satisfies Assumption 2.1, which guarantees that the fast system is open-loop asymptotically stable. If this condition is not satisfied but the fast system is controllable for every $t \in [t_1, t_2]$ and the fast state z is available for feedback, then it is possible to apply a preconditioning state feedback to stabilize the fast system. This result is summarized in the following corollary.

Corollary 7.1

Suppose that all the assumptions of Theorem 7.1 hold except that Assumption 7.2 holds for all $t^* \in [t_1, t_2]$ instead of $t^* = t_2$ only, and $A_{22}(t)$ satisfies $\det[A_{22}(t)] \neq 0$ instead of $\operatorname{Re}\lambda(A_{22}(t)) \leq -c_1$. Moreover, suppose that $z(t)$ is available for feedback. Then there exists an $\varepsilon^* > 0$ such that for all $\varepsilon \in (0, \varepsilon^*)$ the singularly perturbed system (6.1) is controllable on the interval $[t_1, t_2]$.

Proof There exists a continuously differentiable matrix $K(t)$ such that

$$\operatorname{Re}\lambda(A_{22}(t) + B_2(t)K(t)) \leq -c < 0 \quad \forall t \in [t_1, t_2].$$

Application of the state-feedback control law

$$u(t) = K(t)z(t) + v(t) \tag{7.21}$$

results in

$$\left.\begin{aligned} \dot{x} &= A_{11}x + (A_{12} + B_1K)z + B_1v, \\ \varepsilon\dot{z} &= A_{21}x + (A_{22} + B_2K)z + B_2v. \end{aligned}\right\} \tag{7.22}$$

The state-feedback control (7.21) does not affect the controllability of the full singularly perturbed system nor the controllability of the slow system. To see the latter claim notice that the slow system of (7.22) is given by

$$\dot{\hat{x}}_s = \hat{A}_0\hat{x}_s + \hat{B}_0v, \tag{7.23}$$

where

$$\begin{aligned} \hat{A}_0 &= A_{11} - (A_{12} + B_1K)(A_{22} + B_2K)^{-1}A_{21}, \\ \hat{B}_0 &= B_1 - (A_{12} + B_1K)(A_{22} + B_2K)^{-1}B_2. \end{aligned}$$

Using algebraic manipulations similar to those of Chapters 2 and 3, it can be shown (Exercise 5.13) that

$$\hat{A}_0 = A_0 + B_0V, \quad \hat{B}_0 = B_0Y, \tag{7.24}$$

where

$$V = -K(A_{22} + B_2K)^{-1}A_{21},$$
$$Y = I - K(A_{22} + B_2K)^{-1}B_2 = (I + KA_{22}^{-1}B_2)^{-1}.$$

Thus (7.23) is related to (7.1) via the transformation

$$u_s(t) = V(t)x_s(t) + Y(t)v(t), \tag{7.25}$$

where Y is nonsingular. Since controllability of (7.1) is invariant with respect to the transformation (7.25), it follows that (7.1) is controllable if and only if (7.23) is. Therefore the system (7.22) satisfies all the assumptions of Theorem 7.1 and the result follows.

Example 7.2

Consider the system

$$\dot{x} = z,$$
$$\varepsilon\dot{z} = -x + z + u,$$

where $A_{22} = 1$ is not Hurwitz. It is desired to steer the state from $x(0) = z(0) = 0$ to $x(1) = 1$, $z(1) = 0$. The first step is to stabilize the fast system by applying the feedback control

$$u(t) = Kz(t) + v(t),$$

so that the closed-loop system is

$$\dot{x} = z,$$
$$\varepsilon\dot{z} = -x + (1 + K)z + v.$$

Taking $K = -2$, the problem becomes identical with Example 7.1, where v is given by (7.19). Thus

$$u(t) = -2z(t) + 2.313\,e^{-(1-t)} - 2.626\,e^{-(1-t)/\varepsilon}.$$

5.8 Observability

In view of the observability properties studied in Section 2.6 and the controllability properties of time-varying systems established in the previous section, it should come as no surprise that the observability of a time-varying singularly perturbed system is implied, for sufficiently small ε, by the observability of the slow and fast systems. Such a result is given in Theorem 8.1 below. First, let us recall the singularly perturbed time-varying system (6.1) with $u = 0$ and the observed output

$$y(t) = C_1(t)x(t) + C_2(t)z(t), \quad y \in R^p, \tag{8.1}$$

on a finite time interval $[t_0, t_f]$. The slow system is obtained, as usual, by setting $\varepsilon = 0$ and eliminating z:

$$\dot{x}_s(t) = A_0(t)x_s(t), \tag{8.2a}$$

$$y(t) = C_0(t)x_s(t). \tag{8.2b}$$

The fast system at some time instant $t^* \in [t_0, t_f]$ is obtained by expressing (6.1), (8.1) in a fast time scale $\tau = (t - t^*)/\varepsilon$, setting $\varepsilon = 0$ in coefficients (hence x = constant), and eliminating the constant bias from z and y by defining

$$z_f(\tau) = z(t^* + \varepsilon\tau) + A_{22}^{-1}(t^*)A_{21}(t^*)x(t^*), \tag{8.3}$$

$$y_f(\tau) = y(t^* + \varepsilon\tau) - C_0(t^*)x(t^*). \tag{8.4}$$

The resulting fast system is given by

$$\frac{dz_f}{d\tau} = A_{22}(t^*)z_f(\tau), \tag{8.5a}$$

$$y_f(\tau) = C_2(t^*)z_f(\tau), \tag{8.5b}$$

which is a time-invariant system at a frozen time t^*. Let us make the following assumptions on the slow and fast systems.

Assumption 8.1

$$M_s(t_1, t_2) = \int_{t_1}^{t_2} \phi_s^T(t, t_1)C_0^T(t)C_0(t)\phi_s(t, t_1)\, dt > 0,$$

where $\phi_s(\cdot, \cdot)$ is the transition matrix of (8.2) and $M_s(\cdot\ , \cdot)$ is its observability Grammian on $[t_1, t_2]$.

Assumption 8.2

$$\operatorname{rank}[C_2^T(t^*), A_2^T(t^*)C_2^T(t^*), \ldots, A_{22}^{Tm-1}(t^*)C_2^T(t^*)] = m.$$

Theorem 8.1

Suppose that the assumptions of Theorem 3.1 hold on $[t_0, t_f]$, that Assumptions 8.1 and 8.2 (at $t^* = t_1$) are satisfied, and that $C_1(t)$ and $C_2(t)$ are continuously differentiable. Then there exists an $\varepsilon^* > 0$ such that for all

$\varepsilon \in (0, \varepsilon^*)$ the singularly perturbed system (6.1), (8.1) is observable on $[t_1, t_2]$.

Proof The proof is dual to that of Theorem 7.1. Application of the decoupling transformation (3.4) to (6.1), (8.1) yields

$$\dot{\xi} = (A_{11} - A_{12}L)\xi, \tag{8.6}$$

$$\varepsilon\dot{\eta} = (A_{22} + \varepsilon L A_{12})\eta, \tag{8.7}$$

$$y = C_s\xi + C_f\eta, \tag{8.8}$$

where $C_s = C_1 - C_2L$ and $C_f = C_2 + \varepsilon(C_1 - C_2L)H$.

The observability Grammian of (8.6)–(8.8) on $[t_1, t_2]$, as in Section 2.6, takes the scaled form

$$M(t_1, t_2) = \begin{bmatrix} M_{11} & \varepsilon M_{12} \\ \varepsilon M_{12}^{\mathrm{T}} & \varepsilon M_{22} \end{bmatrix}, \tag{8.9}$$

where

$$M_{11}(t_1, t_2) = M_s(t_1, t_2) + O(\varepsilon), \tag{8.10}$$

$$M_{12}(t_1, t_2) = -C_0^{\mathrm{T}}(t_1)C_2(t_1)A_{22}^{-1}(t_1) + O(\varepsilon), \tag{8.11}$$

$$M_{22}(t_1, t_2) = M_f(0, \infty; t_1) + O(\varepsilon) \tag{8.12}$$

and

$$M_f(0, T; t^*) = \int_0^T \mathrm{e}^{A_{22}^{\mathrm{T}}(t^*)\tau} C_2^{\mathrm{T}}(t^*)C_2(t^*)\mathrm{e}^{A_{22}(t^*)\tau}\,\mathrm{d}\tau$$

is the observability Grammian of the fast system (8.7) on $[0, T]$. Hence

$$\det M = \varepsilon^m \det[M_s + O(\varepsilon)]\det[M_f + O(\varepsilon)],$$

which is positive for sufficiently small ε. □

As a consequence of Theorem 8.1, we know that, for sufficiently small ε, the initial states $x(t_1) = x^0$ and $z(t_1) = z^0$ can be obtained from the observations $y(t)$, $t_1 \le t \le t_2$. In particular, using the decoupling transformation (3.4), we have

$$x^0 = \xi^0 + \varepsilon H(t_1)\eta^0, \tag{8.13}$$

$$z^0 = -L(t_1)\xi^0 + [I_m - \varepsilon L(t_1)H(t_1)]\eta^0 \tag{8.14}$$

and (Kwakernaak and Sivan, 1972)

$$\begin{bmatrix} \xi^0 \\ \eta^0 \end{bmatrix} = M^{-1}(t_1, t_2) \int_{t_1}^{t_2} \begin{bmatrix} \phi_1^{\mathrm{T}}(t, t_1) & 0 \\ 0 & \phi_2^{\mathrm{T}}(t, t_1) \end{bmatrix} \begin{bmatrix} C_{\mathrm{s}}^{\mathrm{T}}(t) \\ C_{\mathrm{f}}^{\mathrm{T}}(t) \end{bmatrix} y(t)\,\mathrm{d}t, \qquad (8.15)$$

where $\phi_1(\cdot, \cdot)$ and $\phi_2(\cdot, \cdot)$ are transition matrices of (8.6) and (8.7) respectively. Repeating arguments from Section 5.7, it can be shown that

$$M^{-1}(t_1, t_2) = \begin{bmatrix} N_{11} & N_{12} \\ N_{12}^{\mathrm{T}} & \dfrac{N_{22}}{\varepsilon} \end{bmatrix},$$

$$N_{11} = M_{\mathrm{s}}^{-1}(t_1, t_2) + O(\varepsilon)$$

$$N_{12} = M_{\mathrm{s}}^{-1}(t_1, t_2) C_0^{\mathrm{T}}(t_1) C_2(t_1) A_{22}^{-1}(t_1) M_{\mathrm{f}}^{-1}(0, \infty; t_1) + O(\varepsilon),$$

$$N_{22} = M_{\mathrm{f}}^{-1}(0, \infty; t_1) + O(\varepsilon),$$

$$\xi^0 = M_{\mathrm{s}}^{-1}(t_1, t_2) \int_{t_1}^{t_2} \phi_{\mathrm{s}}^{\mathrm{T}}(t, t_1) C_0^{\mathrm{T}}(t) y(t)\,\mathrm{d}t + O(\varepsilon) \stackrel{\Delta}{=} \tilde{\xi}^0 + O(\varepsilon), \qquad (8.16)$$

$$\eta^0 = M_{\mathrm{f}}^{-1}(0, \infty; t_1) \int_0^{(t_2 - t_1)/\varepsilon} \mathrm{e}^{A_{22}^{\mathrm{T}}(t_1)\tau} C_2^{\mathrm{T}}(t_1) \tilde{y}_{\mathrm{f}}(\tau)\,\mathrm{d}\tau + O(\varepsilon) \stackrel{\Delta}{=} \tilde{\eta}^0 + O(\varepsilon), \qquad (8.17)$$

where $\tilde{y}_{\mathrm{f}}(\tau) = y(t_1 + \varepsilon\tau) - C_0(t_1)\tilde{\xi}^0$.

Defining x_{s}^0 to be the initial state of x_{s}, as determined from the slow model (8.2) using y-observations on $[t_1, t_2]$, we get

$$\tilde{x}_{\mathrm{s}}^0 = M_{\mathrm{s}}^{-1}(t_1, t_2) \int_{t_1}^{t_2} \phi_{\mathrm{s}}^{\mathrm{T}}(t, t_1) C_0^{\mathrm{T}}(t) y(t)\,\mathrm{d}t = \tilde{\xi}^0. \qquad (8.18)$$

Now, replacing $x(t_1)$ in the definition of the fast system (8.5) and of the fast variables (8.3) and (8.4) by its estimate $\tilde{x}_{\mathrm{s}}^0$, and defining $\tilde{z}_{\mathrm{f}}^0$ to be the initial state of z_{f} as determined from the fast model (8.5) and y_{f} observations on $[0, (t_2 - t_1)/\varepsilon]$, we get

$$\begin{aligned} \tilde{z}_{\mathrm{f}}^0 &= M_{\mathrm{f}}^{-1}\left(0, \frac{t_2 - t_1}{\varepsilon}; t_1\right) = \int_0^{(t_2 - t_1)/\varepsilon} \mathrm{e}^{A_{22}^{\mathrm{T}}(t_1)\tau} C_2^{\mathrm{T}}(t_1) \tilde{y}_{\mathrm{f}}(\tau)\,\mathrm{d}\tau \\ &= M_{\mathrm{f}}^{-1}(0, \infty; t_1) \int_0^{(t_2 - t_1)/\varepsilon} \mathrm{e}^{A_{22}^{\mathrm{T}}(t_1)\tau} C_2^{\mathrm{T}}(t_1) \tilde{y}_{\mathrm{f}}(\tau)\,\mathrm{d}\tau + O(\varepsilon) \\ &= \tilde{\eta}^0 + O(\varepsilon). \end{aligned} \qquad (8.19)$$

Thus from (8.13), (8.14) and (8.16)–(8.19) it is obvious that

$$x^0 = \tilde{x}_s^0 + O(\varepsilon), \tag{8.20}$$

$$z^0 = -A_{22}^{-1}(t_1)A_{21}(t_1)\tilde{x}_s^0 + \tilde{z}_f^0 + O(\varepsilon). \tag{8.21}$$

What we have just described is an intuitive procedure to obtain $O(\varepsilon)$ approximations of the initial state x^0 and z^0 from the y-observation on $[t_1, t_2]$. First, to obtain the initial state x^0, we neglect the fast dynamics and use the slow model to obtain $\tilde{x}_s^0$, an $O(\varepsilon)$ approximation of x^0. Second, the fast model is defined by assuming that x is constant with value $\tilde{x}_s^0$, and is used to determine the initial state of the pure fast variable z_f. Finally, an $O(\varepsilon)$ approximation of z^0 is obtained by adding to $\tilde{z}_f^0$ the initial value of the slow component of z. The procedure is illustrated by the following example.

Example 8.1

Let us again return to the familiar DC-motor model

$$\dot{x} = z, \qquad x(0) = x^0,$$

$$\varepsilon\dot{z} = -x - z, \quad z(0) = z^0,$$

$$y = z,$$

where y is observed on $[0, 1]$.

The slow system, obtained by setting $\varepsilon = 0$, is

$$\dot{x}_s = -x_s,$$

$$y = -x_s.$$

The observability Grammian $M_s(0, 1)$ is

$$M_s(0, 1) = \int_0^1 e^{-2t}\, dt = \tfrac{1}{2}(1 - e^{-2}),$$

and an $O(\varepsilon)$ approximation of x^0 is given by

$$\tilde{x}_s^0 = \frac{-2}{1 - e^{-2}} \int_0^1 e^{-t} y(t)\, dt.$$

Using $x(\varepsilon\tau) \approx \tilde{x}_s^0\ \forall\tau$, the fast system is given by

$$\frac{dz_f}{d\tau} = -z_f,$$

$$y_f = z_f,$$

where $z_f = z + \tilde{x}_s^0$ and $y_f = y + \tilde{x}_s^0$. The observability Grammian $M_f(0, \infty; 0)$ is

$$M_f(0, \infty; 0) = \int_0^\infty e^{-2\tau} d\tau = \tfrac{1}{2},$$

and

$$\tilde{z}_f^0 = 2 \int_0^{1/\varepsilon} e^{-\tau} \tilde{y}_f(\tau) d\tau$$

$$= 2 \int_0^{1/\varepsilon} e^{-\tau} [y(\varepsilon\tau) + \tilde{x}_s^0] d\tau.$$

Notice that in calculating $\tilde{z}_f^0$ we have replaced $M_f(0, 1/\varepsilon; 0)$ by $M_f(0, \infty; 0)$, which results in an $O(\varepsilon)$ error. The initial states x^0 and z^0 are approximated by

$$x^0 \approx \tilde{x}_s^0, \quad z^0 \approx -\tilde{x}_s^0 + \tilde{z}_f^0.$$

The errors in calculating x^0 and z^0 for $x^0 = z^0 = 1$ and $\varepsilon = 0.1, 0.05$ and 0.01 are shown in Table 5.2. The numerical results in Table 5.2 show an $O(\varepsilon)$ error (in this case the error $\approx 2\varepsilon$).

TABLE 5.2

ε	0.1	0.05	0.01
$\tilde{x}_s^0$	0.805	0.902	0.980
$x^0 - \tilde{x}_s^0$	0.195	0.098	0.020
$-\tilde{x}_s^0 + \tilde{z}_f^0$	0.805	0.902	0.980
$z^0 - (-\tilde{x}_s^0 + \tilde{z}_f^0)$	0.195	0.098	0.020

5.9 Exercises

Exercise 5.1

Let

$$A(t) = \begin{bmatrix} -2 & e^t \\ e^{-t} & -2 \end{bmatrix}, \quad t \in I = [0, \infty).$$

(a) Verify that $A(t)$ satisfies Assumption 2.1 but not Assumption 2.2.
(b) Calculate $\exp[A(t)\theta]$ as a function of t and show that an inequality like (2.3) does not hold.

Exercise 5.2

Let $\phi(t, s)$, K_1, α_2, γ and ε^* be as in the proof of Lemma 2.2. Show that for all $\varepsilon < \varepsilon^*$

$$\|\phi(t, s)\| \leq K_1\left(1 + \frac{2\gamma}{(1-\gamma)e^2}\right)e^{-\alpha_2(t-s)/\varepsilon}.$$

Exercise 5.3

Let $P(t) = \int_0^\infty [e^{A(t)\tau}]^T[e^{A(t)\tau}]\,d\tau$, and assume that $A(t)$ satisfies Assumptions 2.1–2.3 on $[t_0, \infty)$. Show that for all $t \geq t_0$

(a) $P(t)A(t) + A^T(t)P(t) = -I,$

(b) $\dfrac{1}{2C}x^T x \leq x^T P x \leq \dfrac{K_1^2}{2\alpha_1}x^T x \quad \forall x, \quad C = \sup\|A(t)\|_2,$

(c) $\dot{P}(t) = \displaystyle\int_0^\infty [e^{A(t)\tau}]^T[P(t)\dot{A}(t) + \dot{A}^T(t)P(t)][e^{A(t)\tau}]\,d\tau,$

(d) $\|\dot{P}(t)\| \leq \beta K_1^4/2\alpha_1^2.$

Exercise 5.4

Consider (3.2) and (3.3) and let $L = L_0 + \varepsilon L_1 + \varepsilon^2 R_L$ and $H = H_0 + \varepsilon H_1 + \varepsilon^2 R_H$, where L_0, L_1, H_0 and H_1 are defined by (3.7), (3.8), (3.11) and (3.12) respectively.

(a) Verify that R_L and R_H satisfy equations of the form (3.9) and (3.13). Give expressions for f_1, f_2, g_1 and g_2.

(b) Similarly to the proof of Theorem 3.1, show that the integral equations (3.14) and (3.17) have bounded solutions for sufficiently small ε. Hence

$$L - (L_0 + \varepsilon L_1) = O(\varepsilon^2),$$
$$H - (H_0 + \varepsilon H_1) = O(\varepsilon^2).$$

Exercise 5.5

Verify inequalities (4.4) and (4.5).

Exercise 5.6

This is an extension of Lemma 4.2. Suppose the assumptions of Theorem 4.1

hold. Let $P_s(t) \geq cI_n > 0$ and $P_f(t) \geq cI_m > 0$ satisfy the Lyapunov equations

$$-\dot{P}_s(t) = P_s(t)A_0(t) + A_0^T(t)P_s(t) + Q_s^T Q_s,$$
$$0 = P_f(t)A_{22}(t) + A_{22}^T(t)P_f(t) + Q_f^T Q_f,$$

where Q_s and Q_f are constant nonsingular matrices. Let γ, β_1 and β_2 be defined by

$$\begin{aligned}
[\|Q_f^{-T}P_f(t)Q_f^{-T}\|_2 + 2\|Q_f^{-T}P_f(t)L_0(t)A_{12}(t)Q_f^{-1}\|_2] &\leq \gamma, \\
2\|Q_s^{-T}P_s(t)A_{12}(t)Q_f^{-1}\|_2 &\leq \beta_1, \\
2\|Q_f^{-T}P_f(t)(\dot{L}_0(t) + L_0(t)A_0(t))Q_s^{-1}\|_2 &\leq \beta_2.
\end{aligned}$$

Show that the singularly perturbed system (4.1) is uniformly asymptotically stable for all $\varepsilon < 1/(\gamma + \beta_1\beta_2)$.

Hint: Apply the state transformation

$$\begin{pmatrix} \hat{x} \\ \hat{z} \end{pmatrix} = \begin{pmatrix} Q_s & 0 \\ 0 & Q_f \end{pmatrix}\begin{pmatrix} x \\ z \end{pmatrix}$$

to (4.1), and use Lemma 4.2.

Exercise 5.7

Consider the singularly perturbed (4.1) with

$$A_{11} = \begin{bmatrix} -1 & 0 \\ -\gamma \sin \omega t & -\sigma \end{bmatrix}, \quad A_{12} = \begin{bmatrix} 1 \\ 0 \end{bmatrix},$$
$$A_{21} = [0 \quad \sin \omega t], \qquad A_{22} = -1,$$

where γ, σ and ω are positive constants.

(a) Verify that the assumptions of Theorem 4.1 hold.

(b) Verify that the Lyapunov equations of Exercise 5.6 are satisfied with

$$P_s = \begin{bmatrix} \frac{1}{2} & 0 \\ 0 & \dfrac{1}{2\gamma} \end{bmatrix}, \quad Q_s = \begin{bmatrix} 1 & 0 \\ 0 & \left(\dfrac{\sigma}{\gamma}\right)^{1/2} \end{bmatrix}, \quad P_f = \tfrac{1}{2}, \quad Q_f = 1.$$

(c) Using Exercise 5.6, calculate a bound ε^* such that the system is uniformly asymptotically stable for all $\varepsilon < \varepsilon^*$.

Exercise 5.8

Instead of (5.4), perform the analysis of Sections 5.5 for the adaptation law $\dot{k} = -\gamma(y - y_{\mathrm{m}})r - \sigma k$ and explain the plot of Figs. 2(c), where $\sigma = 0.05$.

Exercise 5.9

The purpose of this exercise is to derive an $O(\varepsilon^2)$ approximation of the solution of (6.1). For convenience set $u = 0$.

(a) Seeking the solutions of (6.2) and (6.3) in the series form

$$\xi(t, \varepsilon) = \xi_0(t) + \varepsilon\xi_1(t) + \ldots,$$
$$\eta(\tau, \varepsilon) = \eta_0(\tau) + \varepsilon\eta_1(\tau) + \ldots,$$

show that ξ_0, ξ_1, η_0 and η_1 are given by

$$\dot{\xi}_0(t) = A_0(t)\xi_0(t), \quad \xi_0(t_0) = x^0,$$
$$\dot{\xi}_1(t) = A_0(t)\xi_1(t) - A_{12}(t)L_1(t)\xi_0(t),$$
$$\xi_1(t_0) = -H_0(t_0)[z^0 + L_0(t_0)x^0],$$
$$\frac{\mathrm{d}\eta_0(\tau)}{\mathrm{d}\tau} = A_{22}(t_0)\eta_0(\tau), \quad \eta_0(0) = L_0(t_0)x^0 + z^0,$$
$$\frac{\mathrm{d}\eta_1(\tau)}{\mathrm{d}\tau} = A_{22}(t_0)\eta_1(\tau) + [\tau\dot{A}_{22}(t_0) + L(t_0)A_{12}(t_0)]\eta_0(\tau),$$
$$\eta_1(0) = L_1(t_0)x^0.$$

(b) Show that, under the assumptions of Theorem 6.1,

$$x(t) = \xi_0(t) + \varepsilon\left[\xi_1(t) + H_0(t)\eta_0\left(\frac{t - t_0}{\varepsilon}\right)\right] + O(\varepsilon^2),$$

$$z(t) = -L_0(t)\xi_0(t) + \eta_0\left(\frac{t - t_0}{\varepsilon}\right) + \varepsilon\left[-L_0(t)\xi_1(t) - L_1(t)\xi_0(t)\right.$$
$$\left. - L_0(t)H_0(t)\eta_0\left(\frac{t - t_0}{\varepsilon}\right) + \eta_1\left(\frac{t - t_0}{\varepsilon}\right)\right] + O(\varepsilon^2).$$

Notice that H_1 need not be calculated; only L_0, L_1 and H_0 are needed to calculate an $O(\varepsilon^2)$ approximation. Will this observation extend to the case $u \neq 0$?

Exercise 5.10

Apply the results of Exercise 5.9 to the singularly perturbed system of Exercise 5.7. Take $x^{0\mathrm{T}} = [1 \quad 1]$, $z^0 = 1$.

Exercise 5.11

Apply the results of Exercise 5.9 to the singularly perturbed system of Example 4.1. Take $x^0 = 1$, $z^{0\mathrm{T}} = [0 \quad 0]$.

Exercise 5.12

Verify expressions (7.14).

Exercise 5.13

Verify expressions (7.24).

Exercise 5.14

Use the results of Section 5.7 to calculate a control that steers that state of the system

$$\begin{aligned}\dot{x} &= x + z_1,\\ \varepsilon\dot{z}_1 &= -z_2,\\ \varepsilon\dot{z}_2 &= z_1 + u\end{aligned}$$

from $x^0 = 0$, $z_1^0 = 0$, $z_2^0 = 0$ to an $O(\varepsilon)$ neighborhood of $x^{\mathrm{f}} = 1$, $z_1^{\mathrm{f}} = 1$, $z_2^{\mathrm{f}} = 1$.

Exercise 5.15

In Lemma 2.2 we saw that $\phi(t, s)$, the transition matrix of (2.1), can be approximated by $\exp[A(s)(t - s)/\varepsilon]$. In this exercise it is shown that $\phi(t, s)$ can also be approximated by $\exp[A(t)(t - s)/\varepsilon]$. Let $\Psi(t, s) = \phi(t, s) - \exp[A(t)(t, s)/\varepsilon]$. Show that $\Psi(t, s)$ satisfies inequality (2.6).

Hint show that $\Psi(t, s)$ is given by

$$\Psi(t, s) = \frac{1}{\varepsilon}\int_s^t \exp\left[A(t)\left(\frac{t-\tau}{\varepsilon}\right)\right][A(\tau) - A(t)]\phi(\tau, s)\mathrm{d}\tau$$

and proceed as in the proof of Lemma 2.2.

5.10 Notes and References

Stability properties of slowly varying systems have been studied in various references. Versions of Lemma 2.2 are given in the books by Coppel (1965), Brockett (1970) and Vidyasagar (1978). Our treatment is patterned after Coppel's, although the contraction mapping proof is different from his. One aspect that is emphasized in our treatment is the calculation of bounds on ε. Slowly varying systems have also been studied under conditional stability assumptions (e.g. Assumption 3.1); for this, the reader may consult Coppel (1967), Chang and Coppel (1969) and Skoog and Lau (1972).

The main sources for the decoupling transformation of Section 5.3 are Chang (1969, 1972). The contraction mapping proof is similar to Chang's although the technical nature of the result is slightly different. Chang employs a conditional stability assumption, while Theorem 3.1 assumes that the eigenvalues of A_{22} have negative real parts. The uniform $O(\varepsilon)$ approximation in Theorem 3.1 is stronger than Chang's results because of stronger differentiability assumptions. A multiparameter version of the decoupling transformation is given in Khalil and Kokotović (1979a, b) and Ladde and Siljak (1983) and Abed (1985a). The uniform asymptotic stability result of Theorem 4.1 is as in Wilde and Kokotović (1972a). It also follows from various results on uniform asymptotic stability of nonautonomous nonlinear singularly perturbed systems, similar to those of Chapter 7. The use of the decoupling transformation results in the present short proof. The stability bound of Lemma 4.1 is due to Javid (1978b), although our derivation is different from his. Also, the stability bound of Lemma 4.2 is taken from Saberi (1983).

The use of singular perturbations to analyze robustness of adaptive systems with respect to unmodeled high-frequency parasitics is treated again in Section 7.4. For a broader perspective, the reader may consult Ioannou and Kokotović (1983, 1984a, b).

The controllability Theorem 7.1 can be found in Sannuti (1977). The calculation of the asymptotic form of the steering control is new, although Kokotović and Haddad (1975a) give similar calculations for time-invariant systems. The feedback control and observer designs for linear time-varying systems, which are not covered here, can be found in O'Reilly (1980).

6 OPTIMAL CONTROL

6.1 Introduction

In optimal control problems on sufficiently long intervals, boundary layer phenomena occur even if the system is not singularly perturbed. In this case, the time scales are due to the different nature of the control tasks near the ends of the interval and over the rest of the interval. In a minimum energy problem, for example, the requirement to satisfy the initial and end conditions dominates at the ends of the interval, while the energy minimization requirement is dominant over the rest of the interval. Optimal trajectories of this kind have a familiar "take-off", "cruise" and "landing" pattern, with the boundary layers being the "take-off" and "landing" parts. For fast subsystems of singularly perturbed systems the intervals are always long compared with their dynamics. Their boundary layers are superimposed on the solution of the slow subsystem.

In this chapter, optimal control problems for singularly perturbed systems are approached by a preliminary study in Section 6.2 of boundary layers appearing in systems that are not singularly perturbed. This section is crucial for the understanding of the rest of the chapter. After a definition of the reduced problem in Section 6.3, we apply the results of Section 6.2 to near-optimal linear and nonlinear control in Sections 6.4 and 6.5, and to "cheap control" and singular arc analysis in Section 6.6.

6.2 Boundary Layers in Optimal Control

When a control problem is formulated over a finite time interval $t \in [0, T]$, the determination of its slow and fast parts depends on the length of the interval T. In this section, we consider that T is long by comparison with the dynamics of the system. By this we mean that the time needed to transfer the state of the system through a distance of $O(1)$, using a control $O(1)$, is small compared with T. In this case, the solutions of many optimal control problems consist of boundary layers, and their properties are similar to those of singularly perturbed systems. Let us display these properties by analyzing a terminal control problem for the system

$$\frac{\partial \tilde{\eta}}{d\tilde{t}} = \tilde{A}(\tilde{t})\tilde{\eta} + \tilde{B}(\tilde{t})\tilde{\nu}, \quad \tilde{\eta}(0) = \eta_0, \quad \tilde{\eta}(T) = \eta_1, \quad \tilde{\eta} \in R^m, \quad \tilde{\nu} \in R^v, \tag{2.1}$$

which is *not* singularly perturbed. A control $\tilde{\nu}(\tilde{t})$ is sought so as to drive the state $\tilde{\eta}(\tilde{t})$ from initial state η_0 at $t = 0$ to a prespecified state η_1, at a prespecified time $t = T$, while minimizing the cost functional

$$\tilde{J} = \frac{1}{2}\int_0^T [\tilde{\eta}^{\mathrm{T}}\tilde{C}^{\mathrm{T}}(\tilde{t})\tilde{C}(\tilde{t})\tilde{\eta} + \tilde{\nu}^{\mathrm{T}}\tilde{R}(\tilde{t})\tilde{\nu}]\,d\tilde{t}, \quad \tilde{R}(\tilde{t}) > 0 \quad \forall \tilde{t} \in [0, T]. \tag{2.2}$$

Treating T as the unit of a new time variable t and denoting the functions of t without the tilde,

$$t = \frac{\tilde{t}}{T}, \quad \eta(t) = \tilde{\eta}(\tilde{t}), \quad \nu(t) = \tilde{\nu}(\tilde{t}). \quad A(t) = \tilde{A}(\tilde{t}), \quad \text{etc.}, \tag{2.3}$$

we rewrite (2.1), (2.2) as

$$\frac{1}{T}\frac{d\eta}{dt} = \frac{1}{T}\dot{\eta} = A(t)\eta + B(t)\nu, \quad \eta(0) = \eta_0, \quad \eta(1) = \eta_1, \tag{2.4}$$

$$J = \frac{T}{2}\int_0^1 [\eta^{\mathrm{T}}C^{\mathrm{T}}(t)C(t)\eta + \nu^{\mathrm{T}}R(t)\nu]\,dt, \quad R(t) > 0 \quad \forall t \in [0, 1]. \tag{2.5}$$

If T is long with respect to the system dynamics, that is, if

$$\varepsilon = 1/T \tag{2.6}$$

is small, then (2.4) is a singularly perturbed system analyzed in Section 5.1 as a "slowly varying system."

In the standard optimality condition for (2.4), (2.5), the Hamiltonian

$$\mathcal{H} = \tfrac{1}{2}T\eta^{\mathrm{T}}C^{\mathrm{T}}(t)C(t)\eta + \tfrac{1}{2}T\nu^{\mathrm{T}}R(t)\nu + T\rho^{\mathrm{T}}[A(t)\eta + B(t)\nu] \tag{2.7}$$

is minimized by the control

$$\nu(t) = -R^{-1}(t)B^{\mathrm{T}}(t)\rho(t), \tag{2.8}$$

where $\rho(t)$ satisfies the adjoint equation

$$\dot{\rho} = -\nabla_\eta \mathcal{H}. \tag{2.9}$$

Substitution of (2.8) into (2.4) and (2.9) results in the boundary value problem

$$\frac{1}{T}\begin{bmatrix}\dot{\eta}\\ \dot{\rho}\end{bmatrix} = \varepsilon\begin{bmatrix}\dot{\eta}\\ \dot{\rho}\end{bmatrix} = \begin{bmatrix}A(t) & -S(t)\\ -Q(t) & -A^{\mathrm{T}}(t)\end{bmatrix}\begin{bmatrix}\eta\\ \rho\end{bmatrix}, \quad \eta(0) = \eta_0, \quad \eta(1) = \eta_1, \tag{2.10}$$

where $S(t) = B(t)R^{-1}(t)B^{\mathrm{T}}(t)$ and $Q(t) = C^{\mathrm{T}}(t)C(t)$. An analysis of this problem is made under the following assumptions.

Assumption 2.1

The derivatives $\dot{A}(t)$, $\dot{B}(t)$, $\dot{C}(t)$ and $\dot{R}(t)$ are continuous for all $t \in [0, 1]$.

Assumption 2.2

The eigenvalues $\lambda^i_{\mathcal{H}}$ of the Hamiltonian matrix,

$$\lambda^i_{\mathcal{H}} \triangleq \lambda^i\begin{bmatrix}A(t) & -S(t)\\ -Q(t) & -A^{\mathrm{T}}(t)\end{bmatrix}, \tag{2.11}$$

lie off the imaginary axis; that is, there exists a constant c such that

$$|\mathrm{Re}\,\lambda^i_{\mathcal{H}}| \geq c > 0, \quad i = 1, \ldots, 2m, \quad \forall t \in [0, 1]. \tag{2.12}$$

Assumption 2.3

The pair $A(t)$, $B(t)$ satisfies the "frozen t" controllability condition, that is,

$$\mathrm{rank}\,[B(t), A(t)B(t), \ldots, A^{m-1}(t)B(t)] = m \quad \forall t \in [0, 1]. \tag{2.13}$$

Under these conditions, the solution of (2.10) has the boundary layer properties established in Theorem 2.1. We approach this theorem by proving a sequence of three lemmas which restate some well known results from linear optimal control theory.

Lemma 2.1

Under Assumptions 2.1–2.3, the time-dependent algebraic Riccati equation

$$KA(t) + A^{\mathrm{T}}(t)K - KS(t)K + Q(t) = 0 \tag{2.14}$$

has a root $K = P(t) \geq 0$ and root $K = N(t) \leq 0$ such that the relationships

$$P(t) - N(t) > 0, \tag{2.15}$$

$$\operatorname{Re} \lambda^i[A(t) - S(t)P(t)] \leq -c < 0, \quad i = 1, \ldots, m, \tag{2.16}$$

$$\lambda_r^i \triangleq \lambda^i[A(t) - S(t)N(t)] = -\lambda_l^i \triangleq -\lambda^i[A(t) - S(t)P(t)], \quad i = 1, \ldots, m, \tag{2.17}$$

$$\lambda_{\mathcal{H}}^i = \begin{cases} \lambda_l^i, & i = 1, \ldots, m, \\ \lambda_r^j, & i = m + j, j = 1, \ldots, m, \end{cases} \tag{2.18}$$

hold for all $t \in [0, 1]$.

Proof Noting that $P(t) \geq 0$ with its property (2.16) is a stablizing solution of (2.14) for each fixed t, the proof follows from established results of linear control theory (e.g. Anderson and Moore, 1971). To see that the same results apply to $N(t) \leq 0$, and to prove (2.15), (2.17) and (2.18), consider the equation

$$-MA(t) - A^{\mathrm{T}}(t)M - MS(t)M + Q(t) = 0, \tag{2.19}$$

which is the replica of (2.14) with $-A(t)$ used instead of $A(t)$. Since (2.13) remains valid for $-A(t)$, there exists a stabilizing solution $P_M(t) \geq 0$ of (2.19). Letting $K = -M$, we identify $N(t) = -P_M(t) \leq 0$ as the matrix satisfying (2.14), (2.15) and (2.17), noting that (2.17) and (2.18) express the well known fact that the eigenvalues $\lambda_{\mathcal{H}}^i$ appear in pairs symmetric with respect to the imaginary axis. □

Lemma 2.2

Under Assumptions 2.1–2.3, there exists an $\varepsilon_1 > 0$ such that for all $\varepsilon \in (0, \varepsilon_1]$ two matrices, $P_\varepsilon(t, \varepsilon)$ and $N_\varepsilon(t, \varepsilon)$, satisfy the differential Riccati equation

$$\varepsilon \dot{K} = -KA(t) - A^{\mathrm{T}}(t)K + KS(t)K - Q(t) \tag{2.20}$$

and are $O(\varepsilon)$ close to the solutions $P(t)$ and $N(t)$ of the algebraic equation (2.14), that is,

$$P_\varepsilon(t, \varepsilon) = P(t) + O(\varepsilon), \quad N_\varepsilon(t, \varepsilon) = N(t) + O(\varepsilon) \quad \forall t \in [0, 1]. \tag{2.21}$$

Proof Substituting

$$N_\varepsilon(t, \varepsilon) - N(t) = \varepsilon \delta N \tag{2.22}$$

into (2.20), we obtain

$$\varepsilon\delta\dot{N} = -\delta N[A(t) - S(t)N(t)] - [A(t) - S(t)N(t)]^{\mathrm{T}}\delta N$$
$$+ \varepsilon\delta N S(t)\delta N - \dot{N}(t), \tag{2.23}$$

where $N(t)$, as a solution of (2.14), is bounded because of Assumption 2.1. The existence of $N_\varepsilon(t, \varepsilon)$ and its property (2.21) are then established using the same argument as in the proof of Theorem 3.1 in Chapter 5. The proof is based on the fact that $-A(t) + S(t)N(t)$ is Hurwitz for each $t \in [0, 1]$. The only difference from the proof of Theorem 3.1, Chapter 5, is that, instead of the state transition matrix for $-A + SN$, in this proof we use the product of the state transition matrices for $-A + SN$ and $(-A + SN)^{\mathrm{T}}$. To prove the existence of $P_\varepsilon(t, \varepsilon)$ and its property (2.21), we repeat the same argument considering that (2.20) is integrated in reverse time $1 - t$ starting at $1 - t = 0$, and making use of the fact that $A(t) - S(t)P(t)$ is Hurwitz for each $t \in [0, 1]$. □

Lemma 2.3

Under Assumptions 2.1–2.3, there exists an $\varepsilon_2 > 0$ such that for all $\varepsilon \in (0, \varepsilon_2]$ the transformation

$$\begin{bmatrix} \eta \\ \rho \end{bmatrix} = \begin{bmatrix} I & I \\ P_\varepsilon(t, \varepsilon) & N_\varepsilon(t, \varepsilon) \end{bmatrix} \begin{bmatrix} l \\ r \end{bmatrix} \tag{2.24}$$

is nonsingular for each $t \in [0, 1]$ and transforms the system (2.10) into two separate systems

$$\varepsilon \dot{l} = [A(t) - S(t)P_\varepsilon(t, \varepsilon)]l, \tag{2.25}$$

$$\varepsilon \dot{r} = [A(t) - S(t)N_\varepsilon(t, \varepsilon)]r, \tag{2.26}$$

which are an expression of the exponential dichotomy of (2.10). Namely, there exist constants $c > 0$ and $k > 0$ such that for all $t \in [0, 1]$, $\varepsilon \in (0, \varepsilon_2]$ the solutions of (2.25) and (2.26) are bounded by

$$\|l(t)\| \leq k\,\mathrm{e}^{-c(t/\varepsilon)}\|l(0)\|, \quad \|r(t)\| \leq k\,\mathrm{e}^{-c(1-t)/\varepsilon}\|r(1)\|. \tag{2.27}$$

Proof The nonsingularity of (2.24) follows from (2.15) and the property of the determinants of partitioned matrices, which, when applied to (2.24), yields

$$\det\begin{bmatrix} I & I \\ P_\varepsilon & N_\varepsilon \end{bmatrix} = \det[N_\varepsilon - P_\varepsilon]. \tag{2.28}$$

In view of (2.15) and (2.21), there exists an $\varepsilon_2 > 0$ such that $\det[N_\varepsilon - P_\varepsilon] \neq 0$ for all $\varepsilon \in (0, \varepsilon_2]$. The fact that (2.24) applied to (2.10) results in (2.25), (2.26) is verified by direct substitution. Finally, replacing $P_\varepsilon(t, \varepsilon)$ by $P(t) + O(\varepsilon)$ and $N_\varepsilon(t, \varepsilon)$ by $N(t) + O(\varepsilon)$ in (2.25) and (2.26) respectively, and using Lemmas 2.1–2.3, Chapter 5, proves (2.27). □

We are now ready to display the boundary layer behavior of the solution η, ρ of (2.10) for ε sufficiently small, that is, for T sufficiently large.

Theorem 2.1

Under Assumptions 2.1 and 2.3, there exists an $\varepsilon_3 > 0$ such that for all $\varepsilon \in (0, \varepsilon_3]$ the solution $\eta(t)$, $\rho(t)$ of (2.10) and the optimal control $\nu(t)$ satisfy

$$\eta(t) = \eta_l(\tau) + \eta_r(\sigma) + O(\varepsilon), \tag{2.29}$$

$$\rho(t) = P(0)\eta_l(\tau) + N(1)\eta_r(\sigma) + O(\varepsilon) \triangleq \rho_l(\tau) + \rho_r(\sigma) + O(\varepsilon), \tag{2.30}$$

$$\begin{aligned}\nu(t) &= -R^{-1}(0)B^{\mathrm{T}}(0)\rho_l(\tau) - R^{-1}(1)B^{\mathrm{T}}(1)\rho_r(\sigma) + O(\varepsilon)\\ &\triangleq \nu_l(\tau) + \nu_r(\sigma) + O(\varepsilon),\end{aligned} \tag{2.31}$$

where τ and σ are the fast time variables

$$\tau = \frac{t}{\varepsilon}, \quad \sigma = \frac{1-t}{\varepsilon}, \tag{2.32}$$

and $\eta_l(\tau)$ and $\eta_r(\sigma)$ are the solutions of the systems

$$\frac{d\eta_l}{d\tau} = [A(0) - S(0)P(0)]\eta_l, \qquad \eta_l(0) = \eta_0, \tag{2.33}$$

$$\frac{d\eta_r}{d\sigma} = -[A(1) - S(1)N(1)]\eta_r, \quad \eta_r(0) = \eta_1, \tag{2.34}$$

referred to as the *initial and terminal boundary-layer systems* respectively.

Proof By Theorem 6.1, Chapter 5, and Lemmas 2.1–2.3, the solution $l(t)$ of (2.25) for $l(0) = \eta_0$ and $\eta_l(\tau)$ defined by (2.33), and the solution $r(t)$ of (2.26) for $r(1) = \eta_1$ and $\eta_r(\sigma)$ defined by (2.34), satisfy for all $t \in [0, 1]$ and all $\varepsilon \in (0, \varepsilon_3]$

$$l(t) = \eta_l(\tau) + O(\varepsilon), \quad r(t) = \eta_\tau(\sigma) + O(\varepsilon). \tag{2.35}$$

Furthermore, $P(t)\eta_l(\tau) = P(0)\eta_l(\tau) + O(\varepsilon)$ and $N(t)\eta_r(\sigma) = N(1)\eta_r(\sigma) + O(\varepsilon)$. Analogous expressions hold for $R^{-1}(t)B^{\mathrm{T}}(t)\rho_l(\tau)$ and

$R^{-1}(t)B^{\mathrm{T}}(t)\rho_r(\sigma)$. In view of (2.24), this proves (2.29) and (2.30), while (2.31) follows by the substitution of (2.30) into (2.8). □

In the time-invariant case, that is, when the matrices A, B, C and R are constant, the roots P and N of the algebraic equation (2.14) satisfy the differential equation (2.20), and (2.21) reduces to

$$P_\varepsilon(t, \varepsilon) \equiv P, \quad N_\varepsilon(t, \varepsilon) \equiv N. \tag{2.36}$$

Consequently, the system (2.25) is identical with the system (2.33), and the system (2.26) is identical with the system (2.34), and it follows from (2.24) that

$$\eta(t) = \mathrm{e}^{(A-SP)t/\varepsilon}\, l(0) + \mathrm{e}^{(A-SN)(t-1)/\varepsilon}\, r(1). \tag{2.37}$$

Using (2.27), we obtain from (2.37) at $t = 0$ and $t = 1$

$$\eta(0) = l(0) + O(\mathrm{e}^{-c/\varepsilon}), \quad \eta(1) = r(1) + O(\mathrm{e}^{-c/\varepsilon}). \tag{2.38}$$

Hence, the error of the approximation $\eta(t) \sim \eta_l(\tau) + \eta_r(\sigma)$ in this case is only $O(\mathrm{e}^{-c/\varepsilon})$, and it is due solely to the approximation of the boundary conditions (2.38).

The importance of Theorem 2.1 is that it approximates a time-varying boundary value problem by two much simpler time-invariant initial value problems, namely the initial and terminal boundary layer regulator problems. The *initial regulator problem* is

$$\frac{\mathrm{d}\eta_l}{\mathrm{d}\tau} = A(0)\eta_l + B(0)\nu_l, \quad \eta_l(0) = \eta_0 \tag{2.39}$$

$$J_l = \frac{1}{2}\int_0^\infty [\eta_l^{\mathrm{T}} C^{\mathrm{T}}(0)C(0)\eta_l + \nu_l^{\mathrm{T}} R(0)\nu_l]\,\mathrm{d}\tau. \tag{2.40}$$

and the optimal state-feedback control

$$\nu_l(\tau) = -R^{-1}(0)B^{\mathrm{T}}(0)P(0)\eta_l(\tau) \tag{2.41}$$

stabilizes the initial boundary layer of the original problem (2.4), (2.5). The *terminal regulator problem* is

$$\frac{\mathrm{d}\eta_r}{\mathrm{d}\sigma} = -A(1)\eta_r - B(1)\nu_r, \quad \eta_r(0) = \eta_1, \tag{2.42}$$

$$J_r = \frac{1}{2}\int_0^\infty [\eta_r^{\mathrm{T}} C^{\mathrm{T}}(1)C(1)\eta_r + \nu_r^{\mathrm{T}} R(1)\nu_r]\,\mathrm{d}\sigma, \tag{2.43}$$

and the optimal state-feedback control

$$\nu_r(\sigma) = -R^{-1}(1)B^{\mathrm{T}}(1)N(1)\eta_r(\sigma) \tag{2.44}$$

stabilizes the terminal boundary layer of the original problem (2.4), (2.5) in reverse time as σ varies from $\sigma = 0$ to $\sigma \to \infty$. This means that in real time t the terminal layer is totally unstable, which is an inherent property of the Hamiltonian boundary value problem (2.10).

Example 2.1

Various aspects of the above analysis are illustrated by way of the scalar problem

$$\varepsilon\dot{\eta} = a(t)\eta + \nu, \quad J = \tfrac{1}{2}T\int_0^1 (q\eta^2 + \nu^2)\,\mathrm{d}t, \tag{2.45}$$

through considering four cases with different values of $a(t)$ and q.

Case 1: $q = 0$, $a(t) = -(1 + \alpha t)$, $|\alpha| < 0.5$. In this case, (2.14) and its solutions $P(t)$ and $N(t)$ are

$$-2(1 + \alpha t)K - K^2 = 0, \quad P(t) = 0, \quad N(t) = -2(1 + \alpha t). \tag{2.46}$$

Thus $\nu_l(\tau) = 0$ and the initial layer is $\eta_l(\tau) = \mathrm{e}^{-\tau}\eta_0$, because the open-loop system (2.45) is asymptotically stable and $q = 0$, that is, η is not observable from J. The terminal layer $\eta_r(\sigma)$ and the corresponding control $\nu_r(\sigma)$ are

$$\eta_r(\sigma) = \mathrm{e}^{-(1+\alpha)\sigma}\,\eta_1, \quad \nu_r(\sigma) = 2(1 + \alpha)\eta_r(\sigma), \tag{2.47}$$

and hence the approximate optimal trajectory (2.29) is

$$\eta(t) = \mathrm{e}^{-t/\varepsilon}\,\eta_0 + \mathrm{e}^{(1+\alpha)(t-1)/\varepsilon}\,\eta_1 + O(\varepsilon). \tag{2.48}$$

In this case it is possible to obtain the exact solution and to compare it with (2.48). Approximate controls and trajectories are sketched in Fig. 6.1(1).

Case 2: $q = 1$, $a = 0$. In this case, the open-loop system (2.45) is stable, but not asymptotically stable, while η is observable in J, so that (2.14) and its solutions P and N are

$$-K^2 + 1 = 0, \quad P = 1, \quad N = -1. \tag{2.49}$$

Both initial and terminal layers are controlled,

$$\nu_l(\tau) = -\eta_l(\tau), \quad \nu_r(\sigma) = \eta_r(\sigma), \tag{2.50}$$

and the approximate trajectory (2.37) is

$$\eta(t) = \mathrm{e}^{-t/\varepsilon}\,\eta_0 + \mathrm{e}^{(t-1)/\varepsilon}\,\eta_1 + O(\mathrm{e}^{-c/\varepsilon}), \tag{2.51}$$

and is shown in Fig. 6.1(2).

Case 3: $q = 0$, $a = 1$. The open-loop system is unstable and η is unobservable from J, so that (2.14) and its solutions P and N are

$$2K - K^2 = 0, \quad P = 2, \quad N = 0. \tag{2.52}$$

In this case, only the initial layer is controlled by $\nu_l(\tau) = -2\eta_l(\tau)$, while the terminal layer is the solution of (2.42) with $\nu_r(\sigma) = 0$. The expression for the approximate trajectory is the same as in (2.51); see also Fig. 6.1(3).

Case 4: $q = 0$, $a = 0$. The system is as in Case 2, but now η is unobservable in J and both eigenvalues of the Hamiltonian matrix (2.11) are zero:

$$\lambda_{\mathcal{H}}^{1,2} = \lambda^{1,2}\begin{bmatrix} 0 & -1 \\ 0 & 0 \end{bmatrix} = 0. \tag{2.53}$$

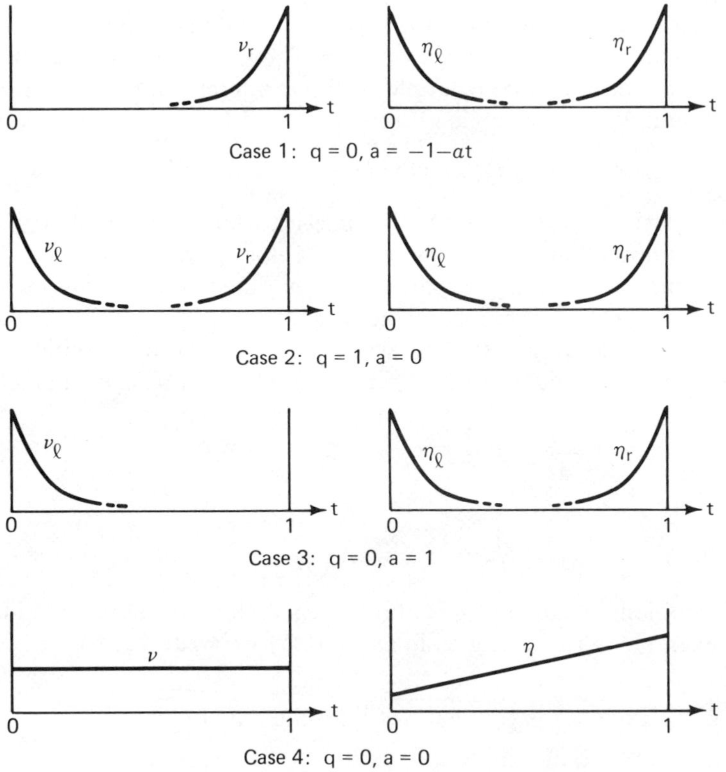

Fig. 6.1. The boundary layer behavior of the solution to problem (2.45) for $\varepsilon = 0.2$ in cases 1, 2, 3. In case 4, the solution does not exhibit boundary layer behavior.

The boundary layer approximation does not apply because the condition (2.11) is violated. In fact, the exact optimal solution

$$\nu(t) = \varepsilon(\eta_1 - \eta_0) = \text{const}, \quad \eta(t) = \eta_0 + (\eta_1 - \eta_0)t, \tag{2.54}$$

shown in Fig. 6.1(4), does not exhibit a boundary layer behavior.

It is clear from this example that the presence of boundary layers depends on the observability of individual eigenvalues and, in particular, on the observability of the zero eigenvalues. In a higher-order problem the part of the solution corresponding to the unobservable zero eigenvalues will not possess boundary layers, but the remaining part will be in the boundary layer form. For unobservable eigenvalues away from the imaginary axis, both layers will appear, but only one of them will be controlled: the initial layer will be controlled if $\operatorname{Re}\lambda > 0$, while the terminal layer will be controlled if $\operatorname{Re}\lambda < 0$. For observable eigenvalues both layers will be controlled.

Thus far, we have discussed the problem with the terminal state η fixed. Similar conclusions can be drawn from Theorem 2.1 for other types of boundary conditions. For example, if the terminal state $\eta(1)$ is free, then $\rho(1) = 0$ and we see from (2.30) that

$$\rho(1) = 0 = N(1)\eta(1) + O(\varepsilon) = N\eta(1) + O(\mathrm{e}^{-c/\varepsilon}), \tag{2.55}$$

where the last equality applies to the time-invariant case, along with (2.37), (2.38). It follows that $\eta(1)$ will be regulated close to zero if N is nonsingular, which is the case when the pair (A, C) is detectable, that is, when the unstable eigenvalues are observable in J.

Let us conclude this section with a glimpse at the boundary layer phenomena occurring in a more general nonlinear problem, namely,

$$\frac{\mathrm{d}\tilde{\eta}}{\mathrm{d}\tilde{t}} = g(\tilde{\eta}, \tilde{\nu}), \quad \tilde{\eta}(0) = \eta_0, \quad \tilde{\eta}(T) = \eta_1, \tag{2.56}$$

$$\tilde{J} = \int_0^T V(\tilde{\eta}, \tilde{\nu})\,\mathrm{d}\tilde{t}. \tag{2.57}$$

This problem, a current topic of research, is sketched here and illustrated by an example. Proceeding as in (2.1)–(2.5) we write (2.56), (2.57) as

$$\frac{1}{T}\dot{\eta} = \varepsilon\dot{\eta} = g(\eta, \nu), \quad \eta(0) = \eta_0, \quad \eta(1) = \eta_1, \tag{2.58}$$

$$J = T\int_0^1 V(\eta, \nu)\,\mathrm{d}t. \tag{2.59}$$

Motivated by the state-feedback interpretation of the boundary layer regulators (2.39)–(2.44), we examine the Hamilton–Jacobi optimality condition for (2.56), (2.57), that is,

$$-\frac{\partial \tilde{J}^*}{\partial \tilde{t}} = \min_{\tilde{\nu}} \left[V(\tilde{\eta}, \tilde{\nu}) + \frac{\partial \tilde{J}^*}{\partial \tilde{\eta}} g(\tilde{\eta}, \tilde{\nu}) \right], \tag{2.60}$$

which, rescaled for (2.58), (2.59), becomes

$$-\varepsilon \frac{\partial J^*}{\partial t} = \min_{\nu} \left[V(\eta, \nu) + \frac{\partial J^*}{\partial \eta} g(\eta, \nu) \right], \tag{2.61}$$

where $J^*(\eta, t)$ is Bellman's optimal value function (Athans and Falb, 1966). A nonlinear generalization of Theorem 2.1 would establish that the condition (2.61) can be approximately satisfied by two boundary layer regulator solutions obtained from the simplified condition

$$0 = \min_{\nu} \left[V(\eta, \nu) + \frac{\partial J^*}{\partial \eta} g(\eta, \nu) \right]. \tag{2.62}$$

As in the linear problem, the initial-layer feedback control $\nu_l = \nu_l(\eta_l)$ would stabilize the system (2.58) in real time, while the terminal-layer feedback control $\nu_r = \nu_r(\eta_r)$ would stabilize it in reverse time.

Example 2.2

For the nonlinear problem

$$\varepsilon \dot{\eta} = \eta^3 + \nu, \quad \eta(0) = \eta_0, \quad \eta(1) = \eta_1, \tag{2.63}$$

$$J = \tfrac{1}{2} T \int_0^1 (q\eta^2 + \nu^2)\,dt, \quad \varepsilon = \frac{1}{T}, \quad q = 1, \tag{2.64}$$

the minimizing control for (2.62) is $\nu = -\partial J^*/\partial \eta$, which, when substituted back into (2.62), yields

$$\nu^2 + 2\eta^3 \nu - \eta^2 = 0. \tag{2.65}$$

The roots of this equation, which is a nonlinear counterpart of (2.14), are

$$\nu_{1,2} = -\eta^3 \pm \eta(1 + \eta^4)^{1/2}, \tag{2.66}$$

and the decision is to be made as to which of them is ν_l and which of them is ν_r. Since ν_1 is stabilizing in real time and ν_2 is stabilizing in reverse time, the correct choice is

$$\nu_l = -\eta_l^3 - \eta_l(1 + \eta_l^4)^{1/2}, \quad \nu_r = -\eta_r^3 + \eta_r(1 + \eta_r^4)^{1/2}. \tag{2.67}$$

A nonlinear analog of Theorem 2.1 would then establish that the solution of (2.58)–(2.59) is approximated by

$$\eta(t) = \eta_l(t) + \eta_r(t) + O(\varepsilon), \quad \nu(t) = \nu_l(t) + \nu_r(t) + O(\varepsilon), \tag{2.68}$$

where the corresponding terms are obtained from the boundary layer systems of (2.63) controlled respectively by ν_l and ν_r given by (2.67).

6.3 The Reduced Problem

Let us now consider the problem of finding a control $u(t)$ that steers the state x, z of the singularly perturbed system

$$\dot{x} = f(x, z, t, \varepsilon, u), \quad x \in R^n, \quad u \in R^r, \tag{3.1}$$

$$\varepsilon\dot{z} = g(x, z, t, \varepsilon, u), \quad z \in R^m, \tag{3.2}$$

from x_0, z_0 at $t = 0$ to x_1, z_1 at $t = 1$, that is,

$$x(0) = x_0, \quad z(0) = z_0, \quad x(1) = x_1, \quad z(1) = z_1, \tag{3.3}$$

while minimizing the cost functional

$$J = \int_0^1 V(x, z, t, \varepsilon, u)\, dt. \tag{3.4}$$

This problem can be simplified by neglecting ε in two different ways. First, an optimality condition can be formulated for the exact problem and simplified by setting $\varepsilon = 0$. The result will be a *reduced optimality condition*. Second, by neglecting ε in the system (3.1), (3.2) the same type of optimality condition can be formulated for the *reduced system*. When the obtained optimality conditions are identical, it is said that the *reduced problem is formally correct*. Since this may not be obvious in a Hamiltonian formulation, we first formulate a necessary optimality condition for the exact problem, using the Lagrangian

$$\mathcal{L} = V + p^{\mathrm{T}}(\dot{x} - f) + q^{\mathrm{T}}(\varepsilon\dot{z} - g), \tag{3.5}$$

where p and q are the multipliers (costates or adjoint variables) associated with x and z respectively. The Lagrangian necessary condition is

$$\frac{\partial \mathcal{L}}{\partial p} = 0 \Rightarrow \dot{x} = f, \quad x(0) = x_0, \quad x(1) = x_1, \tag{3.6}$$

$$\frac{\partial \mathcal{L}}{\partial q} = 0 \Rightarrow \varepsilon\dot{z} = g, \quad z(0) = z_0, \quad z(1) = z_1; \tag{3.7}$$

$$\frac{\partial \mathcal{L}}{\partial u} = 0 \Rightarrow \nabla_u V - f_u^{\mathrm{T}} p - g_u^{\mathrm{T}} q = 0, \tag{3.8}$$

$$\frac{\partial \mathcal{L}}{\partial x} - \frac{\mathrm{d}}{\mathrm{d}t}\frac{\partial \mathcal{L}}{\partial \dot{x}} = 0 \Rightarrow \dot{p} = \nabla_x \mathcal{L}, \tag{3.9}$$

$$\frac{\partial \mathcal{L}}{\partial z} - \frac{\mathrm{d}}{\mathrm{d}t}\frac{\partial \mathcal{L}}{\partial \dot{z}} = 0 \Rightarrow \varepsilon \dot{q} = \nabla_z \mathcal{L}. \tag{3.10}$$

The *reduced necessary condition* is now obtained by setting $\varepsilon = 0$ in (3.6)–(3.10) and disregarding the requirement that $z(0) = z_0$, $z(1) = z_1$, which must be dropped because (3.7) and (3.10) have degenerated to $0 = g$ and $0 = \nabla_z \mathcal{L}$, and are not free to satisfy arbitrary boundary conditions.

On the other hand, the *reduced problem* is obtained by setting $\varepsilon = 0$ in (3.1), (3.2) and (3.4), and dropping the requirement that $z(0) = z_0$, $z(1) = z_1$; that is, the reduced problem is defined as

$$\dot{x}_s = f(x_s, z_s, t, 0, u_s), \quad x_s(0) = x_0, \tag{3.11}$$

$$0 = g(x_s, z_s, t, 0, u_s), \quad x_s(1) = x_1, \tag{3.12}$$

$$J_s = \int_0^1 V(x_s, z_s, t, 0, u_s)\, \mathrm{d}t. \tag{3.13}$$

(The subscript "s" intimates our intention to use the solution of the reduced problem for an approximation of the slow ("outer") part of the optimal solution.) The question of formal correctness is answered by comparing the Lagrangian necessary condition for the reduced problem (3.11)–(3.13) with the reduced necessary condition (3.6)–(3.10).

The Lagrangian for the reduced problem (3.11)–(3.13) is

$$\mathcal{L}_s = V(x_s, t_s, t, 0, u_s) + p_s^{\mathrm{T}}[\dot{x}_s - f(x_s, z_s, t, 0, u_s)] \\ - q_s^{\mathrm{T}} g(x_s, z_s, t, 0, u_s), \tag{3.14}$$

and the corresponding necessary condition is

$$\frac{\partial \mathcal{L}_s}{\partial p_s} = 0 \Rightarrow \dot{x}_s = f(x_s, z_s, t, 0, u_s), \tag{3.15}$$

$$\frac{\partial \mathcal{L}_s}{\partial q_s} = 0 \Rightarrow 0 = g(x_s, z_s, t, 0, u_s), \tag{3.16}$$

$$\frac{\partial \mathcal{L}_s}{\partial u_s} = 0 \Rightarrow \nabla_{u_s} V(x_s, z_s, t, 0, u) - f_{u_s}^{\mathrm{T}}(x_s, z_s, t, 0, u_s) p_s \\ - g_{u_s}^{\mathrm{T}}(x_s, z_s, t, 0, u_s) q_s = 0, \tag{3.17}$$

$$\frac{\partial \mathcal{L}_s}{\partial x_s} - \frac{d}{dt}\frac{\partial \mathcal{L}_s}{\partial \dot{x}_s} = 0 \Rightarrow \dot{p}_s = \nabla_{x_s}\mathcal{L}_s, \tag{3.18}$$

$$\frac{\partial \mathcal{L}_s}{\partial z_s} = 0 \Rightarrow 0 = \nabla_{z_s}\mathcal{L}_s, \tag{3.19}$$

subject to $x_s(0) = x_0$, $x_s(1) = x_1$. By setting $\varepsilon = 0$ in (3.6)–(3.10), and comparing the result with (3.15)–(3.19), we come to the following conclusion.

Lemma 3.1

The reduced problem (3.11)–(3.13) is formally correct.

We have used the Lagrangian necessary condition because it conveniently incorporates both differential and algebraic constraints. The more common Hamiltonian form is obtained from (3.6)–(3.10) via the Hamiltonian function

$$\mathcal{H} = V + p^T f + q^T g. \tag{3.20}$$

Rewriting (3.6)–(3.10) in terms of $\mathcal{H}$, we get

$$\dot{x} = \nabla_p \mathcal{H} = f, \quad x(0) = x_0, \quad x(1) = x_1, \tag{3.21}$$

$$\varepsilon\dot{z} = \nabla_g \mathcal{H} = g, \quad z(0) = z_0, \quad z(1) = z_1, \tag{3.22}$$

$$0 = \nabla_u \mathcal{H}, \tag{3.23}$$

$$\dot{p} = -\nabla_x \mathcal{H}, \tag{3.24}$$

$$\varepsilon\dot{q} = -\nabla_z \mathcal{H}, \tag{3.25}$$

It is important to observe that in the definition of $\mathcal{H}$ use is made of g rather than of g/ε. This means that the Hamiltonian adjoint (costate) variable associated with z is εq. Let us apply the Hamiltonian condition to the problem

$$\dot{x} = A_{11}x + A_{12}z + B_1 u, \quad x(0) = x_0, \quad x(1) = x_1, \tag{3.26}$$

$$\varepsilon\dot{z} = A_{21}x + A_{22}z + B_2 u, \quad z(0) = z_0, \quad z(1) = z_1, \tag{3.27}$$

$$y = C_1 x + C_2 z, \tag{3.28}$$

$$J = \frac{1}{2}\int_0^1 (y^T y + u^T R u)\, dt, \quad R > 0. \tag{3.29}$$

Using the Hamiltonian

$$\mathcal{H} = \tfrac{1}{2}y^T y + \tfrac{1}{2}u^T R u + p^T(A_{11}x + A_{12}z + B_1 u) + q^T(A_{21}x + A_{22}z + B_2 u), \tag{3.30}$$

we obtain from (3.23), (3.24) and (3.25)

$$u = -R^{-1}(B_1^{\mathrm{T}}p + B_2^{\mathrm{T}}q), \tag{3.31}$$

$$\dot{p} = -A_{11}^{\mathrm{T}}p - A_{21}^{\mathrm{T}}q - C_1^{\mathrm{T}}C_1x - C_1^{\mathrm{T}}C_2z. \tag{3.32}$$

$$\varepsilon\dot{q} = -A_{12}^{\mathrm{T}}p - A_{22}^{\mathrm{T}}q - C_2^{\mathrm{T}}C_1x - C_2^{\mathrm{T}}C_2z. \tag{3.33}$$

The substitution of (3.31) into (3.26), (3.27) results in the boundary value problem consisting of the system

$$\begin{bmatrix}\dot{x}\\ \dot{p}\end{bmatrix} = \begin{bmatrix}A_{11} & -S_{11}\\ -Q_{11} & -A_{11}^{\mathrm{T}}\end{bmatrix}\begin{bmatrix}x\\ p\end{bmatrix} + \begin{bmatrix}A_{12} & -S_{12}\\ -Q_{12} & -A_{21}^{\mathrm{T}}\end{bmatrix}\begin{bmatrix}z\\ q\end{bmatrix}, \tag{3.34}$$

$$\varepsilon\begin{bmatrix}\dot{z}\\ \dot{q}\end{bmatrix} = \begin{bmatrix}A_{21} & -S_{12}^{\mathrm{T}}\\ -Q_{12}^{\mathrm{T}} & -A_{12}^{\mathrm{T}}\end{bmatrix}\begin{bmatrix}x\\ p\end{bmatrix} + \begin{bmatrix}A_{22} & -S_{22}\\ -Q_{22} & -A_{22}^{\mathrm{T}}\end{bmatrix}\begin{bmatrix}z\\ q\end{bmatrix}, \tag{3.35}$$

and the boundary conditions

$$x(0) = x_0, \quad x(1) = x_1, \tag{3.36}$$

$$z(0) = z_0, \quad z(1) = z_1, \tag{3.37}$$

where $Q_{ij} = C_i^{\mathrm{T}}C_j$ and $S_{ij} = B_iR^{-1}B_j^{\mathrm{T}}$, $i, j = 1, 2$. Wherever convenient, the system (3.34), (3.35) will be rewritten in the more compact notation

$$\begin{bmatrix}\dot{x}\\ \dot{p}\end{bmatrix} = F_{11}\begin{bmatrix}x\\ p\end{bmatrix} + F_{12}\begin{bmatrix}z\\ q\end{bmatrix}, \tag{3.38}$$

$$\varepsilon\begin{bmatrix}\dot{z}\\ \dot{q}\end{bmatrix} = F_{21}\begin{bmatrix}x\\ p\end{bmatrix} + F_{22}\begin{bmatrix}z\\ q\end{bmatrix}, \tag{3.39}$$

with the obvious definitions of F_{11}, F_{12}, F_{21} and F_{22}. This system is in the familiar singularly perturbed form analyzed in Chapters 2 and 5. To define its reduced system, we need F_{22}^{-1}. However, anticipating that F_{22} will be the Hamiltonian matrix for a fast subproblem, we make an assumption analogous to Assumption 2.2.

Assumption 3.1

The eigenvalues of $F_{22}(t)$ lie off the imaginary axis for all $t \in [0, 1]$; that is, there exists a constant c such that

$$|\mathrm{Re}\,\lambda(F_{22})| \geq c > 0 \quad \forall t \in [0, 1]. \tag{3.40}$$

Under this assumption F_{22}^{-1} exists and the reduced necessary condition is

$$\begin{bmatrix} \dot{x}_s \\ \dot{p}_s \end{bmatrix} = (F_{11} - F_{12}F_{22}^{-1}F_{21})\begin{bmatrix} x_s \\ p_s \end{bmatrix}, \quad \begin{bmatrix} z_s \\ q_s \end{bmatrix} = -F_{22}^{-1}F_{21}\begin{bmatrix} x_s \\ p_s \end{bmatrix}, \tag{3.41}$$

subject to

$$x_s(0) = x_0, \quad x_s(1) = x_1. \tag{3.42}$$

From Lemma 3.1, we know that the same necessary condition is obtained by setting $\varepsilon = 0$ in (3.26)–(3.29), even if A_{22}^{-1} does not exist. Accepting a loss of generality, we assume that A_{22}^{-1} exists, and define the reduced problem as

$$\dot{x}_s = A_0x_s + B_0u_s, \quad x_s(0) = x_0, \quad x_s(1) = x_1, \tag{3.43}$$

$$y_s = C_1x_s + C_2z_s = C_0x_s + D_0u_s, \tag{3.44}$$

$$J_s = \frac{1}{2}\int_0^1 (x_s^T C_0^T C_0 x_s + 2u_s^T D_0^T C_0 x_s + u_s^T R_0 u_s)\, dt, \tag{3.45}$$

where z_s is obtained from (3.27) with $\varepsilon\dot{z} = 0$ and substituted into (3.26), (3.28) and (3.29), that is,

$$z_s = -A_{22}^{-1}(A_{21}x_s + B_2u_s), \tag{3.46}$$

whence

$$\begin{aligned} A_0 &= A_{11} - A_{12}A_{22}^{-1}A_{21}, \quad B_0 = B_1 - A_{12}A_{22}^{-1}B_2, \\ C_0 &= C_1 - C_2A_{22}^{-1}A_{21}, \quad D_0 = -C_2A_{22}^{-1}B_2, \quad R_0 = R + D_0^T D_0. \end{aligned} \tag{3.47}$$

Application of the necessary condition $0 = \nabla u_s \mathscr{H}_s$, $\dot{p}_s = -\nabla x_s \mathscr{H}_s$ to the Hamiltonian for (3.43)–(3.45),

$$\mathscr{H}_s = \tfrac{1}{2}(x_s^T C_0^T C_0 x_s + 2u_s^T D_0^T C_0 x_s + u_s^T R_0 u_s) + p_s^T(A_0x_s + B_0u_s), \tag{3.48}$$

and substitution of the control u_s minimizing $\mathscr{H}_s$,

$$u_s = -R_0^{-1}(D_0^T C_0 x_s + B_0^T p_s), \tag{3.49}$$

into (3.43) and $\dot{p}_s = -\nabla x_s \mathscr{H}_s$, results in the boundary value problem

$$\begin{bmatrix} \dot{x}_s \\ \dot{p}_s \end{bmatrix} = \begin{bmatrix} A_0 - B_0R_0^{-1}D_0^T C_0 & -B_0R_0^{-1}B_0^T \\ -C_0^T C_0 + C_0^T D_0 R_0^{-1} D_0^T C_0 & -(A_0 - B_0R_0^{-1}D_0^T C_0)^T \end{bmatrix} \begin{bmatrix} x_s \\ p_s \end{bmatrix}$$

$$x_s(0) = x_0, \qquad x_s(1) = x_1. \tag{3.50}$$

Specialized to this case, Lemma 3.1 reads as follows.

Lemma 3.2

The matrix $F_{11} - F_{12}F_{22}^{-1}F_{21}$ in (3.41) and the Hamiltonian matrix in (3.50) are identical, that is,

$$F_{11} - F_{12}F_{22}^{-1}F_{21} \equiv \begin{bmatrix} A_0 - B_0R_0^{-1}D_0^TC_0 & -B_0R_0^{-1}B_0^T \\ -C_0^TC_0 + C_0^TD_0R_0^{-1}D_0^TC_0 & -(A_0 - B_0R_0^{-1}D_0^TC_0) \end{bmatrix} \overset{\Delta}{=} F_0. \quad (3.51)$$

If one were to prove this fact without the help of Lemma 3.1, one would have to go through manipulations as in Chapter 3, involving an expression for F_{22}^{-1} in terms of A_{22}^{-1}. While in the Lagrangian formulation, it is almost obvious that the reduced problem is formally correct, this property is obscured in situations in which algebraic constraints are explicitly solved and eliminated from the problem. A special case in which (3.51) can readily be verified by algebraic manipulations is when $C_2 = 0$. In this case, $D_0 = 0$, $R_0 = R$, $C_0 = C_1$, and

$$-B_0R_0^{-1}B_0^T = -S_{11} + A_{12}A_{22}^{-1}S_{12}^T + S_{12}(A_{22}^T)^{-1}A_{12}^T - A_{12}A_{22}^{-1}S_{22}(A_{22}^T)^{-1}A_{12}^T. \quad (3.52)$$

On the other hand, the substitution of

$$F_{22}^{-1} = \begin{bmatrix} A_{22} & -S_{22} \\ 0 & -A_{22}^T \end{bmatrix}^{-1} = \begin{bmatrix} A_{22}^{-1} & -A_{22}^{-1}S_{22}(A_{22}^T)^{-1} \\ 0 & -(A_{22}^T)^{-1} \end{bmatrix}. \quad (3.53)$$

into (3.41) gives

$$F_0 = \begin{bmatrix} A_{11} - A_{12}A_{22}^{-1}A_{21} & -S_{11} + A_{12}A_{22}^{-1}S_{12}^T + S_{12}(A_{22}^T)^{-1}A_{12}^T - A_{12}A_{22}^{-1}S_{22}(A_{22}^T)^{-1}A_{12}^T \\ -Q_{11} & -(A_{11} - A_{12}A_{22}^{-1}A_{21})^T \end{bmatrix} \quad (3.54)$$

and confirms (3.51). Fortunately, Lemma 3.1 and its consequence, Lemma 3.2, render such calculations superflous.

All the results in the remaining part of this chapter will be obtained under the following assumption.

Assumption 3.2

The solution of the reduced problem exists and is unique.

In other words, our analysis assumes that the reduced solution is given and uses it to construct two-time-scale approximations of the exact solution. Depending on the particular problem, this assumption implies several other assumptions discussed in detail in standard texts on optimal control. For example, in a fixed-end-point problem x_1 must be reachable from x_0, which in the linear case implies the controllability of the reduced system (3.43). Another example is a convexity assumption frequently made to avoid the so-called conjugate points in the Hamiltonian problem (3.50). Since the matrices appearing in (3.50) have already been examined in Chapter 3, when dealing with the composite control, we know that this assumption is automatically satisfied because the (2,1)-block of F_0 in (3.51) is nonpositive,

$$-C_0^{\mathrm{T}}(I - D_0 R_0^{-1} D_0^{\mathrm{T}})C_0 = -C_0^{\mathrm{T}}(I + D_0 R D_0^{\mathrm{T}})^{-1} C_0 \leqslant 0, \tag{3.55}$$

owing to $R > 0$; see (4.19) of Chapter 3. Let us conclude this section with a discussion of the reduced problems for a linear example and a nonlinear example.

Example 3.1

Suppose that a DC motor without load, modeled by

$$\dot{x} = z, \quad \varepsilon\dot{z} = -x - z + u, \tag{3.56}$$

where x is its angular velocity and z and u are the armature current, and voltage respectively, is to be accelerated from $x_0 = 0$ to x_1, while the initial and terminal values of the current are to be kept at $z_0 = 0$, $z_1 = 0$. The control accomplishing this task should also minimize the energy function

$$J = \frac{1}{2}\int_0^1 u^2 \, \mathrm{d}t. \tag{3.57}$$

The Hamiltonian $\mathcal{H}$ for (3.56), (3.57) and the result of $\nabla_u \mathcal{H} = 0$ are

$$\mathcal{H} = \tfrac{1}{2}u^2 + pz + q(-x - z + u), \quad \nabla_u \mathcal{H} = 0 \Rightarrow u = -q, \tag{3.58}$$

which, when substituted into (3.24), (3.25) and (3.56), give

$$\left.\begin{aligned} &\dot{x} = z, \quad \varepsilon\dot{z} = -x - z - q, \quad x(0) = x_0 \quad x(1) = x_1, \\ &\dot{p} = q, \quad \varepsilon\dot{q} = -p + q, \qquad z(0) = z_0, \quad z(1) = z_1. \end{aligned}\right\} \tag{3.59}$$

Letting $\varepsilon = 0$ in (3.59), one obtains the reduced necessary condition

$$\left.\begin{aligned} &\dot{x}_s = -x_s - p_s, \quad z_s = -x_s - p_s, \quad x_s(0) = x_0, \quad x_s(1) = x_1, \\ &\dot{p}_s = p_s, \quad q_s = p_s, \quad u_s = -p_s, \end{aligned}\right\} \tag{3.60}$$

which is also the necessary condition for the reduced problem

$$\dot{x}_s = -x_s + u_s; \quad J_s = \frac{1}{2}\int_0^1 u_s^2 \, dt, \quad x_s(0) = x_0, \quad x_s(1) = x_1, \quad (3.61)$$

as predicted by Lemma 3.2. We will continue the analysis of this problem in the next section.

Example 3.2

The nonlinear optimal control problem

$$\begin{aligned} \dot{x} &= xz, & x(0) &= \frac{1}{\sqrt{2}}, & x(1) &= \tfrac{1}{2}, \\ \varepsilon \dot{z} &= -z + u, & z(0) &= 0, & z(1) &= 0, \end{aligned} \quad (3.62)$$

$$J = \int_0^1 (x^4 + \tfrac{1}{2}z^2 + \tfrac{1}{2}u^2) \, dt \quad (3.63)$$

occurs in chemical processes, where x represents the concentration of an ingredient whose rate of change is proportional to the flow z. The flow z is controlled by u through a first-order delay with the time constant ε. The penalty x^4 in (3.63) attempts to reduce large increases in the concentration. The Hamiltonian $\mathcal{H}$ and the result of $\nabla_u \mathcal{H} = 0$ are

$$\mathcal{H} = x^4 + \tfrac{1}{2}z^2 + \tfrac{1}{2}u^2 + pxz + q(-z + u), \quad \nabla_u \mathcal{H} = 0 \Rightarrow u = -q, \quad (3.64)$$

and their substitution into (3.24), (3.25) and (3.62) yields the necessary condition in the form of the nonlinear boundary value problem

$$\left.\begin{aligned} \dot{x} &= xz, & \varepsilon \dot{z} &= -z - q, & x(0) &= \frac{1}{\sqrt{2}}, & x(1) &= \tfrac{1}{2}, \\ \dot{p} &= -4x^3 - pz, & \varepsilon \dot{q} &= -px - z + q, & z(0) &= 0, & z(1) &= 0. \end{aligned}\right\} \quad (3.65)$$

The reduced problem obtained by setting $\varepsilon = 0$ in (3.62), (3.63) is

$$\dot{x}_s = x_s u_s, \quad J_s = \int_0^1 (x_s^4 + u_s^2) \, dt, \quad x_s(0) = \frac{1}{\sqrt{2}}, \quad x_s(1) = \tfrac{1}{2}, \quad (3.66)$$

and its necessary condition is

$$\left.\begin{aligned} \dot{x}_s &= -\tfrac{1}{2}x_s^2 p_s, \quad u_s = -\tfrac{1}{2}x_s p_s, \quad x_s(0) = \frac{1}{\sqrt{2}}, \quad x_s(1) = \tfrac{1}{2}, \\ \dot{p}_s &= -4x_s^3 + \tfrac{1}{2}x_s p_s^2, \quad z_s = u_s. \end{aligned}\right\} \quad (3.67)$$

The same reduced necessary condition is obtained by setting $\varepsilon = 0$ in (3.65). We shall return to this problem in Section 6.5 in the context of near-optimal nonlinear control.

6.4 Near-Optimal Linear Control

Our understanding of boundary-layer phenomena and the relationship between the exact and the reduced necessary conditions serve as a basis for a two-time-scale approximation of the exact solution. To begin with, let us consider the linear boundary value problem (3.34)–(3.37), assuming that all of its coefficient matrices are constant. Using the more compact notation, we apply our block-diagonalizing transformation, developed in Chapter 2, to this singularly perturbed system; namely, we let

$$\begin{bmatrix} x \\ p \end{bmatrix} = \begin{bmatrix} \xi \\ \pi \end{bmatrix} + \varepsilon H \begin{bmatrix} \eta \\ \rho \end{bmatrix}, \quad \begin{bmatrix} z \\ q \end{bmatrix} = -L \begin{bmatrix} \xi \\ \pi \end{bmatrix} + (I_{2m} - \varepsilon L H) \begin{bmatrix} \eta \\ \rho \end{bmatrix}, \tag{4.1}$$

where L and H are respectively the solutions of the matrix equations (2.8) and (4.3) of Chapter 2, with F_{ij} appearing instead of A_{ij}, $i, j = 1, 2$. For our purposes it is sufficient to remember from (2.13) and (4.7) of Chapter 2 that L and H can be approximated by

$$L = F_{22}^{-1} F_{21} + \varepsilon F_{22}^{-2} F_{21} F_0 + O(\varepsilon^2), \quad H = F_{12} F_{22}^{-1} + O(\varepsilon), \tag{4.2}$$

where $F_0 = F_{11} - F_{12} F_{22}^{-1} F_{21}$, as in (3.41). Using (4.1), the system (3.38), (3.39) is transformed into two separate subsystems; namely, the *slow subsystem*

$$\begin{bmatrix} \dot{\xi} \\ \dot{\pi} \end{bmatrix} = (F_{11} - F_{12} L) \begin{bmatrix} \xi \\ \pi \end{bmatrix}, \tag{4.3}$$

and the *fast subsystem*

$$\varepsilon \begin{bmatrix} \dot{\eta} \\ \dot{\rho} \end{bmatrix} = (F_{22} + \varepsilon L F_{12}) \begin{bmatrix} \eta \\ \rho \end{bmatrix}. \tag{4.4}$$

To obtain the boundary conditions for the new variables, we apply the inverse transformation to the boundary conditions (3.36), (3.37). For the slow subproblem we get

$$\begin{bmatrix} \xi \\ \pi \end{bmatrix} = (I_{2n} - \varepsilon H L) \begin{bmatrix} x \\ p \end{bmatrix} - \varepsilon H \begin{bmatrix} \eta \\ \rho \end{bmatrix} = \begin{bmatrix} x \\ p \end{bmatrix} + O(\varepsilon). \tag{4.5}$$

It is important that, allowing an error of $O(\varepsilon)$, the conditions for ξ can be taken to be equal to those for x. For the fast subproblem, we obtain from

$$\begin{bmatrix} \eta \\ \rho \end{bmatrix} = L \begin{bmatrix} x \\ p \end{bmatrix} + \begin{bmatrix} z \\ q \end{bmatrix} = F_{22}^{-1} F_{21} \begin{bmatrix} x \\ p \end{bmatrix} + \begin{bmatrix} z \\ q \end{bmatrix} + O(\varepsilon) \tag{4.6}$$

that the boundary conditions for η depend on the values of all four original variables x, z, p, q at $t = 0$ and $t = 1$. This suggests that we should first consider the slow subproblem, which, taking into account (4.2) and (4.5), can be written as

$$\begin{bmatrix} \dot{\xi} \\ \dot{\pi} \end{bmatrix} = (F_0 - \varepsilon F_{12} F_{22}^{-2} F_{21} F_0 + O(\varepsilon^2)) \begin{bmatrix} \xi \\ \pi \end{bmatrix}, \quad \begin{cases} \xi_0 = x_0 + O(\varepsilon), \\ \xi_1 = x_1 + O(\varepsilon). \end{cases} \tag{4.7}$$

The limit of this problem as $\varepsilon \to 0$ is the reduced problem (3.41), (3.42) for which in Assumption 3.2 we have assumed the existence of a unique solution. In (4.7) this reduced solution is subjected to both an $O(\varepsilon)$ perturbation of the system matrix F_0 and an $O(\varepsilon)$ perturbation of its boundary conditions. In theory such perturbations will produce only an $O(\varepsilon)$ perturbation of the solution, but in practice care must be exercised in implementing this result because F_0, being a Hamiltonian matrix, has n unstable eigenvalues. Whenever the problem (4.7) is solved by a method involving numerical integration, it must be ensured that its unstable part is integrated in reverse time. With this caveat the solution to the reduced problem will be an $O(\varepsilon)$ approximation of the exact solution $\xi(t)$, $\pi(t)$ of the slow subproblem (4.7), that is,

$$\xi(t) = x_s(t) + O(\varepsilon), \quad \pi(t) = p_s(t) + O(\varepsilon) \quad \forall t \in [0, 1]. \tag{4.8}$$

This result furnishes the data for an approximate solution of the fast subproblem, owing to the approximation of $p(t)$, $q(t)$ in (4.6) by $p_s(t)$, $q_s(t)$. At $t = 0$ and $t = 1$ this gives

$$\begin{aligned} \begin{bmatrix} \eta_0 \\ \rho_0 \end{bmatrix} &= F_{22}^{-1} F_{21} \begin{bmatrix} x_0 \\ p_s(0) \end{bmatrix} + \begin{bmatrix} z_0 \\ q_s(0) \end{bmatrix} + O(\varepsilon), \\ \begin{bmatrix} \eta_1 \\ \rho_1 \end{bmatrix} &= F_{22}^{-1} F_{21} \begin{bmatrix} x_1 \\ p_s(1) \end{bmatrix} + \begin{bmatrix} z_1 \\ q_s(1) \end{bmatrix} + O(\varepsilon), \end{aligned} \tag{4.9}$$

and the data to solve (4.4) as a pure initial value problem at either end of the interval seem to be complete. However, if we were to do so, we would be disregarding our knowledge of the boundary layer phenomena and ignoring the caveat about the unstable eigenvalues. Remembering (2.27), it is, however, expected that the unstable modes of (4.4) diverge at an exponential rate $O(e^{ct/\varepsilon})$. At this rate an $O(\varepsilon)$ error at $t = 0$ would become $O(\varepsilon e^{c/\varepsilon})$ at $t = 1$, when evaluated analytically, and would certainly be much worse numerically. Therefore one must make sure that the boundary values for the unstable part of the fast subsystem (4.4) are prescribed at the end of the interval, $t = 1$, to allow the unstable modes to be integrated in their stable direction, that is in reverse time $1 - t$ from $1 - t = 0$ to $1 - t = 1$. This can be accomplished by using the Riccati machinery of Section 6.2 to exploit the exponential dichotomy of (4.4). With the obvious change of notation, $A \to A_{22}$, $B \to B_2$, $C \to C_2$, $S \to S_{22}$, etc., the results of Section 6.2 are now applied to the fast subproblem (4.4)–(4.9).

Lemma 4.1

If (A_{22}, B_2) is a controllable pair and Assumption 3.1 holds, then

$$\lambda^i(F_{22} + \varepsilon L F_{12}) = \begin{cases} \lambda_l^i + \Delta(\varepsilon), & \lambda_l^i \overset{\Delta}{=} \lambda^i(A_{22} - S_{22}P), \quad i = 1, \ldots, m, \\ \lambda_r^j + \Delta(\varepsilon), & \lambda_r^j \overset{\Delta}{=} \lambda^j(A_{22} - S_{22}N), \quad i = m + j, \quad j = 1, \ldots, m, \end{cases} \tag{4.10}$$

where $P \geq 0$ and $N \leq 0$ are as in Lemma 2.1 and $\Delta(\varepsilon) \to 0$ as $\varepsilon \to 0$. Therefore there exist constants $c_1 > 0$ and $\varepsilon_4 > 0$ such that for all $\varepsilon \in [0, \varepsilon_4]$

$$\left.\begin{aligned} &\operatorname{Re} \lambda^i(F_{22} + \varepsilon L F_{12}) \leq -c_1 < 0, \quad i = 1, \ldots, m, \\ &\operatorname{Re} \lambda^i(F_{22} + \varepsilon L F_{12}) \geq c_1 > 0, \quad i = m + j, \quad j = 1, \ldots, m. \end{aligned}\right\} \tag{4.11}$$

This lemma is a direct consequence of Lemma 2.1 and the continuous dependence of the eigenvalues on matrix coefficients. Then, by Lemma 2.3, a dichotomy transformation for (4.4) exists and is $O(\varepsilon)$ close to (2.24); that is,

$$\eta = l + r + O(\varepsilon), \quad \rho = Pl + Nr + O(\varepsilon), \tag{4.12}$$

$$\varepsilon \dot{l} = [A_{22} - S_{22}P + O(\varepsilon)]l, \quad \varepsilon \dot{r} = [A_{22} - S_{22}N + O(\varepsilon)]r, \tag{4.13}$$

and constants $c_1 > 0$ and $k_1 > 0$ exist such that for all $t \in [0, 1]$ and $\varepsilon \in (0, \varepsilon_4]$

$$\|l(t)\| \leq k_1 \, e^{-c_1(t/\varepsilon)} \|l(0)\|, \quad \|r(t)\| \leq k_1 \, e^{-c_1(1-t)/\varepsilon} \|r(1)\|. \tag{4.14}$$

A conclusion analogous to that of Theorem 2.1 is that the solution to the fast subproblem (4.4), (4.9) is approximated by the sum of the boundary layers

$$\eta_l = e^{(A_{22} - S_{22}P)t/\varepsilon}\,\eta_0, \quad \eta_r = e^{-(A_{22} - S_{22}N)(1-t)/\varepsilon}\,\eta_1, \tag{4.15}$$

$$\eta = \eta_l + \eta_r + O(\varepsilon), \quad \rho = P\eta_l + N\eta_r + O(\varepsilon), \tag{4.16}$$

where $\eta_l(\tau)$ and $\eta_r(\sigma)$ can be interpreted as the solutions of the boundary layer regulator problems (2.39)–(2.41) and (2.42)–(2.44) respectively.

So far our discussion in this section has been restricted to the case when all the coefficient matrices are constant. The only new step to be made in order to state the analogous results for the time-varying problem is to treat (4.1) as a time-varying transformation, allowing for the fact that m of the $2m$ eigenvalues of $F_{22}(t)$ have positive real parts for each $t \in [0, 1]$. This problem is addressed in Chapter 5, where Theorem 3.1 and its Corollary 3.1 establish the existence of matrices $L(t, \varepsilon)$ and $H(t, \varepsilon)$ with the same properties as in (4.2) for each $t \in [0, 1]$ and all $\varepsilon \in (0, \varepsilon_5]$, where $\varepsilon_5 > 0$ is a constant. The results of the two preceding sections in this chapter are also valid for the time-varying case. In this way, we obtain the following approximation result.

Theorem 4.1

Consider the optimal control problem (3.26)–(3.29) in which the time derivatives of all the matrices are continuous and bounded and $R(t) > 0$ for all $t \in [0, 1]$. If Assumptions 3.1 and 3.2 hold and if

$$\operatorname{rank}\,[B_2(t), A_{22}(t)B_2(t), \ldots, A_{22}^{m-1}(t)B_2(t)] = m \quad \forall t \in [0, 1] \tag{4.17}$$

then there exists an $\varepsilon_5 > 0$ such that for all $\varepsilon \in (0, \varepsilon_5]$ and all $t \in [0, 1]$ the optimal trajectory $x(t, \varepsilon)$, $z(t, \varepsilon)$ and the corresponding optimal control $u(t, \varepsilon)$ satisfy

$$x(t, \varepsilon) = x_s(t) + O(\varepsilon), \tag{4.18}$$

$$z(t, \varepsilon) = z_s(t) + \eta_l\left(\frac{t}{\varepsilon}\right) + \eta_r\left(\frac{1-t}{\varepsilon}\right) + O(\varepsilon), \tag{4.19}$$

$$u(t, \varepsilon) = u_s(t) + \nu_l\left(\frac{t}{\varepsilon}\right) + \nu_r\left(\frac{1-t}{\varepsilon}\right) + O(\varepsilon), \tag{4.20}$$

where $x_s(t)$, $z_s(t)$, $u_s(t)$ is the optimal solution of the reduced problem (3.43)–(3.50), while $\nu_l(\tau)$ and $\nu_r(\sigma)$, solve the boundary layer regulator problems (2.39)–(2.41) and (2.42)–(2.44) respectively, for η_0 and η_1 given by (4.9), and with A, B, C replaced by A_{22}, B_2, C_2.

Proof The result (4.18) follows from (4.5) and (4.7), (4.8). To prove (4.19), we note from (3.39) with $\varepsilon = 0$ that the reduced solution satisfies

$$0 = F_{21}\begin{bmatrix} x_s \\ p_s \end{bmatrix} + F_{22}\begin{bmatrix} z_s \\ q_s \end{bmatrix} \quad \forall t \in [0, 1]. \tag{4.21}$$

Subtracting this zero quantity from (4.6) and taking (4.8) into account, that is, the fact that $x - x_s$ and $p - p_s$ are $O(\varepsilon)$, we obtain

$$\begin{bmatrix} \eta \\ \rho \end{bmatrix} = \begin{bmatrix} z(t, \varepsilon) - z_s(t) \\ q(t, \varepsilon) - q_s(t) \end{bmatrix} + O(\varepsilon) \quad \forall t \in [0, 1], \tag{4.22}$$

which, in view of our previous discussion on boundary layers, proves (4.19) and (4.20). □

This theorem is extremely simple to apply. First, we solve the reduced problem (3.43)–(3.50). Next, we determine the initial conditions $\eta_l(0)$ and $\eta_r(0)$ from (4.19) at $t = 0$ and $t = 1$ as

$$\eta_l(0) = z_0 - z_s(0), \quad \eta_r(0) = z_1 - z_s(1). \tag{4.23}$$

Their meaning as the correctors of the discrepancies in the boundary conditions for z caused by the reduced solution $z_s(t)$ is now clear. Using these initial conditions, the *optimal stabilizing solutions* of the layer regulator problems (2.39)–(2.41) and (2.42)–(2.44) are evaluated with, of course, A, B, C replaced by A_{22}, B_2, C_2.

Example 4.1

Continuing the discussion of Example 3.1, the solution of the reduced problem (3.60) is

$$x_s(t) = \mathrm{e}^{-t}[x_0 + g(\mathrm{e}^{2t} - 1)], \quad g = \frac{\mathrm{e}x_1 - x_0}{\mathrm{e}^2 - 1}, \tag{4.24}$$

$$u_s(t) = -p_s(t) = 2g\mathrm{e}^t, \quad z_s(t) = -x_s(t) + u_s(t), \tag{4.25}$$

and hence the corrective initial conditions are

$$\eta_0 = \eta_l(0) = z_0 + x_0 - 2g, \quad \eta_1 = \eta_r(0) = z_1 + x_1 - 2g\mathrm{e}. \tag{4.26}$$

In this case $A_{22} = -1$, $B_2 = 1$, $C_2 = 0$, $R = 1$, $S_{22} = 1$. The Riccati equation (2.14) and its solutions $P \geq 0$ and $N \leq 0$ are

$$2K + K^2 = 0, \quad P = 0, \quad N = -2, \quad A_{22} - S_{22}P = -1, \quad A_{22} - S_{22}N = 1, \tag{4.27}$$

and hence the stabilizing solutions of the boundary layer regulators are

$$\eta_l = e^{-t/\varepsilon}(z_0 + x_0 - 2g), \quad \eta_r = e^{-(1-t)/\varepsilon}(z_1 + x_1 - 2ge), \tag{4.28}$$

$$\nu_l = -R^{-1}B_2P\eta_l = 0, \qquad \nu_r = -R^{-1}B_2N\eta_r = 2\eta_r. \tag{4.29}$$

The approximation to the solution of the optimal control problem for the DC motor (3.56), (3.57) can now be composed as in (4.18)–(4.20). Only the terminal layer is controlled by the control ν_r, whose task is to regulate

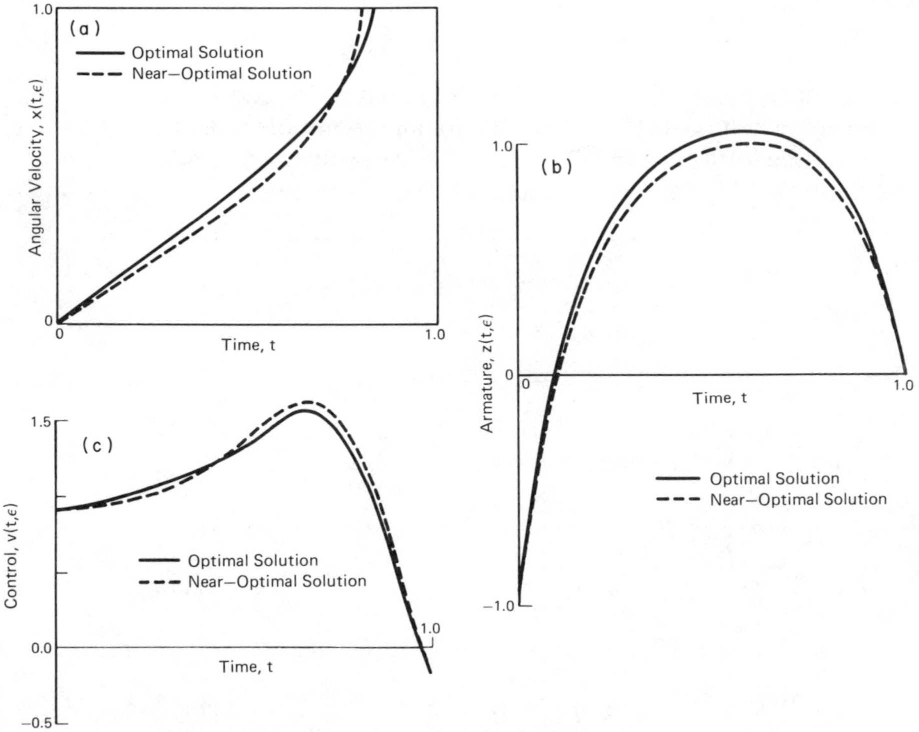

Fig. 6.2. (a), (b), (c) Near-optimal control of a DC motor.

the current from $z_s(1)$ to zero, $z_2 = 0$. In practice, this would prepare the unloaded DC motor for a constant speed operation at $x(t) = x_1$, after $t = 1$. Exact and approximate solutions are plotted in Fig. 6.2. The reader should note that the same problem was solved using the controllability Grammian in Chapter 5, Example 7.1.

6.5 Nonlinear and Constrained Control

The linear results of the preceding sections are readily extended to the nonlinear problem of the form

$$\dot{x} = a_1(x) + A_1(x)z + B_1(x)u, \quad x(0) = x_0, \quad x(1) = x_1, \tag{5.1}$$

$$\varepsilon\dot{z} = a_2(x) + A_2(x)z + B_2(x)u, \quad z(0) = z_0, \quad z(1) = z_1, \tag{5.2}$$

$$J = \int_0^1 [V_1(x) + z^{\mathrm{T}}V_2(x)z + u^{\mathrm{T}}R(x)u]\,\mathrm{d}t;$$

$$V_1(x), \quad V_2(x), \quad R(x) > 0 \quad \forall x \in R^n, \tag{5.3}$$

where all the functions of x possess continuous and bounded derivatives with respect to x and $A_2(x)$ is nonsingular for all x. Some of these assumptions, as well as the form of the problem, are unnecessarily restrictive, and are made only to simplify the presentation. The Hamiltonian

$$\begin{aligned}\mathcal{H} = V_1 + z^{\mathrm{T}}V_2z + u^{\mathrm{T}}Ru + p^{\mathrm{T}}(a_1 + A_1z + B_1u)\\ + q^{\mathrm{T}}(a_2 + A_2z + B_2u),\end{aligned} \tag{5.4}$$

is minimized by the control

$$u = -\tfrac{1}{2}R^{-1}(B_1^{\mathrm{T}}p + B_2^{\mathrm{T}}q), \tag{5.5}$$

and when this control is substituted into (5.1), (5.2) and (3.24), (3.25), the resulting nonlinear boundary value problem is

$$\dot{x} = a_1 + A_1z - \tfrac{1}{2}B_1R^{-1}(B_1^{\mathrm{T}}p + B_2^{\mathrm{T}}q) \stackrel{\triangle}{=} g_1(x, p, z, q), \quad x(0) = x_0, \tag{5.6}$$

$$\dot{p} = -\nabla_x\mathcal{H} \stackrel{\triangle}{=} g_2(x, p, z, q), \qquad x(1) = x_1, \tag{5.7}$$

$$\varepsilon\dot{z} = a_2 + A_2z - \tfrac{1}{2}B_2R^{-1}(B_1^{\mathrm{T}}p + B_2^{\mathrm{T}}q) \stackrel{\triangle}{=} g_3(x, p, z, q), \quad z(0) = z_0, \tag{5.8}$$

$$\varepsilon\dot{q} = -2V_2z - A_1^{\mathrm{T}}p - A_2^{\mathrm{T}}q \stackrel{\triangle}{=} g_4(x, p, z, q), \qquad z(1) = z_1. \tag{5.9}$$

The reduced problem obtained by setting $\varepsilon = 0$ in (5.2) and substituting

$$z_s = -A_2^{-1}(a_2 + B_2u_s) \tag{5.10}$$

into (5.1) and (5.3) is

$$\dot{x}_s = a_0(x_s) + B_0(x_s)u_s, \quad x_s(0) = x_0, \quad x_s(1) = x_1, \tag{5.11}$$

$$J_s = \int_0^1 [Q_0(x_s) + 2D_0^{\mathrm{T}}(x_s)u_s + u_s^{\mathrm{T}}R_0(x_s)u_s]\,\mathrm{d}t, \tag{5.12}$$

where

$$a_0 = a_1 - A_1A_2^{-1}a_2, \quad B_0 = B_1 - A_1A_2^{-1}B_2, \tag{5.13}$$

$$\begin{aligned} Q_0 &= V_1 + a_2^{\mathrm{T}}A_2^{\mathrm{T}-1}V_2A_2^{-1}a_2, \quad D_0 = B_2^{\mathrm{T}}A_2^{\mathrm{T}-1}V_2A_2^{-1}a_2, \\ R_0 &= R + B_2^{\mathrm{T}}A_2^{\mathrm{T}-1}V_2A_2^{-1}B_2. \end{aligned} \tag{5.14}$$

The Hamiltonian

$$\mathcal{H}_{\mathrm{s}} = Q_0 + 2D_0^{\mathrm{T}}u_{\mathrm{s}} + u_{\mathrm{s}}^{\mathrm{T}}R_0u_{\mathrm{s}} + p_{\mathrm{s}}^{\mathrm{T}}(a_0 + B_0u_{\mathrm{s}}) \tag{5.15}$$

is minimized by

$$u_{\mathrm{s}} = -\tfrac{1}{2}R_0^{-1}(2D_0 + B_0^{\mathrm{T}}p_{\mathrm{s}}), \tag{5.16}$$

and the corresponding reduced boundary value problem is

$$\begin{aligned} \dot{x}_{\mathrm{s}} &= a_0 - B_0R_0^{-1}D_0 - \tfrac{1}{2}B_0R_0^{-1}B_0^{\mathrm{T}}p_{\mathrm{s}} \\ &\overset{\Delta}{=} f_1(x_{\mathrm{s}}, p_{\mathrm{s}}), \quad x_{\mathrm{s}}(0) = x_0, \quad x_{\mathrm{s}}(1) = x_1, \end{aligned} \tag{5.17}$$

$$\dot{p}_s = -\nabla x_{\mathrm{s}}\mathcal{H}_{\mathrm{s}} \overset{\Delta}{=} f_2(x_{\mathrm{s}}, p_{\mathrm{s}}). \tag{5.18}$$

It is assumed that this problem has a unique solution. Anticipating that $x_{\mathrm{s}}(t)$ will be $O(\varepsilon)$ close to the optimal solution $x(t, \varepsilon)$, we introduce the quantities

$$\begin{aligned} \varepsilon\delta a_2 &= a_2(x) - a_2(x_{\mathrm{s}}), \\ \varepsilon\delta A_2 &= A_2(x) - A_2(x_{\mathrm{s}}), \\ \varepsilon\delta B_2 &= B_2(x) - B_2(x_{\mathrm{s}}) \end{aligned} \tag{5.19}$$

and substitute them into (5.2), taking into account that z_{s} is defined by (5.10). In this way, one obtains

$$\varepsilon(\dot{z} - \dot{z}_{\mathrm{s}}) = A_2(x_{\mathrm{s}})(z - z_{\mathrm{s}}) + B_2(x_{\mathrm{s}})(u - u_{\mathrm{s}}) + \varepsilon\Delta, \tag{5.20}$$

where Δ incorporates the terms δa_2, $\delta A_2 z$, $\delta B_2 u$ and $\dot{z}_{\mathrm{s}}$. By analogy with the linear problem, we treat

$$\eta = z - z_{\mathrm{s}} + O(\varepsilon), \quad \nu = u - u_{\mathrm{s}} + O(\varepsilon) \tag{5.21}$$

as the variables for the boundary layer regulator problems, at $t = 0$, in forward time τ,

$$\left.\begin{aligned} \frac{\mathrm{d}\eta_l}{\mathrm{d}\tau} &= A_2(x_0)\eta_l + B_2(x_0)\nu_l, \quad \eta_l(0) = z_0 - z_{\mathrm{s}}(0), \\ J_l &= \int_0^\infty [\eta_l^{\mathrm{T}}V_2(x_0)\eta_l + \nu_l^{\mathrm{T}}R(x_0)\nu_l]\,\mathrm{d}\tau, \end{aligned}\right\} \tag{5.22}$$

and at $t = 1$, in reverse time σ,

$$\frac{\mathrm{d}\eta_r}{\mathrm{d}\sigma} = -A_2(x_1)\eta_r - B_2(x_1)\nu_r, \quad \eta_r(0) = z_1 - z_s(1), \tag{5.23}$$

$$J_r = \int_0^\infty [\eta_r^{\mathrm{T}} V_2(x_1)\eta_r + \nu_r^{\mathrm{T}} R(x_1)\nu_r]\,\mathrm{d}\sigma. \tag{5.24}$$

Then, under conditions analogous to those in Theorem 4.1, the same approximation (4.18)–(4.20) holds for the nonlinear problem (5.1)–(5.3). The proof of this result makes use of the fact that the linearization of (5.6)–(5.9), along the approximate solution, has the form of the linear necessary conditions in (3.34), (3.35). For details of the proof and higher-order asymptotic expansions the reader is referred to Chow (1979).

Example 5.1

Proceeding with the chemical process problem of Example 3.2, it is observed that (3.62), (3.63) is in the form (5.1)–(5.3). The reduced problem (3.67), although nonlinear, can be solved analytically:

$$\left.\begin{aligned} x_s(t) &= [2(t+1)]^{-1/2}, \quad & u_s(t) &= -x_s^2(t), \\ p_s(t) &= 2x_s(t), & z_s(t) &= u_s(t) = -x_s^2(t), \end{aligned}\right\} \tag{5.25}$$

and the initial conditions for the boundary layer problems are

$$\begin{aligned} \eta_0 &= \eta_l(0) = z_0 - z_s(0) = 0 - (-\tfrac{1}{2}) = \tfrac{1}{2}, \\ \eta_1 &= \eta_r(0) = z_1 - z_s(1) = 0 - (-\tfrac{1}{4}) = \tfrac{1}{4}. \end{aligned} \tag{5.26}$$

The Riccati equation and its solutions $P \geq 0$ and $N \leq 0$ are

$$0 = 2K + K^2 - 1, \quad P = -1 + \sqrt{2}, \quad N = -1 - \sqrt{2}. \tag{5.27}$$

Hence the boundary layer regulators are

$$\frac{\mathrm{d}\eta_l}{\mathrm{d}\tau} = -\sqrt{2}\eta_l, \quad \nu_l = -P\eta_l; \quad \frac{\mathrm{d}\eta_r}{\mathrm{d}\sigma} = -\sqrt{2}\eta_r, \quad \nu_r = -N\eta_r. \tag{5.28}$$

When the solutions for (5.28) are combined with the reduced solution (5.25), the near-optimal solution (4.18)–(4.20) for this problem is

$$x(t, \varepsilon) = [2(t+1)]^{-1/2} + O(\varepsilon), \tag{5.29}$$

$$z(t, \varepsilon) = -[2(t+1)]^{-1} + \tfrac{1}{2}\mathrm{e}^{-\sqrt{2}t/\varepsilon} + \tfrac{1}{4}\mathrm{e}^{\sqrt{2}(t-1)/\varepsilon} + O(\varepsilon), \tag{5.30}$$

$$u(t, \varepsilon) = -[2(t+1)]^{-1} - \tfrac{1}{2}(\sqrt{2} - 1)\mathrm{e}^{-\sqrt{2}t/\varepsilon} + \tfrac{1}{4}(\sqrt{2} + 1)\mathrm{e}^{\sqrt{2}(t-1)/\varepsilon} + O(\varepsilon). \tag{5.31}$$

These approximations are compared with the optimal solutions obtained for $\varepsilon = 0.1$ numerically in Fig. 6.3. A satisfactory agreement is achieved in particular for $x(t, \varepsilon)$, while the outer approximation of $z(t, \varepsilon)$ can be further improved by a first-order correction of the slow part of the solution.

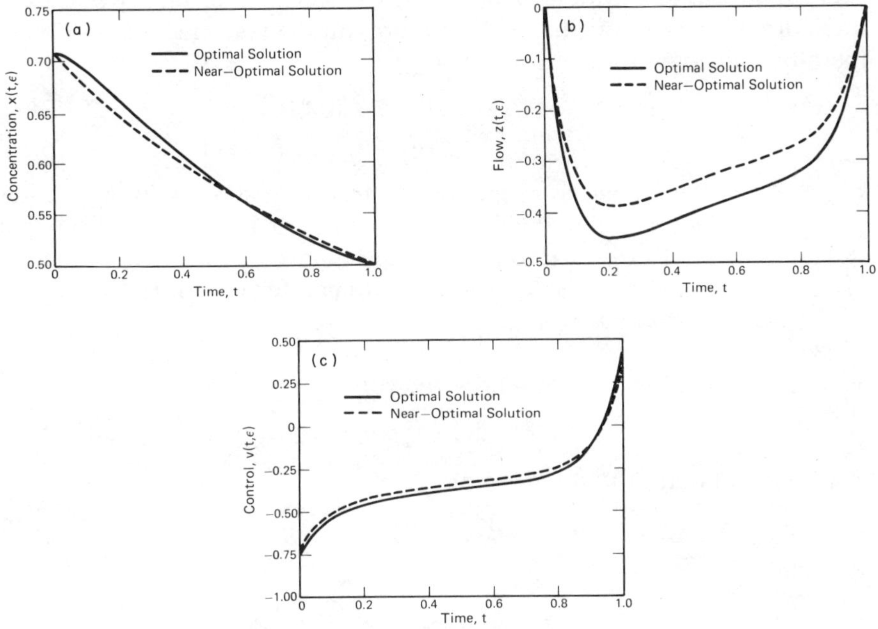

Fig. 6.3. (a), (b), (c) Near-optimal control of nonlinear chemical process example.

All the optimal control problems considered thus far have been without control constraints. The additional difficulty with constrained problems is that both the slow (reduced) control $u_s(t)$ and the fast controls $\nu_l(\tau)$ and $\nu_r(\sigma)$ share the same constraints. Nevertheless, the separation of time scales and the insight gained by the analysis of the boundary layer phenomena is helpful in constructing asymptotic approximations similar to those of Theorem 4.1. As an illustration, consider the time-optimal problem for the linear time-invariant system

$$\dot{x} = A_{11}x + A_{12}z + b_1 u, \quad x(0) = x_0, \quad x(T) = 0, \tag{5.32}$$

$$\varepsilon\dot{z} = A_{21}x + A_{22}z + b_2 u, \quad z(0) = z_0, \quad z(T) = 0, \tag{5.33}$$

in which the scalar control is constrained by

$$|u| \leq 1 \quad \forall t \in [0, T], \tag{5.34}$$

and is to steer $x(t)$, $z(t)$ from x_0, z_0 at $t = 0$ to zero in minimum time T. The results for this special case are readily extended to the time-optimal control problem for the nonlinear system (5.1), (5.2) with a constrained control vector u.

The solution of the time-optimal control problem (5.32)–(5.34) is an exercise in the application of the results of Chapter 2. First we use the transformation (4.8) of Chapter 2 to represent the system (5.32)–(5.34) in the block-diagonal form

$$\dot{\xi} = A_s \xi + b_s u, \quad \xi(0) = \xi_0, \quad \xi(T) = 0, \tag{5.35}$$

$$\varepsilon \dot{\eta} = A_f \eta + b_f u, \quad \eta(0) = \eta_0, \quad \eta(T) = 0, \tag{5.36}$$

where ξ_0, η_0 are known from the inverse transformation (4.9) of Chapter 2, and u and its constraint (5.34) are unaltered. Obviously, a control u that steers ξ, η from ξ_0, η_0 to zero in minimum time T will steer x_0, z_0 to zero in the same minimum time T. The Hamiltonian for (5.35), (5.36)

$$\mathcal{H} = 1 + p^{\mathrm{T}}(A_s \xi + b_s u) + q^{\mathrm{T}}(A_f \eta + b_f u) \tag{5.37}$$

is minimized, subject to (5.34), by the control

$$u = -\operatorname{sign}(b_s^{\mathrm{T}} p + b_f^{\mathrm{T}} q), \tag{5.38}$$

where p and q are the solutions of

$$\dot{p} = -A_s^{\mathrm{T}} p, \quad \varepsilon \dot{q} = -A_f^{\mathrm{T}} q. \tag{5.39}$$

Remembering that

$$\left.\begin{aligned} A_s &= A_0 - \varepsilon A_{12} A_{22}^{-2} A_{21} A_0 + O(\varepsilon^2), \\ A_f &= A_{22} + \varepsilon A_{22}^{-1} A_{21} A_{12} + O(\varepsilon^2), \\ b_s &= b_0 - \varepsilon A_{12} A_{22}^{-2} A_{21} b_1 + O(\varepsilon^2), \\ b_f &= b_2 + \varepsilon A_{22}^{-1} A_{21} b_1 + O(\varepsilon^2), \end{aligned}\right\} \tag{5.40}$$

let us assume that A_{22} is a Hurwitz matrix, then so is A_f, for ε sufficiently small; that is,

$$\operatorname{Re} \lambda(A_{22}) < 0 \Rightarrow \operatorname{Re} \lambda(A_f) < 0 \quad \forall \varepsilon \in [0, \varepsilon_6]. \tag{5.41}$$

Substituting into (5.38) the solution of (5.39) for some end condition, yet to be found,

$$p(T) = p_1, \quad q(T) = q_1, \tag{5.42}$$

we arrive at the conclusion that the switching law (5.38) has a slow-plus-fast behavior, namely,

$$u = -\operatorname{sign}(b_s^{\mathrm{T}} \mathrm{e}^{-A_s^{\mathrm{T}}(t-T)} p_1 + b_f^{\mathrm{T}} \mathrm{e}^{-A_f^{\mathrm{T}}(t-T)/\varepsilon} q_1). \tag{5.43}$$

The switches due to the second term in parentheses occur only near $t = T$ because of (5.41). They are of a high frequency, while the switches due to the first term are of a low frequency and occur throughout the interval $[0, T]$. If we approximate A_s and b_s by $A_s \approx A_0$ and $b_s \approx b_0$, we will discover that the first term is the solution of the time-optimal control problem for the reduced system. Suppose the minimal time for the reduced problem is T_0, and hence $x_s(T_0) = 0$. However, $z(T_0) \approx z_s(T_0)$ is in general different from zero, and the task of the high-frequency switches near the end of the interval is to steer z to zero. Again we see that, because A_{22} is Hurwitz, only the terminal layer is controlled. One way to use (5.43) is to separately calculate the slow and the fast switching instants. Let us illustrate this by an example.

Example 5.2

Suppose that a double-integrator plant is controlled through an actuator with two small time constants ε and 0.5ε (see Fig. 6.4). In this case, $\dot{x}_1 = x_2$, $\dot{x}_2 = z_1$, $\varepsilon\dot{z}_1 = -2z_1 + 2z_2$, $\varepsilon\dot{z}_2 = -z_2 + u$, and the reduced system is $\dot{x}_{1s} = x_{2s}$, $\dot{x}_{2s} = u_s$, that is,

$$A_0 = \begin{bmatrix} 0 & 1 \\ 0 & 0 \end{bmatrix}, \quad b_0 = \begin{bmatrix} 0 \\ 1 \end{bmatrix}, \tag{5.44}$$

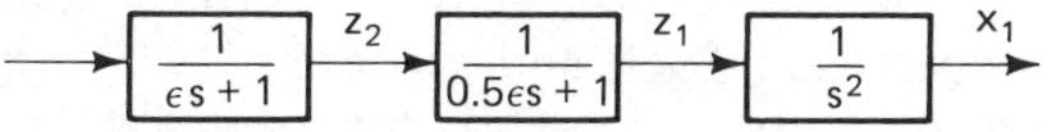

Fig. 6.4. Time-optimal control of a double-integrator plant with two parasitic time constants.

The reduced solution is well known,

$$u_s(t) = -\text{sign}\,(2x_{10} + x_{20}|x_{20}|), \tag{5.45}$$

and the one switch time is $t_s = -u_s x_{20} + [0.5x_{20}^2 - u_s x_{10}]^{1/2}$, while the reduced minimum final time is $T_s = -u_s x_{20} + 2[0.5x_{20}^2 - u_s x_{10}]^{1/2}$. At $t = T_s$ the fast variables are not zero:

$$z_1(T_s) \approx -u_s(T_s), \quad z_2(T_s) \approx -u_s(T_s). \tag{5.46}$$

The terminal layer control v_r is to steer $z(T_s)$ to zero in minimum time. This is also a textbook example, for which the solution is

$$v_r(\sigma) = \begin{cases} u_s & (0 \leq \sigma = (t - T_s)/\varepsilon < \ln 3), \\ -u_s & (\ln 3 \leq \sigma \leq \ln 4). \end{cases} \tag{5.47}$$

The concatenation of $u_s(t)$ and $v_r((t - T_s)/\varepsilon)$ will steer x, z to an $O(\varepsilon)$ neighborhood of the origin. This result is obtained with $A_s \approx A_0$, $b_s \approx b_0$ and

$$A_f \sim A_{22} = \begin{bmatrix} -2 & 2 \\ 0 & -1 \end{bmatrix}, \quad b_f \sim b_2 = \begin{bmatrix} 0 \\ 1 \end{bmatrix}. \tag{5.48}$$

A further improvement is possible by a closer approximation of A_s, b_s, A_f and b_f.

6.6 Cheap Control and Singular Arcs

An unconstrained optimal control that satisfies the necessary optimality condition

$$\nabla_u \mathcal{H} = 0, \quad \frac{\partial^2 \mathcal{H}}{\partial u^2} \geq 0 \tag{6.1}$$

but does not satisfy the strengthened Legendre–Clebsch condition $\partial^2\mathcal{H}/\partial u_0^2$, that is, such that

$$\frac{\partial^2 \mathcal{H}}{\partial u^2} \quad \text{is singular,} \tag{6.2}$$

is referred to as a *singular control* and the corresponding part of the trajectory as a *singular arc* (Bell and Jacobson, 1975; Bryson and Ho, 1975). A well known necessary condition for optimality in this case is

$$(-1)^k \frac{\partial}{\partial u}\left[\frac{\mathrm{d}^{2k}}{\mathrm{d}t^{2k}} \nabla_u \mathcal{H}\right] \geq 0, \quad k = 0, 1, 2, \ldots. \tag{6.3}$$

A direct use of this condition in optimal control computations is complicated by the need to connect a singular arc with boundary points which may not be on the arc.

Some singular control problems can be treated as the limits of simpler "cheap control" problems, such as the following class of problems analyzed by O'Malley and Jameson (1975):

$$\dot{x} = A_{11}x + A_{12}z, \qquad x(0) = x_0, x(T) = x_1, \tag{6.4}$$

$$\dot{z} = A_{21}x + A_{22}z + u, \quad z(0) = z_0, z(T) = z_1, \tag{6.5}$$

$$J = \frac{1}{2}\int_0^T \left(\begin{bmatrix} x \\ z \end{bmatrix}^{\mathrm{T}} Q \begin{bmatrix} x \\ z \end{bmatrix} + \varepsilon^2 u^{\mathrm{T}} R u\right) \mathrm{d}t. \tag{6.6}$$

The term "cheap" indicates that ε is small, and hence the cost of control is low. There is no loss of generality in considering (6.4), (6.5), because every linear system with a full-rank control input matrix B can be represented in this form after a similarity transformation. Although the system (6.4), (6.5) is not singularly perturbed, our (x, z) notation anticipates that a large u will force the z-variable to act as a fast variable. The Hamiltonian function for this problem is

$$\mathcal{H} = \frac{1}{2}\begin{bmatrix} x \\ z \end{bmatrix}^{\mathrm{T}} Q \begin{bmatrix} x \\ z \end{bmatrix} + \tfrac{1}{2}\varepsilon^2 u^{\mathrm{T}} R u + \begin{bmatrix} p \\ \tilde{q} \end{bmatrix}^{\mathrm{T}} \left(A \begin{bmatrix} x \\ z \end{bmatrix} + \begin{bmatrix} 0 \\ I \end{bmatrix} u \right), \tag{6.7}$$

and it is obvious that for $\varepsilon = 0$ the problem is singular, since then $\mathcal{H}$ is linear in u and (6.2) holds. For $\varepsilon > 0$ and $R > 0$, the control minimizing it is

$$u = -\frac{1}{\varepsilon^2} R^{-1} \tilde{q}. \tag{6.8}$$

Since its substitution into (6.5) would lead to a multiplication of $\dot{z}$ by ε^2, we rescale the adjoint variable $\tilde{q}$ as $q = \tilde{q}/\varepsilon$ and write the adjoint equations in the form

$$\dot{p} = -\nabla_x \mathcal{H}, \quad \varepsilon \dot{q} = -\nabla_z \mathcal{H}. \tag{6.9}$$

Combining (6.4), (6.5), (6.8) and (6.9), we obtain the singularly perturbed boundary value problem

$$\left.\begin{aligned} \begin{bmatrix} \dot{x} \\ \dot{p} \end{bmatrix} &= \begin{bmatrix} A_{11} & 0 \\ -Q_{11} & -A_{11}^{\mathrm{T}} \end{bmatrix} \begin{bmatrix} x \\ p \end{bmatrix} + \begin{bmatrix} A_{12} & 0 \\ -Q_{12} & -\varepsilon A_{21}^{\mathrm{T}} \end{bmatrix} \begin{bmatrix} z \\ q \end{bmatrix}, \\ x(0) &= x_0, \quad x(T) = x_1, \end{aligned}\right\} \tag{6.10}$$

$$\left.\begin{aligned} \varepsilon \begin{bmatrix} \dot{z} \\ \dot{q} \end{bmatrix} &= \begin{bmatrix} \varepsilon A_{21} & 0 \\ -Q_{12}^{\mathrm{T}} & -A_{12}^{\mathrm{T}} \end{bmatrix} \begin{bmatrix} x \\ p \end{bmatrix} + \begin{bmatrix} \varepsilon A_{22} & -R^{-1} \\ -Q_{22} & -\varepsilon A_{22}^{\mathrm{T}} \end{bmatrix} \begin{bmatrix} z \\ q \end{bmatrix}, \\ z(0) &= z_0, \quad z(T) = z_1, \end{aligned}\right\} \tag{6.11}$$

which is a special case of the problem (3.34)–(3.37). The analysis of Sections 6.3 and 6.4 is applicable if $Q_{22} > 0$. Under this assumption we set $\varepsilon = 0$, solve (6.11) for

$$z_s = -Q_{22}^{-1}(Q_{12}^{\mathrm{T}} x_s + A_{12}^{\mathrm{T}} p_s) \tag{6.12}$$

$$q_s = 0, \tag{6.13}$$

and, substituting into (6.10), obtain the reduced slow system

$$\begin{bmatrix} \dot{x}_s \\ \dot{p}_s \end{bmatrix} = \begin{bmatrix} A_{11} - A_{12}Q_{22}^{-1}Q_{12}^{T} & -A_{12}Q_{22}^{-1}A_{12}^{T} \\ -Q_{11} + Q_{12}Q_{22}^{-1}Q_{12}^{T} & -A_{11}^{T} + Q_{12}Q_{22}^{-1}A_{12}^{T} \end{bmatrix} \begin{bmatrix} x_s \\ p_s \end{bmatrix}, \tag{6.14}$$

with the boundary conditions

$$x_s(0) = x_0, \quad x_s(T) = x_1. \tag{6.15}$$

The expressions (6.12)–(6.15) define the same singular arc $x_s(t)$, $z_s(t)$ that would have been obtained using the condition (6.3), if such an arc exists. Conditions for its existence can be established by an analysis of (6.14), (6.15) as a linear boundary value problem. To prove that this problem is the limit, as $\varepsilon \to 0$, of the necessary condition for optimality (6.10), (6.11), we need to show what happens with the fast transients which occur near $t = 0$ and $t = T$ because, in general,

$$z_s(0) \neq z_0, \quad z_s(t) \neq z_1. \tag{6.16}$$

The analysis in Section 6.4 shows that these transients are governed by the boundary layer regulators (4.13). In the case of (6.11), the Hamiltonian matrix of these regulators is

$$\begin{bmatrix} 0 & -R^{-1} \\ -Q_{22} & 0 \end{bmatrix}, \tag{6.17}$$

and the corresponding Riccati equation is

$$KR^{-1}K - Q_{22} = 0. \tag{6.18}$$

Its positive- and negative-definite solutions P and N are respectively

$$P = R^{1/2}(R^{-1/2}Q_{22}R^{-1/2})^{1/2}R^{1/2}, \quad N = -P. \tag{6.19}$$

The boundary-layer regulators (4.13) are therefore

$$\varepsilon\dot{l} = -R^{-1}Pl, \qquad l(0) = z_0 - z_s(0), \tag{6.20}$$

$$\varepsilon\dot{r} = -R^{-1}Nr, \quad r(T) = z_1 - z_s(T), \tag{6.21}$$

Thus, by Theorem 4.1, a near-optimal solution to the "cheap control" problem (6.4)–(6.6) is

$$x(t, \varepsilon) = x_s(t) + O(\varepsilon) \tag{6.22}$$

$$z(t, \varepsilon) = z_s(t) + e^{-R^{-1}Pt/\varepsilon}(z_0 - z_s(0)) + e^{R^{-1}P(t-T)/\varepsilon}(z_1 - z_s(T)) + O(\varepsilon). \tag{6.23}$$

The singular arc $x_s(t)$, $z_s(t)$ and the fast transients away from it are clearly exhibited.

Example 6.1

In order to compare the singular perturbation approach to singular control problems with a more common use of the necessary condition (6.3), we consider the problem

$$\dot{y}_1 = y_2 + u, \quad y_1(0) = a, \quad y_1(T) = 0, \tag{6.24}$$

$$\dot{y}_2 = -u, \quad y_2(0) = b, \quad y_2(T) = 0, \tag{6.25}$$

$$J = \frac{1}{2}\int_0^T y_1^2 \, dt; \quad a, b \text{ and } T \text{ given.} \tag{6.26}$$

Using (6.3), it is shown by Bryson and Ho (1975) that a singular arc for this problem lies on the curve

$$\tfrac{1}{2}y_1^2 + y_1 y_2 = \text{const} \tag{6.27}$$

and that impulsive transients occur at both ends of the arc. However, an analysis based on (6.3) does not reveal the nature of these transients.

To apply the methodology of this section we first use $x = y_1 + y_2$, $z = y_2$ to transform (6.24), (6.25) into

$$\dot{x} = -z, \quad x(0) = a + b, \quad x(T) = 0, \tag{6.28}$$

$$\dot{z} = u, \quad z(0) = -b, \quad z(T) = 0, \tag{6.29}$$

We also add a "cheap control" term to (6.26) as

$$J = \frac{1}{2}\int_0^T [(x + z)^2 + \varepsilon^2 u^2] \, dt. \tag{6.30}$$

The boundary value problem (6.10), (6.11) is in this case

$$\dot{x} = -z, \quad x(0) = a + b, \tag{6.31}$$

$$\dot{p} = -x - z, \quad x(T) = 0, \tag{6.32}$$

$$\varepsilon\dot{z} = -q, \quad z(0) = -b, \tag{6.34}$$

$$\varepsilon\dot{q} = -x + p - z, \quad z(T) = 0. \tag{6.35}$$

Since $Q_{22} = 1 > 0$, the reduced slow system (6.14) is obtained by setting $\varepsilon = 0$ and solving (6.34), (6.35) for

$$z_s = -x_s + p_s, \quad q_s = 0, \tag{6.36}$$

which, substituted into (6.31), (6.32), give

$$\dot{x}_s = x_s - p_s, \quad x_s(0) = a + b, \tag{6.37}$$

$$\dot{p}_s = -p_s, \quad x_s(T) = 0. \tag{6.38}$$

The singular arc $x_s(t)$, $z_s(t)$ is thus obtained by solving (6.37), (6.38) for $x_s(t)$, $p_s(t)$ and then evaluating $z_s(t)$ from (6.36). To compare this result with (6.27), we note from (6.36)–(6.38) that $\dot{z}_s = -\dot{x}_s + \dot{p}_s = x_s$ and $\dot{x}_s = -z_s$. Therefore

$$\tfrac{1}{2}x_s^2 - \tfrac{1}{2}z_s^2 = \text{const}, \tag{6.39}$$

which, in view of $x = y_1 + y_2$, $z = -y_2$, is the same as (6.27).

In order to complete the approximation (6.22), (6.23), we calculate the boundary layers. Using the Riccati solution

$$K^2 - 1 = 0, \quad P = 1, \quad N = -1, \tag{6.40}$$

we obtain

$$z(t, \varepsilon) = z_s(t) + e^{-t/\varepsilon}(b - z_s(0)) - e^{(t-T)/\varepsilon} z_s(T) + O(\varepsilon). \tag{6.41}$$

As can be seen, this analysis identifies not only the curve on which a singular arc can be found, but also both ends of the arc. Moreover, it characterizes the singular behavior near these ends.

6.7 Exercises

Exercise 6.1

Consider the optimal control problem

$$\dot{x}_1 = x_2, \quad x_1(0) = a_1, \quad x_1(T) = b_1$$

$$\dot{x}_2 = u, \quad x_2(0) = a_2, \quad x_2(T) = b_2$$

$$J = \frac{1}{2}\int_0^T (x_1^2 + u^2)\,dt,$$

where constants a_1, a_2, b_1 and b_2 are given and $T = 1/\varepsilon \gg 1$. Following the methodology of Section 6.2, find an $O(\varepsilon)$ approximation to the solution of this problem and compare it with the exact solution when $\varepsilon = 1/T = 0.1$.

Exercise 6.2

(a) Verify the formal correctness of the singularly perturbed optimal control problem

$$\dot{x} = -x + z^3 + u, \qquad x(0) = a_0, \quad x(1) \text{ is free},$$

$$\varepsilon\dot{z} = z^3 + u, \qquad z(0) = b_0, \quad z(1) = b_1,$$

$$J = \frac{1}{2}\int_0^1 (x^2 + z^6 + u^2)\,dt,$$

where a_0, b_0 and b_1 are given constants. First, write the Hamiltonian necessary optimality condition and obtain its reduced form by setting $\varepsilon = 0$. Next, set $\varepsilon = 0$ in the problem statement and write the Hamiltonian necessary condition for the reduced problem thus obtained. Compare the two necessary conditions.

(b) Paralleling the derivations in Example 2.2, find an $O(\varepsilon)$ approximation of the optimal solution. Explain the form of its slow part by an uncontrollability property.

Exercise 6.3

(a) Verify that the optimal control problem

$$\begin{aligned}
\dot{x}_0 &= z_1, & x(0) &= a_0, & x(1) &= b_0,\\
\varepsilon \dot{z}_1 &= z_2, & z_1(0) &= a_1, & z_1(1) &= b_1,\\
\varepsilon \dot{z}_2 &= u, & z_2(0) &= a_2, & z_2(1) &= b_2,
\end{aligned}$$

$$J = \frac{1}{2}\int_0^1 (x^2 + z_1^2 + u^2)\,dt,$$

where a_0, a_1, a_2, b_0, b_1 and b_2 are given constants, satisfies Assumptions 3.1 and 3.2 and that the matrices P and N of Lemma 4.1 are the same as those found in Exercise 6.1.

(b) Using Theorem 4.1, find an $O(\varepsilon)$ approximation to the optimal solution.

Exercise 6.4

Using Theorem 4.1, find an $O(\varepsilon)$ approximation to the solution of the time-varying optimal control problem

$$\dot{x} = z,$$

$$\varepsilon \dot{z} = tz + u,$$

$$J = \frac{1}{2}\int_1^2 [x^2 + (q - t^2)z^2 + u^2]\,dt,$$

in which $x(t)$ and $z(t)$ are fixed at both ends,

$$x(1) = a_1, \quad x(2) = a_2, \quad z(1) = b_1, \quad z(2) = b_2,$$

where a_1, a_2, b_1 and b_2 are given constants. Discuss the implementation of this approximate solution.

Exercise 6.5

Applying the procedure (5.37)–(5.47), find a near-optimal control for the minimum-time problem

$$\dot{x} = -x + u, \quad x(0) = x_0, \quad x(T) = 0,$$

$$\varepsilon\dot{z} = az + u, \quad z(0) = z_0, \quad z(T) = 0,$$

$$|u(t)| \leqslant 1,$$

where $a = -1$. Observe that the fast switch occurs near the end of the interval. Can a near-optimal control be found and implemented if $a = 1$?

6.8 Notes and References

Trajectory optimization problems were among the first problems to attract control engineers to singular perturbations. Kokotović and Sannuti (1968) introduced the formulation presented in Section 6.3, and proposed a correction of the reduced solution. Kelley and Edelbaum (1970) and Kelley (1970a, b, 1971a, b) demonstrated the applicability of reduced-order modeling and boundary layer corrections to aircraft manoeuver optimization. With this impetus, earlier works on singularly perturbed boundary value problems became relevant. Among them are results of Levin (1957) on stable initial manifolds, Tupchiev's two-point boundary value analysis reported by Vasil'eva (1963), and the dichotomy decomposition of Chang (1972).

Our presentation of the exponential dichotomy as a characteristic of Hamiltonian boundary value problems in Section 6.2 and its use in Section 6.4 are patterned on the exposition of Wilde and Kokotović (1972b, 1973). Their advantage is that a boundary value problem is treated as a composition of initial-value problems. The nonlinear extension of this approach in Section 6.5 is due to Chow (1979) and is related to the nonlinear stability and regulator problems of Chapter 7.

Asymptotic expansions of more general boundary value problems, not covered in this book, are found in Bagirova, Vasil'eva and Imanaliev (1967), Hadlock (1970), O'Malley (1972a, b, 1974a, b). Sannuti (1974a, b, 1975), Freedman and Kaplan (1976), Vasil'eva and Dmitriev (1980) and others. The relationship of the assumptions made by the two approaches was investigated by O'Malley (1975). The assumptions differ in how they state the "matching" conditions, that is, the conditions that guarantee the stability of the initial layer in forward time and of the terminal layer in reverse time. Geometrically, they require a saddle point property, that is the existence of two manifolds, one stable and the other totally unstable. To avoid "turning points", this saddle point property must hold throughout the interval. While for general boundary value problems this requirement is restrictive, further research may show that most optimal control problems satisfy it owing to their Hamiltonian structure.

Asymptotic analysis of problems with constraints and singular arcs has been much more difficult. Only for linear time-optimal problems of the type considered in Section 6.5 have simple near-optimal solutions been proposed—by Collins (1973), Kokotović and Haddad (1975a, b), Javid and Kokotović (1977), Javid (1978a) and Halanay and Mirica (1979). Approximate solutions to nonlinear constrained problems have been obtained by application-specific techniques. They are particularly promising in flight dynamics and guidance, where results of Ardema (1976, 1979), Calise (1976, 1978, 1979, 1980, 1981), Sridhar and Gupta (1980) and Shinar (1983) extend earlier results of Kelley (1973). Analytical elucidation of some heuristic concepts emerging from these applications is expected to broaden the applicability of the time-scale methodology to optimal control problems. The use of a "cheap control" problem for providing a limiting analysis of a more difficult singular optimal control problem in Section 6.6 is discussed in O'Malley and Jameson (1975), and, for time-varying systems, in O'Malley and Jameson (1977).

7 NONLINEAR SYSTEMS

7.1 Introduction

An advantage of the two-time-scale approach to system modeling and control systems design is that it is not restricted to linear systems. In this chapter we exploit this advantage in our study of the stability properties and the composite state feedback control of nonlinear systems.

First, Section 7.2 addresses a fundamental question of this book, and one that has long interested singular perturbation researchers: assuming that the associated slow and fast systems are each asymptotically stable, what additional conditions will guarantee the asymptotic stability of the original nonlinear singularly perturbed system for a sufficiently small singular perturbation parameter ε? Using Lyapunov stability methods, a composite Lyapunov function is constructed in a step-by-step manner, starting with the stability of a slowly varying system of the type considered in Section 5.2. It is shown that under mild conditions, any weighted sum of quadratic-type Lyapunov functions for the slow and fast systems is a quadratic-type Lyapunov function for the singularly perturbed system when ε is sufficiently small. The choice of the weights of the composite Lyapunov function involves a trade off between obtaining a large estimate of the domain of attraction and a large upper bound on the singular perturbation parameter ε. Analysis developed in Section 7.2 for autonomous systems is extended in Section 7.5 to nonautonomous systems.

The application of this method is illustrated in two case studies presented in Sections 7.3 and 7.4. The method is applied in Section 7.3 to estimate the domain of attraction of the stable equilibrium point of a synchronous generator connected to an infinite bus. The second case study, Section 7.4, illustrates an important application to the analysis of robustness of adaptive control systems with respect to parasitics. This is illustrated in Section 7.4 by a second-order example.

In Sections 7.6 and 7.7 we extend the composite state feedback control method of Chapter 3 to nonlinear system models including those that cannot be stabilized through linearization. The design is sequential in general, since the fast control design depends on the slow control design. However, should the system be linear in the fast states and control inputs, as in the nonlinear regulator problem of Section 7.7, the fast–slow designs are separated as before. The near-optimal regulator design is radically different from the finite-interval trajectory optimization results of Chapter 6 because of the stability and boundedness requirements fundamental in infinite-time problems which require feedback solutions.

7.2 Stability Analysis: Autonomous Systems

We consider a nonlinear autonomous singularly perturbed system

$$\left.\begin{aligned} \dot{x} &= f(x, z), \quad x \in R^n, \\ \varepsilon\dot{z} &= g(x, z), \quad z \in R^m, \end{aligned}\right\} \tag{2.1}$$

which has an isolated equilibrium at the origin $x = 0$, $z = 0$. It is assumed throughout that f and g are smooth enough to ensure that, for specified initial conditions, (2.1) has a unique solution. Stability of the equilibrium is investigated by examining the reduced (slow) system

$$\dot{x} = f(x, h(x)), \tag{2.2}$$

where $z = h(x)$ is an isolated root of $0 = g(x, z)$, and the boundary-layer (fast) system

$$\frac{dz}{d\tau} = g(x, z(\tau)), \quad \tau = \frac{t}{\varepsilon}, \tag{2.3}$$

where x is treated as a fixed parameter. Theorem 2.1 states, essentially, that if $x = 0$ is an asymptotically stable equilibrium of the reduced system (2.2), $z = h(x)$ is an asymptotically stable equilibrium of the boundary-layer system (2.3), uniformly† in x, and $f(\cdot,\cdot)$ and $g(\cdot,\cdot)$ satisfy certain growth conditions, then the origin is an asymptotically stable equilibrium of the singularly perturbed system (2.1).

In Theorem 2.1, asymptotic stability requirements on the reduced and boundary-layer systems are expressed by requiring the existence of Lyapunov functions for each system. The growth requirements on $f(\cdot,\cdot)$ and $g(\cdot,\cdot)$ take the form of inequalities satisfied by the Lyapunov functions.

† That is, the ε-δ definition of Lyapunov stability and the convergence $z \to h(x)$ are uniform in x (Vidyasagar, 1978).

Most of the analysis in this section is devoted to motivating the conditions of Theorem 2.1 and relating them to the previous analysis of Section 5.2. We start by studying the slowly varying system

$$\varepsilon \dot{z}(t) = A(x(t))z(t), \quad z \in R^m, \tag{2.4}$$

where $x(t)$ satisfies

$$\dot{x}(t) = f(x(t)), \quad x \in R^n. \tag{2.5}$$

It is assumed that for all $t \geq 0$, $x(t) \in S$, where S is a closed bounded subset of R^n. This assumption does not require the trajectory $x(t)$ to converge to any limit as $t \to \infty$. It merely requires that $x(t)$ remains bounded as $t \to \infty$. System (2.4) is a special case of the linear slowly varying system studied in Section 5.2. Instead of having A as an explicit function of t, A now depends on x, which depends on t. Therefore, in view of Lemma 2.2, it is seen that

$$\operatorname{Re} \lambda(A(x)) \leq -c_1, \quad \|A(x)\| \leq c_2, \quad \left\| \frac{\partial a_{ij}}{\partial x} f(x) \right\| \leq c_{ij} \quad \forall x \in S \tag{2.6}$$

imply that $z(t) \to 0$ as $t \to \infty$. The same conclusion can be reached by employing Lyapunov methods. In view of (2.6), for every $x \in S$ there exists a symmetric positive-definite matrix $P(x)$, bounded on S, satisfying the Lyapunov equation

$$P(x)A(x) + A^{\mathrm{T}}(x)P(x) = -I. \tag{2.7}$$

Moreover, $\dot{P}(x)$, the derivative of $P(x)$ along the trajectory of (2.5), is bounded on S. This follows from the explicit form of the solution of (2.7), given by

$$P(x) = \int_0^\infty \mathrm{e}^{A^{\mathrm{T}}(x)\sigma} \mathrm{e}^{A(x)\sigma}\, \mathrm{d}\sigma.$$

With $P(x)$ we can form a Lyapunov function candidate† for (2.4) as

$$W(x, z) = z^{\mathrm{T}} P(x) z. \tag{2.8}$$

The function $W(x, z)$ can be viewed as a Lyapunov function of (2.4) when x is treated as a fixed parameter which may take any value in S. If x were fixed, the derivative of W along the trajectory of (2.4) would have been given by $-(1/\varepsilon)z^{\mathrm{T}}z$. But rather than being a constant, x is slowly varying according to the solution of (2.5), and hence the derivative of $W(x, z)$ along

† Following the terminology of Vidyasagar (1978), a function W is called a Lyapunov function candidate if it satisfies the usual differentiability and positive definiteness requirements. If, for a particular system, $\dot{W} \leqslant 0$ then W is referred to as a Lyapunov function.

the trajectory of (2.4), (2.5) is given by

$$\dot{W} = -\frac{1}{\varepsilon} z^{\mathrm{T}} z + z^{\mathrm{T}} \dot{P}(x) z$$

$$\le -\frac{1}{\varepsilon}\|z\|^2 + \gamma\|z\|^2 = -\frac{1}{\varepsilon}(1 - \varepsilon\gamma)\|z\|^2 < 0 \quad \text{for } \varepsilon < \frac{1}{\gamma}.$$

Thus $W(x(t), z(t)) \to 0$ as $t \to \infty$, implying that $z(t) \to 0$ as $t \to \infty$. In the above inequality, and throughout this chapter, $\|\cdot\|$ denotes the Euclidean norm $\|\cdot\|_2$.

Let us point out that the Lyapunov function $W(x, z)$ has two properties

$$\frac{\partial W}{\partial z} A(x) z \le -\|z\|^2, \tag{2.9}$$

$$\frac{\partial W}{\partial x} f(x) \le \gamma\|z\|^2. \tag{2.10}$$

These two properties will carry over to Lyapunov functions of the more general system (2.1).

Now consider the system

$$\varepsilon\dot{z}(t) = A(x(t))z(t) + b(x(t)). \tag{2.11}$$

When x is treated as a fixed parameter (2.11) has an equilibrium $h(x) \triangleq -A^{-1}(x)b(x)$ instead of the zero equilibrium in (2.4). The existence of $A^{-1}(x)$ is guaranteed by (2.6). In addition, we assume that $\partial h/\partial x$ exists in S. A natural Lyapunov function candidate for (2.11) is

$$W(x, z) = (z - h(x))^{\mathrm{T}} P(x)(z - h(x)).$$

This function has the following properties for every $x \in S$:

$$\text{(i)} \quad \frac{\partial W}{\partial z}(A(x)z + b(x)) \le -\|z - h(x)\|^2, \tag{2.12}$$

$$\text{(ii)} \quad \frac{\partial W}{\partial x} f(x) \le \gamma\|z - h(x)\|^2 + \beta_2\left\|\frac{\partial h}{\partial x} f(x)\right\|\|z - h(x)\|. \tag{2.13}$$

Comparing (2.12), (2.13) with (2.9), (2.10) shows that $\|z - h(x)\|$ replaces $\|z\|$, reflecting the fact that now the equilibrium is $h(x)$ instead of zero. More significantly, there is an extra term in (2.13) not appearing in (2.10) which is due to the slow motion of the equilibrium. When the derivative of W is computed along the trajectories of (2.11) and (2.5) we get

$$\dot{W} = \frac{1}{\varepsilon}\frac{\partial W}{\partial z}(Ax + b) + \frac{\partial W}{\partial x} f(x)$$
$$\le -\frac{1}{\varepsilon}(1 - \varepsilon\gamma)\|z - h(x)\|^2 + \beta_2 \left\| \frac{\partial h}{\partial x} f(x) \right\| \|z - h(x)\|. \quad (2.14)$$

It is important to see that now we cannot assure that $z(t) \to h(x(t))$ as $t \to \infty$ without additional requirements on the asymptotic behavior of $x(t)$ as $t \to \infty$. This brings us to the next step of our study, where we consider the full singularly perturbed system

$$\dot{x} = f(x, z), \quad (2.15)$$

$$\varepsilon \dot{z} = g(x, z). \quad (2.16)$$

Let $B_x \subset R^n$ and $B_z \subset R^m$ denote closed sets.

Assumption 2.1

The origin is an isolated equilibrium of (2.15), (2.16) in $B_x \times B_z$, i.e.

$$\left.\begin{aligned} 0 &= f(0, 0), \\ 0 &= g(0, 0), \end{aligned}\right\} \quad (2.17)$$

and $z = h(x)$ is the unique root of

$$0 = g(x, z) \quad (2.18)$$

in $B_x \times B_z$, i.e.

$$0 = g(x, h(x)). \quad (2.19)$$

We study the asymptotic stability of the origin using Lyapunov methods. To construct a Lyapunov function candidate for the singularly perturbed system (2.15), (2.16) we consider each of the two equations separately. Starting with (2.16), we make the following assumption.

Assumption 2.2

There exists a Lyapunov function candidate $W(x, z)$ such that for all $(x, z) \in B_x \times B_z$

(i) $W(x, z) > 0 \quad \forall z \neq h(x)$ and $W(x, h(x)) = 0$,

(ii) $$\frac{\partial W}{\partial z} g(x, z) \le -\alpha_2 \phi^2(z - h(x)), \quad \alpha_2 > 0, \quad (2.20)$$

(iii) $$\frac{\partial W}{\partial x} f(x, z) \le \gamma \phi^2(z - h(x)) + \beta_2 \psi(x) \phi(z - h(x)), \quad (2.21)$$

where $\psi(\cdot)$ and $\phi(\cdot)$ are scalar functions of vector arguments which vanish only when their arguments are zero, e.g. $\psi(x) = 0$ if and only if $x = 0$. They will be referred to as comparison functions.

Conditions (2.20) and (2.21) look exactly like (2.12) and (2.13) with $\|z - h(x)\|$ replaced by $\phi(z - h(x))$ and $\|(\partial h/\partial x) f(x)\|$ replaced by $\psi(x)$. Working with comparison functions more general than norms allows us to handle a wider class of nonlinearities, as will be shown later by examples. The function $W(x, z)$ is a Lyapunov function of the boundary layer system (2.3) in which x is treated as a fixed parameter.

We turn now to (2.15). Adding and subtracting $f(x, h(x))$ to the right-hand side of (2.15) yields

$$\dot{x} = f(x, h(x)) + f(x, z) - f(x, h(x)). \tag{2.22}$$

The term $f(x, z) - f(x, h(x))$ can be viewed as a perturbation of the reduced system

$$\dot{x} = f(x, h(x)). \tag{2.23}$$

It is natural first to look for a Lyapunov function candidate for (2.23) and then to consider the effect of the perturbation term $f(x, z) - f(x, h(x))$. In this spirit, we assume the existence of a Lyapunov function $V(x)$ that satisfies

$$\frac{\partial V}{\partial x} f(x, h(x)) \leq -\alpha_1 \psi^2(x), \quad \alpha_1 > 0. \tag{2.24}$$

Condition (2.24) guarantees that $x = 0$ is an asymptotically stable equilibrium of (2.23). It is important to observe here that the comparison function in (2.24) is the same as in (2.21). Computing the derivative of V along the solution of (2.22) yields

$$\begin{aligned} \dot{V} &= \frac{\partial V}{\partial x} f(x, h(x)) + \frac{\partial V}{\partial x} [f(x, z) - f(x, h(x))] \\ &\leq -\alpha_1 \psi^2(x) + \frac{\partial V}{\partial x} [f(x, z) - f(x, h(x))]. \end{aligned}$$

We assume that

$$\frac{\partial V}{\partial x} [f(x, z) - f(x, h(x))] \leq \beta_1 \psi(x) \phi(z - h(x)), \tag{2.25}$$

so that

$$\dot{V} \leq -\alpha_1 \psi^2(x) + \beta_1 \psi(x) \phi(z - h(x)). \tag{2.26}$$

Condition (2.25), in a sense, determines the allowed growth of f in z. In typical problems, verifying (2.25) reduces to verifying the inequality

$$\|f(x, z) - f(x, h(x))\| \le \phi(z - h(x)),$$

which means that the rate of growth of f in z cannot be faster than the rate of growth of $\phi(\cdot)$.

The conditions on $V(x)$ are put together in the following assumption.

Assumption 2.3

There exists a Lyapunov function $V(x)$ that satisfies inequalities (2.24) and (2.25) for all $x \in B_x$.

With the Lyapunov functions $V(x)$ and $W(x, z)$ to hand, we consider $\nu(x, z)$, defined by a weighted sum of V and W,

$$\nu(x, z) = (1 - d)V(x) + dW(x, z), \quad 0 < d < 1, \tag{2.27}$$

as a Lyapunov function candidate for the singularly perturbed system (2.15), (2.16). To explore the freedom we have in choosing the weights, we take d as an unspecified parameter in (0, 1). The derivative of ν along the trajectories of (2.15), (2.16) is given by

$$\begin{aligned}\dot{\nu} &= (1 - d)\frac{\partial V}{\partial x} f(x, z) + \frac{d}{\varepsilon}\frac{\partial W}{\partial z} g(x, z) + d\frac{\partial W}{\partial x} f(x, z)\\ &= (1 - d)\frac{\partial V}{\partial x} f(x, h(x)) + (1 - d)\frac{\partial V}{\partial x}[f(x, z) - f(x, h(x))]\\ &\quad + \frac{d}{\varepsilon}\frac{\partial W}{\partial z} g(x, z) + d\frac{\partial W}{\partial x} f(x, z).\end{aligned}$$

Using inequalities (2.20), (2.21), (2.24) and (2.25), we get

$$\begin{aligned}\dot{\nu} &\le -(1 - d)\alpha_1\psi^2(x) + (1 - d)\beta_1\psi(x)\phi(z - h(x))\\ &\quad - \frac{d}{\varepsilon}\alpha_2\phi^2(z - h(x)) + d\gamma\phi^2(z - h(x)) + d\beta_2\psi(x)\phi(z - h(x))\\ &= -\begin{bmatrix}\psi(x)\\ \phi(z - h(x))\end{bmatrix}^{\mathrm{T}}\begin{bmatrix}(1 - d)\alpha_1 & -\frac{1}{2}(1 - d)\beta_1 - \frac{1}{2}d\beta_2\\ -\frac{1}{2}(1 - d)\beta_1 - \frac{1}{2}d\beta_2 & d\left(\frac{\alpha_2}{\varepsilon} - \gamma\right)\end{bmatrix}\\ &\quad \times\begin{bmatrix}\psi(x)\\ \phi(z - h(x))\end{bmatrix}.\end{aligned}$$

The right-hand side of the above inequality is a quadratic form in $\psi(\cdot)$ and $\phi(\cdot)$. This quadratic form is negative-definite when

$$d(1-d)\alpha_1\left(\frac{\alpha_2}{\varepsilon}-\gamma\right) > \tfrac{1}{4}[(1-d)\beta_1 + d\beta_2]^2,$$

which is equivalent to

$$\frac{1}{\varepsilon} > \frac{1}{\alpha_1\alpha_2}\left[\alpha_1\gamma + \frac{1}{4(1-d)d}((1-d)\beta_1 + d\beta_2)^2\right]. \tag{2.28}$$

The inequality (2.28) shows that for any choice of d the corresponding ν is a Lyapunov function for the singularly perturbed system (2.15), (2.16) for all ε satisfying (2.28). Notice that in the above development only α_1 and α_2 are required by definition to be positive. The other three parameters β_1, β_2 and γ could, in general, be positive, negative or zero. In most problems, however, one arrives at inequalities (2.21) and (2.25) using norm inequalities leading automatically to nonnegative values for β_1, β_2 and γ. Therefore we continue our discussion assuming $\beta_1 \geq 0$, $\beta_2 \geq 0$ and $\gamma \geq 0$. The inequality (2.28) can be written as

$$\varepsilon < \frac{\alpha_1\alpha_2}{\alpha_1\gamma + \dfrac{1}{4(1-d)d}[(1-d)\beta_1 + d\beta_2]^2} \triangleq \varepsilon_d. \tag{2.29}$$

The dependence of the right-hand side of (2.29) on the unspecified parameter d is sketched in Fig. 7.1.

It can easily be seen that the maximum value of ε_d occurs at $d^* =$

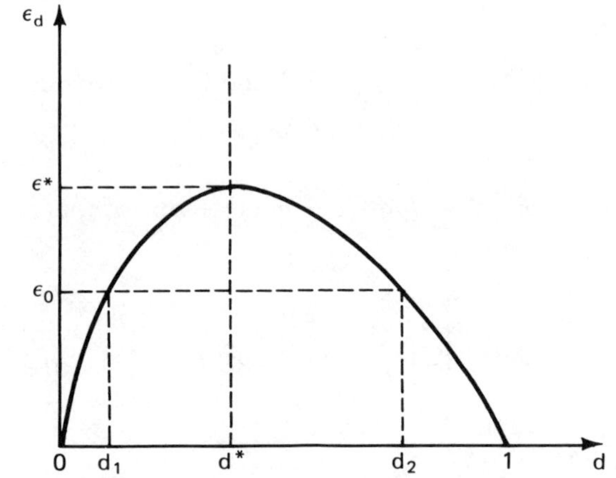

Fig. 7.1. Stability upper bounds on ε.

$\beta_1/(\beta_1+\beta_2)$ and is given by

$$\varepsilon^* = \frac{\alpha_1\alpha_2}{\alpha_1\gamma+\beta_1\beta_2}. \tag{2.30}$$

It follows that the equilibrium point of (2.1) is asymptotically stable for all $\varepsilon < \varepsilon^*$. The number ε^* is the best upper bound on ε that can be provided by the above stability analysis. Theorem 2.1 summarizes our findings.

Theorem 2.1

Let Assumptions 2.1–2.3 be satisfied. Then the origin is an asymptotically stable equilibrium of the singularly perturbed system (2.1) for all $\varepsilon \in (0, \varepsilon^*)$, where ε^* is given by (2.30). Moreover, for every number $d \in (0, 1)$

$$\nu(x, z) = (1-d)V(x) + dW(x, z)$$

is a Lyapunov function for all $\varepsilon \in (0, \varepsilon_d)$, where $\varepsilon_d \le \varepsilon^*$ is given by (2.29).

Corollary 2.1

Let the assumptions of Theorem 2.1 hold for all $x, z \in R^n \times R^m$ and let $V(x)$ and $W(x, z)$ be radially unbounded (i.e. $V(x) \to \infty$ as $\|x\| \to \infty$ and $W(x, z) \to \infty$ as $\|z - h(x)\| \to \infty$). Then, the equilibrium $(x = 0,\ z = 0)$ is globally asymptotically stable for all $\varepsilon < \varepsilon^*$.

Corollary 2.2

Let all the assumptions of Theorem 2.1 hold and suppose, in addition, that $V(x)$ and $W(x, z)$ satisfy the inequalities

$$e_1\psi^2(x) \le V(x) \le e_2\psi^2(x) \qquad \forall x \in B_x,$$

$$e_3\phi^2(z-h(x)) \le W(x, z) \le e_4\phi^2(z-h(x)) \quad \forall (x, z) \in B_x \times B_z,$$

where $e_1, \ldots, e_4$ denote positive constants. Then, the conclusions of Theorem 2.1 hold, with exponential stability replacing asymptotic stability.

Corollary 2.3

Let f, g and h be continuously differentiable. Suppose that $x = 0$ is an exponentially stable equilibrium of the reduced system (2.2) and $z = h(x)$ is an exponentially stable equilibrium of the boundary layer system (2.3), uniformly in x, i.e.

$$\|z(\tau) - h(x)\| \leq K e^{-\alpha\tau} \|z(0) - h(x)\|,$$

where α and K are independent of x. Then, the origin is an exponentially stable equilibrium of the singularly perturbed system (2.1), for sufficiently small ε.

Proof This follows from Corollary 2.2 with the conceptual Lyapunov functions

$$V(x) = \int_0^T \|X(t, x)\|^2 \, dt,$$

$$W(x, z) = \int_0^T \|Z(\tau, z; x) - h(x)\|^2 \, d\tau,$$

where $X(t, x)$ is the trajectory of the reduced system starting at initial point x at $t = 0$ and $Z(\tau, z; x)$ is the trajectory of the boundary layer system starting at initial point z at $\tau = 0$. It is left for the reader (Exercise 7.3) to verify that V and W satisfy all the conditions of Corollary 2.2 □

The stability analysis that led to Theorem 2.1 delineates a procedure for constructing Lyapunov function candidates for the singularly perturbed system (2.1). One starts by studying the equilibria of the reduced and boundary layer systems, searching for Lyapunov functions $V(x)$ and $W(x, z)$ that satisfy (2.24) and (2.20) respectively. Then inequalities (2.21) and (2.25), which may be called *interconnection conditions*, are checked. Several choices of V and W may be tried before one arrives at a choice for which (2.21) and (2.25) hold. At this point, the asymptotic stability of the equilibrium ($x = 0$, $z = 0$) is guaranteed for ε sufficiently small. However, for a given value of ε, say ε_0, one wants to improve the bound ε^* by exploiting the freedom that usually exists in verifying inequalities (2.21) and (2.25) (e.g. freedom in defining boundaries of the sets B_x and B_z). The Lyapunov function $\nu(x, z)$ is known to be valid when $\varepsilon_0 < \varepsilon^*$. If so, one uses the fact, illustrated in Fig. 7.1, that for a given ε_0 there is a range (d_1, d_2) such that any $d \in (d_1, d_2)$ results in a valid Lyapunov function. In other words, $\nu(x, z) = (1 - d)V(x) + d(W(x, z)$, $d \in (d_1, d_2)$, defines a parameterized family of Lyapunov functions for (2.1). The freedom in choosing d can be used to achieve other objectives like improving estimates of the domain of attraction. To illustrate this point, let c be any positive number such that

$$L \stackrel{\triangle}{=} \{(x, z) : \nu(x, z) \leq c\} \subset B_x \times B_z. \tag{2.31}$$

Then L is included in the domain of attraction of the equilibrium. If L_s and $L_f(x)$ are estimates of the domains of attraction of the equilibria of the reduced and boundary layer systems, defined by

$$L_s = \{x : V(x) \le c_s\} \subset B_x,$$

$$L_f(x) = \{z : W(x, z) \le c_f\} \subset B_z \quad \text{for } x \in L_s,$$

then c can be chosen as

$$c = \min\{(1 - d)c_s, dc_f\},$$

which is maximum if d can be chosen as $d = c_s/(c_s + c_f)$.

Example 2.1

The second-order system

$$\dot{x} = x - x^3 + z = f(x, z), \tag{2.32a}$$

$$\varepsilon\dot{z} = -x - z = g(x, z) \tag{2.32b}$$

has a unique equilibrium at the origin. We shall investigate the stability of the equilibrium in three steps.

Step 1: Reduced system Setting $\varepsilon = 0$ in (2.32b), we get

$$z = -x \triangleq h(x),$$

which when substituted in (2.32a) yields the reduced system

$$\dot{x} = -x^3 = f(x, h(x)). \tag{2.33}$$

Taking $V(x) = \frac{1}{4}x^4$, we have $\partial V/\partial x = x^3$ and

$$\frac{\partial V}{\partial x} f(x, h(x)) = -x^6.$$

Hence inequality (2.24) is satisfied with $\alpha_1 = 1$ and $\psi(x) = |x|^3$.

Step 2: Boundary layer system Treating x on the right-hand side of (2.32b) as a constant and expressing (2.32b) in the fast time scale $\tau = t/\varepsilon$, we obtain the boundary layer system

$$\frac{dz}{d\tau} = -(x + z).$$

We take $W(x, z) = \frac{1}{2}(x + z)^2$ so that inequality (2.20) is satisfied with $\alpha_2 = 1$ and $\phi(x + z) = |x + z|$.

Step 3: Interconnection conditions It remains now to verify inequalities (2.21) and (2.25):

$$\left|\frac{\partial V}{\partial x}[f(x,z) - f(x,h(x))]\right| = |x^3(x+z)| = \psi\phi,$$

$$\left|\frac{\partial W}{\partial x}f(x,z)\right| \le \left|\frac{\partial W}{\partial x}f(x,h(x))\right| + \left|\frac{\partial W}{\partial x}[f(x,z) - f(x,h(x))]\right|$$

$$= |x+z||x|^3 + |x+z|^2 = \psi\phi + \phi^2.$$

Hence, inequalities (2.21) and (2.25) are satisfied with $\beta_1 = \beta_2 = \gamma = 1$.

All the inequalities hold globally and V and W are radially unbounded. Therefore, by Corollary 2.3, the equilibrium is globally asymptotically stable for all $\varepsilon < \varepsilon^* = 0.5$. To see how conservative this bound is, let us note that the characteristic equation of the linearized system is

$$\lambda^2 + \left(\frac{1}{\varepsilon} - 1\right)\lambda = 0,$$

which shows that the equilibrium is unstable for $\varepsilon > 1$.

In Step 1 above, we could have chosen $V(x) = \frac{1}{2}x^2$, which would satisfy (2.24) with $\alpha_1 = 1$ and $\psi(x) = x^2$. This choice, however, would not satisfy inequality (2.25) since $|\partial V/\partial x| = |x|$. This observation serves as a guide in our search for Lyapunov functions that would satisfy the interconnection conditions. We should search for Lyapunov functions that satisfy (2.20), (2.24) and the inequalities $|\partial V/\partial x| \le K_1\psi$, $|f(x,h(x))| \le K_2\psi$ and $|\partial W/\partial x| \le K_3\phi$. We conclude the example by the following remark.

Remark 2.1 Since all the conditions of Theorem 2.1 are satisfied with $V(x) = \frac{1}{4}x^4$ and $W(x,z) = \frac{1}{2}(x+z)^2$, we know that

$$\nu(x,z) = (1-d)V(x) + dW(x,z), \quad d \in (0,1),$$

is a Lyapunov function for the singularly perturbed system (2.32) for $\varepsilon < \varepsilon_d^*$. Since our example is a simple second-order system, we may try to compute the derivative of ν with respect to (2.32) and see if we can get an upper bound on ε less conservative than the one provided by Theorem 2.1:

$$\dot{\nu} = (1-d)x^3(x - x^3 + z) + d(x+z)\left[x - x^3 + z - \frac{1}{\varepsilon}(x+z)\right]$$

$$= -(1-d)x^6 + (1-d)x^3(x+z) - dx^3(x+z)$$

$$-d\left(\frac{1}{\varepsilon} - 1\right)(x+z)^2.$$

It is apparent that the choice $d = \frac{1}{2}$ cancels the cross-product terms and yields

$$\dot{\nu} = -\frac{1}{2}x^6 - \frac{1}{2}\left(\frac{1}{\varepsilon} - 1\right)(x + z)^2,$$

which is negative-definite for all $\varepsilon < 1$ and is the actual range of ε for which the origin is asymptotically stable.

Motivated by Remark 2.1, let us explore another angle of using singular perturbation decompositions in Lyapunov stability analysis. In the stability analysis of Theorem 2.1, singular perturbation decompositions have been employed for two different purposes. First, they were used to define separate reduced and boundary layer systems whose Lyapunov functions were used to form a Lyapunov function candidate for the full system. This process simplifies the search for Lyapunov functions since it is done at the level of the lower-order systems. Second, the singularly perturbed form of the system together with inequalities (2.20), (2.21), (2.24) and (2.25) were used to show that the derivative of the Lyapunov function candidate is bounded by a negative upper bound. However, even if some of the abovementioned inequalities are not satisfied, the same Lyapunov function candidate may still be used to investigate the stability of the full system by calculating its derivative along the trajectory of the full system.

The stability analysis of this section will be illustrated by two case studies presented in subsequent sections. In the synchronous machine case study of Section 7.3, we start by choosing functions V and W that satisfy all the conditions for the application of Theorem 2.1. Then, in the spirit of Remark 2.1, we try another choice of W whose derivative with respect to the boundary layer system is only negative-semidefinite, so that (2.20) is not satisfied. Nonetheless, we form $\nu = (1 - d)V + dW$, calculate its derivative with respect to the full singularly perturbed system and show that d can be chosen such that $\dot{\nu}$ is negative-semidefinite.

The adaptive control case study of Section 7.4 is particularly interesting because it illustrates an extension of the method when the system has an equilibrium set rather than an equilibrium point. Subsequently, one tries to establish stability rather than asymptotic stability. A stability analysis similar to the one preceding Theorem 2.1 is performed to construct a Lyapunov function candidate and show that its derivative is negative-semidefinite.

7.3 Case Study: Stability of a Synchronous Machine

The stability analysis of Section 7.2 will be employed to estimate the domain of attraction of the stable equilibrium of a synchronous generator connected to an infinite bus through a transmission line. The generator is represented by the one-axis model (Anderson and Fouad, 1977; Pai, 1981) given by

$$\tau'_{d0}\frac{dE'_q}{d\tau} = -\frac{x_d + x_e}{x'_d + x_e}E'_q + \frac{x_d - x'_d}{x'_d + x_e}V\cos\delta + E_{FD}, \tag{3.1a}$$

$$\frac{d\delta}{d\tau} = \omega \tag{3.1b}$$

$$M\frac{d\omega}{d\tau} = -D\omega + P_m - \frac{E'_q V}{x'_d + x_e}\sin\delta, \tag{3.1c}$$

where

τ'_{d0} = direct-axis open-circuit transient time constant,
x_d = direct-axis synchronous reactance of the generator,
x'_d = direct-axis transient reactance of the generator,
E'_q = instantaneous voltage proportional to field-flux linkage,
δ = angle between voltage of infinite bus and E'_q,
D = damping coefficients of the generator,
E_{FD} = field voltage (assumed constant),
P_m = mechanical input to the generator (assumed constant),
ω = slip velocity of the generator,
x_e = reactance of the transmission line,
V = voltage of infinite bus.

The variable E'_q is usually slower than δ and ω as a result of a relatively large value of the time constant τ'_{d0}. Viewing E'_q as a slow variable and δ and ω as fast variables, we study the domain of attraction of the stable equilibrium using time-scale decomposition. The first task is to bring (3.1) into the singularly perturbed form. Let $\bar{E}'_q$, $\bar{\delta}$ and $\bar{\omega}$ denote the stable equilibrium of (3.1). The state variables are taken as $x = (E'_q - \bar{E}'_q)/\bar{E}'_q$, $z_1 = \delta - \bar{\delta}$ and $z_2 = \omega - \bar{\omega}$, so that x will be the slow variable. In order to have the singularly perturbed form (2.1), we define ε as $\varepsilon = 1/\tau'_{d0}$ and change the time scale from τ to $t = \tau/\tau'_{d0}$ to obtain

$$\dot{x} = -ax + b[\cos(z_1 + \bar{\delta}) - \cos\bar{\delta}] \triangleq f(x, z), \tag{3.2a}$$

$$\varepsilon\dot{z}_1 = z_2 \triangleq g_1(x, z), \tag{3.2b}$$

$$\varepsilon\dot{z}_2 = -\lambda z_2 - c[(1 + x)\sin(z_1 + \bar{\delta}) - \sin\bar{\delta}] \triangleq g_2(x, z), \tag{3.2c}$$

where

$$a = \frac{x_e + x_d}{x_e + x'_d}, \quad b = \frac{x_d - x'_d}{x_e + x'_d}\cdot\frac{V}{\bar{E}'_q}, \quad c = \frac{\bar{E}'_q V}{M(x_e + x'_d)}, \quad \lambda = \frac{D}{M}.$$

The reduced system is obtained by setting $\varepsilon = 0$ in (3.2) to get

$$z = h(x) = \begin{pmatrix} h_1(x) \\ h_2(x) \end{pmatrix} = \begin{pmatrix} \sin^{-1}\left(\dfrac{\sin \bar{\delta}}{1+x}\right) - \bar{\delta} \\ 0 \end{pmatrix}, \tag{3.3}$$

which when substituted in (3.2a) yields

$$\dot{x} = -ax + b[\cos(h_1(x) + \bar{\delta}) - \cos \bar{\delta}] \stackrel{\Delta}{=} -ax + bN(x). \tag{3.4}$$

The boundary layer system is defined by

$$\frac{\mathrm{d}z_1}{\mathrm{d}\tau} = z_2, \tag{3.5a}$$

$$\frac{\mathrm{d}z_2}{\mathrm{d}\tau} = -\lambda z_2 - c[(1+x)\sin(z_1 + \bar{\delta}) - \sin \bar{\delta}]. \tag{3.5b}$$

In order for the reduced and boundary layer systems to be well defined, as well as for other reasons that will be apparent later, we restrict x and z_1 to a region defined by the following inequalities:

$$x \geq -\theta, \tag{3.6a}$$

$$-(\pi + h_1(x) + 2\bar{\delta}) \leq z_1 \leq \pi - (h_1(x) + 2\bar{\delta}), \tag{3.6b}$$

$$[z_1 - h_1(x)][\sin(z_1 + \bar{\delta}) - \sin(h_1(x) + \bar{\delta})] \geq \eta |z_1 - h_1(x)|^2, \tag{3.6c}$$

where the positive numbers η and θ will be specified later; θ should satisfy

$$\left|\frac{\sin \bar{\delta}}{1-\theta}\right| < 1.$$

To study the stability of the equilibrium of the reduced system (3.4), we choose

$$V(x) = \int_0^x [a\sigma - bN(\sigma)]\,\mathrm{d}\sigma \tag{3.7}$$

as a Lyapunov function candidate. From the definition of $N(x)$ it follows that $N(0) = 0$, $xN(x) > 0$ and

$$\left|\frac{\partial N}{\partial x}\right| = \left|\frac{1}{[(1+x)^2 - (\sin \bar{\delta})^2]^{1/2}}\left(\frac{\sin \bar{\delta}}{1+x}\right)^2\right| \leq K_N,$$

where

$$K_N = \frac{1}{[(1-\theta)^2 - (\sin \bar{\delta})^2]^{1/2}}\left(\frac{\sin \bar{\delta}}{1-\theta}\right)^2.$$

We assume that the positive number θ can be chosen such that $K_N < a/b$. It follows that $V(x)$ is positive-definite. The derivative of $V(x)$ along the trajectory of (3.4) is given by

$$\dot{V} = \frac{\partial V}{\partial x}(-ax + bN(x)) = -(ax - bN(x))^2.$$

Therefore inequality (2.24) is satisfied with $\psi(x) = |(ax - bN(x))|$ and $\alpha_1 = 1$. Next, we consider the boundary layer system (3.5). This system has an equilibrium at $z_1 = h_1(x)$ and $z_2 = 0$. We choose a Lur'e-type Lyapunov function candidate $W(x, z)$ given by

$$W(x, z) = \frac{\lambda}{2}(z_1 - h_1(x))^2 + z_2(z_1 - h_1(x)) + \frac{q}{2\lambda} z_2^2$$
$$+ \frac{qc}{\lambda}(1 + x)\int_{h_1(x)}^{z_1} M(x, \sigma)\,d\sigma, \tag{3.8}$$

where $q > 1$ and

$$M(x, \sigma) = \sin(\sigma + \bar{\delta}) - \sin(h_1(x) + \bar{\delta}).$$

Inequalities (3.6a, b) guarantee that $W(x, z)$ is positive-definite. The derivative of $W(x, z)$ along the trajectory of (3.5) is given by

$$\frac{dW}{d\tau} = (\nabla_z W(x, z))^{\mathrm{T}} g(x, z) = -(q - 1)z_2^2 - c(1 + x)(z_1 - h_1(x))M(x, z_1).$$

Using inequality (3.6c), we get

$$\frac{dW}{d\tau} \leq -(q - 1)z_2^2 - c(1 - \theta)\eta|z_1 - h_1(x)|^2.$$

A comparison function for the boundary-layer system (3.5) is taken as

$$\phi(z - h(x)) = \left\| \begin{pmatrix} |z_1 - h_1(x)| \\ \xi|z_2| \end{pmatrix} \right\|,$$

where ξ is an arbitrary positive number to be chosen later to improve the upper bound on ε. With the above choice of the comparison function and with the choice $q = 1 + c(1 - \theta)\eta\xi^2$, inequality (2.20) is satisfied with $\alpha_2 = c(1 - \theta)\eta$. It remains now to verify the interconnection conditions, inequalities (2.21) and (2.25). For (2.21) we have

$$\frac{\partial}{\partial x} W(x, z) f(x, z) = [-(ax - bN(x)) + b(\cos(z_1 + \bar{\delta}) - \cos(h_1(x) + \bar{\delta}))]$$
$$\times \left[-\lambda \frac{\partial h_1}{\partial x}(z_1 - h_1(x)) - z_2 \frac{\partial h_1}{\partial x} - \frac{qc}{\lambda}(\cos(z_1 + \bar{\delta}) - \cos(h_1(x) + \bar{\delta})) \right].$$

Using the inequality $|\cos(z_1 + \bar{\delta}) - \cos(h_1(x) + \bar{\delta})| \le |z_1 - h_1(x)|$, we get

$$\begin{aligned} \frac{\partial W}{\partial x} f(x, z) &\le \lambda K_1 \psi(x)|z_1 - h_1(x)| + K_1 \psi(x)|z_2| + \frac{qc}{\lambda}\psi(x)|z_1 - h_1(x)| \\ &\quad + bK_1|z_1 - h_1(x)|\,|z_2| + bK_1\lambda|z_1 - h_1(x)|^2 \\ &\le \gamma\phi^2(z - h(x)) + \beta_2\psi(x)\phi(z - h(x)), \end{aligned}$$

where

$$\gamma = \frac{1}{\xi}\left(\tfrac{1}{2}bK_1 + bK_1\lambda\xi\right),$$

$$\beta_2 = \frac{1}{\xi}\left[K_1^2 + \xi^2\left(\lambda K_1 + \frac{qc}{\lambda}\right)^2\right]^{1/2}$$

and K_1 is an upper bound on $\partial h_1/\partial x$ (i.e. $|\partial h_1/\partial x| \le K_1$), which is given by

$$K_1 = \frac{\sin\bar{\delta}}{(1 - \theta)[(1 - \theta)^2 - (\sin\bar{\delta})^2]^{1/2}}.$$

So, inequality (2.21) holds. Finally, for (2.25),

$$\begin{aligned} \frac{\partial V}{\partial x}(f(x, z) - f(x, h(x))) &= b(ax - bN(x)) \\ &\quad \times (\cos(z_1 + \bar{\delta}) - \cos(h_1(x) + \bar{\delta})) \\ &\le b\psi(x)|z_1 - h_1(x)| \\ &\le b\psi(x)\phi(z - h(x)). \end{aligned}$$

Thus all the conditions of Theorem 2.1 hold and we conclude that the equilibrium of (3.2) is asymptotically stable for all $\varepsilon < \varepsilon_d$ for any $d \in (0, 1)$, where

$$\varepsilon_d = \frac{c(1 - \theta)\eta\xi}{\tfrac{1}{2}bK_1 + bK_1\lambda\xi + \dfrac{1}{4d(1 - d)\xi}\left\{b(1 - d)\xi + d\left[K_1^2 + \xi^2\left(\lambda K_1 + \dfrac{qc}{\lambda}\right)^2\right]^{1/2}\right\}^2}. \tag{3.9}$$

Moreover, $\nu(x, z)$, given by

$$\nu(x, z) = (1 - d) \int_0^x [a\sigma - bN(\sigma)]\,d\sigma + d\left[\frac{\lambda}{2}(z_1 - h_1(x))^2 + z_2(z_1 - h_1(x)) + \frac{q}{2\lambda} z_2^2 + \frac{qc}{\lambda}(1 + x)\int_{h_1(x)}^{z_1} M(x, \sigma)\,d\sigma\right], \quad (3.10)$$

is a Lyapunov function of (3.2).

In (3.9) and (3.10) there are four unspecified positive parameters, namely, d, θ, η and ξ. The first three should be chosen to satisfy certain restrictions that were imposed throughout the derivation, while the fourth parameter ξ is arbitrary. In making these choices, we are guided by the requirement that ε_d (given by (3.9)) should be greater than ε, and by our desire to obtain a large estimate of the domain of attraction. Let us illustrate that by a numerical example. Consider the typical numerical data (Siddiqee, 1968): $M = 147 \times 10^{-4}$, $x'_{\mathrm{d}} = 0.3$, $x_{\mathrm{e}} = 0.2$, $x_{\mathrm{d}} = 1.15$, $\tau'_{\mathrm{d0}} = 6.6$, $P_{\mathrm{m}} = 0.815$, $V = 1$ and $E_{\mathrm{FD}} = 1.22$, where M is in (per unit power) $\times$ second^2/radian, τ'_{d0} is in seconds and all parameters are per unit. These data yield $\bar{\delta} = 0.42$ rad, $a = 2.7$, $b = 1.7$ and $c = 2 \times 10^4/147$. In addition, we take $\lambda = D/M = 4\,\mathrm{s}^{-1}$. Choosing $d = 0.01$, $\theta = 0.42$, $\eta = 0.35$ and $\xi = 0.1$ yields $\varepsilon_d = 0.1844$, which is acceptable since $1/\tau'_{\mathrm{d0}} = 0.1515$. Hence $\nu(x, z)$ is given by

$$\nu(x, z) = 0.99\, V(x) + 0.01\, W(x, z).$$

An estimate of the domain of attraction can be obtained by means of closed Lyapunov surfaces as in (2.31). However, a better estimate can be obtained by means of open Lyapunov surfaces (Willems, 1969). Let C and D be two points on the line $x = -\theta$ such that for any point on the segment CD the derivative $\dot{x}$ is positive and let $C_{\max} = \min\{\nu(x_{\mathrm{C}}, z_{\mathrm{C}}), \nu(x_{\mathrm{D}}, z_{\mathrm{D}})\}$. It can be shown that the set

$$\mathscr{L} = \{x, z : \nu(x, z) \le C_{\max} \text{ and } x \ge -\theta\}$$

is included in the domain of attraction. A sketch of the set $\mathscr{L}$ is shown in Fig. 7.2.

Let us now use this example to elaborate on Remark 2.1 and the discussion thereafter. Another Lyapunov function candidate for the boundary layer system (3.5) is the energy-type function given by

$$\tilde{W}(x, z) = \tfrac{1}{2} z_2^2 + c(1 + x)\int_{h_1(x)}^{z_1} M(x, \sigma)\,d\sigma. \quad (3.11)$$

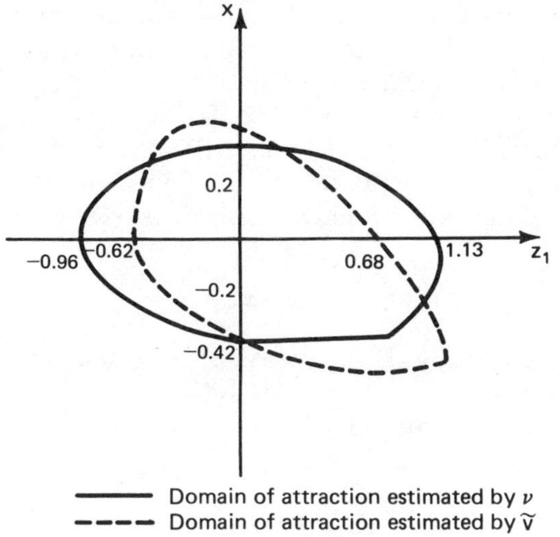

Fig. 7.2. Estimates of the domain of attraction.

This function, though, does not satisfy the conditions of Theorem 2.1, since its derivative along the solution of boundary layer system is only negative-semidefinite, so that inequality (2.20) is not satisfied. However, we may still consider

$$\tilde{\nu}(x, z) = (1 - d)V(x) + d\tilde{W}(x, z) \tag{3.12}$$

as a Lyapunov function candidate. It is interesting to note that computing the derivative of $\tilde{\nu}$ along the trajectory of the full system (3.2) shows that there is a unique choice of d, namely $d = b/(b + c)$, that makes $\dot{\tilde{\nu}}$ negative-semidefinite. Moreover, it can be checked that the trivial solution is the only solution of $\dot{\tilde{\nu}} = 0$ in the neighborhood of the origin. Thus $\tilde{\nu}$ is a Lyapunov function for all $\varepsilon > 0$. In comparing the two Lyapunov functions ν and ν, we note that in $\tilde{\nu}$ there is no freedom in forming the composite Lyapunov function, while in ν the parameter d is free and can be used to generate a family of composite Lyapunov functions. The freedom in choosing d was used to enlarge the domain of attraction estimated by ν in the direction of the δ-axis (z_1-axis), at the expense of its extension in the direction of the x-axis, compared with the domain of attraction obtained by $\tilde{\nu}$ (see Fig. 7.2). It is usually more beneficial to extend the domain of attraction in the direction of the δ-axis (Saberi and Khalil, 1984). Other estimates of the domain of attraction can be obtained by different choices of the parameters d, θ, η and ξ. The price paid for the freedom in choosing

d is the restriction of the range of ε to $(0, \varepsilon_d)$. Getting an acceptable ε_d hinges on having $\lambda = D/M$ not too small. With the numerical values we used here we need $\lambda \geqslant 1$, which is not unrealistic. Although for uncontrolled machines when D accounts only for field damping a typical value of λ could be as low as 0.2 (Anderson and Fouad, 1977), for a feedback-controlled machine in which D accounts for damping introduced by power system stabilizers typical values of λ could be as high as 10 (Demello and Concordia, 1969).

7.4 Case Study: Robustness of an Adaptive System

We consider a second-order plant

$$\dot{y} = ay + 2z - u, \tag{4.1}$$

$$\varepsilon\dot{z} = -z + u, \tag{4.2}$$

with unknown constant parameter a and a "parasitic" time constant ε. It is required to design an adaptive control law to regulate the output y (i.e. for $y(t) \to 0$ as $t \to \infty$). The design problem can be simplified by neglecting the "parasitic" time constant ε to obtain a reduced-order model

$$\dot{y} = ay + u. \tag{4.3}$$

An adaptive control for (4.3) can be taken as

$$u = -Ky, \tag{4.4}$$

with the adaptive law proposed in Narendra *et al.* (1980):

$$\dot{K} = by^2(t), \quad b > 0. \tag{4.5}$$

Application of the adaptive control (4.4) to the reduced-order model (4.3) results in

$$\dot{y} = -(K - a)y. \tag{4.6}$$

To show that $y(t) \to 0$ as $t \to \infty$, consider a Lyapunov function candidate

$$V(y, K) = \tfrac{1}{2}y^2 + \frac{1}{2b}(K - K^*)^2, \quad K^* > a. \tag{4.7}$$

The derivative of V along the trajectory of (4.5), (4.6) is given by

$$\dot{V} = y\dot{y} + \frac{1}{b}(K - K^*)\dot{K} = -(K^* - a)y^2 \leq 0. \tag{4.8}$$

Moreover, $\dot{V} = 0$ only at the equilibrium set $y = 0$, $K =$ constant. It is easily seen that as $t \to \infty$, V converges to a finite value and $\dot{V}$ converges to zero. Therefore $y(t) \to 0$ and $K(t) \to$ constant as $t \to \infty$. This convergence holds for all bounded initial conditions $y(0)$, $K(0)$, and $\lim_{t\to\infty} K(t)$ depends on the initial conditions.

We employ singular perturbation stability analysis to study the effect of the model–plant mismatch due to the parasitic time constant ε. Application of the control law (4.4), (4.5) to the plant with parasitics (4.1), (4.2) results in the closed-loop system

$$\dot{y} = (a + K)y + 2z, \tag{4.9}$$

$$\dot{K} = by^2, \tag{4.10}$$

$$\varepsilon\dot{z} = -z - Ky, \tag{4.11}$$

which takes the form (2.1) with $x = (y, K)^{\mathrm{T}}$. Following the procedure of Section 7.2, we start by analyzing the reduced and boundary-layer systems and then investigate interconnections. The reduced system, obtained by setting $\varepsilon = 0$ in (4.11), is given by

$$\dot{y} = -(K - a)y, \tag{4.12}$$

$$\dot{K} = by^2, \tag{4.13}$$

which is the closed-loop system obtained in the design stage by neglecting parasitics. We have already seen that $V(x) = V(y, K) = \frac{1}{2}y^2 + (1/2b)(K - K^*)^2$ is a Lyapunov function for this system. We rewrite (4.8) as

$$\dot{V} = -\alpha_1 \psi^2(x), \quad \alpha_1 = K^* - a > 0, \quad \psi(x) = |y|. \tag{4.14}$$

The boundary-layer system is given by

$$\frac{\mathrm{d}z}{\mathrm{d}\tau} = -Ky - z, \tag{4.15}$$

where Ky is treated as a constant parameter. This linear system has a quadratic Lyapunov function $W(x, z) = \frac{1}{2}(z + Ky)^2$. Furthermore,

$$\frac{\mathrm{d}W}{\mathrm{d}\tau} = -(z + Ky)^2 = -\alpha_2\phi^2(z + Ky), \quad \alpha_2 = 1, \quad \phi(\cdot) = |\cdot|. \tag{4.16}$$

Finally, to study interconnections, we analyze, in the notation of Section

7.2, $(\partial V/\partial x)[f(x, z) - f(x, Ky)]$ and $(\partial W/\partial x) f(x, z)$:

$$
\begin{aligned}
\frac{\partial V}{\partial x}[f(x, z) - f(x, Ky)] &= [y \quad b^{-1}(K - K^*)] \begin{bmatrix} 2(z + Ky) \\ 0 \end{bmatrix} \\
&= 2y(z + Ky) \\
&\leq \beta_1 \psi(x) \phi(z + Ky), \quad \beta_1 = 2,
\end{aligned} \tag{4.17}
$$

$$
\begin{aligned}
\frac{\partial W}{\partial x} f(x, z) &= [(z + Ky)K \quad (z + Ky)y] \begin{bmatrix} -(K - a)y + 2(Ky + z) \\ by^2 \end{bmatrix} \\
&= -(K - a)Ky(z + Ky) + 2K(z + Ky)^2 + b(z + Ky)y^3 \\
&\leq \gamma \phi^2(z + Ky) + \beta_2 \psi(x) \phi(z + Ky),
\end{aligned} \tag{4.18}
$$

where γ and β_2 are positive constants such that for all values of interest of y and K the following inequalities hold:

$$2K \leq \gamma, \tag{4.19}$$

$$|by^2 - K(K - a)| \leq \beta_2. \tag{4.20}$$

Thus the Lyapunov functions $V(x)$ and $W(x, z)$ satisfy inequalities (2.20), (2.21), (2.24) and (2.25) of Theorem 2.1. The only assumption of Theorem 2.1 that does not hold is the requirement that $\psi(x) = 0$ if and only if $x = 0$. In the current problem $\psi(x) = |y|$, which vanishes at $y = 0$ for all K. The asymptotic stability conclusion of Theorem 2.1 does not hold, but taking $\nu = (1 - d)V + dW$ as a Lyapunov function candidate, we obtain

$$
\begin{aligned}
\dot{\nu} &\leq -\begin{bmatrix} \psi \\ \phi \end{bmatrix}^{\mathrm{T}} \begin{bmatrix} (1 - d)\alpha_1 & -\frac{1}{2}(1 - d)\beta_1 - \frac{1}{2}d\beta_2 \\ -\frac{1}{2}(1 - d)\beta_1 - \frac{1}{2}d\beta_2 & d\left(\dfrac{\alpha_2}{\varepsilon} - \gamma\right) \end{bmatrix} \begin{bmatrix} \psi \\ \phi \end{bmatrix} \\
&\leq 0 \quad \text{for } \varepsilon < \varepsilon_d,
\end{aligned} \tag{4.21}
$$

where

$$
\begin{aligned}
\varepsilon_d &= \frac{\alpha_1 \alpha_2}{\alpha_1 \gamma + \dfrac{1}{4(1 - d)d}[(1 - d)\beta_1 + d\beta_2]^2} \\
&= \frac{K^* - a}{(K^* - a)\gamma + \dfrac{1}{4(1 - d)d}[2(1 - d) + d\beta_2]^2}.
\end{aligned} \tag{4.22}
$$

Consider now the set $\mathcal{L}(d, c) = \{(y, K, z) : \nu(y, K, z) \leq \frac{1}{2}c(1 - d)\}$ for some

$c > 0$. It can be easily verified that in $\hat{\mathscr{L}}(d, c)$, inequalities (4.20) and (4.21) are satisfied with

$$\gamma = 2[K^* + (cb)^{1/2}], \tag{4.23}$$

$$\beta_2 = \max\{cb + (\tfrac{1}{2}a)^2, \quad (K^* + (cb)^{1/2})(K^* - a + (cb)^{1/2})\}. \tag{4.24}$$

Choosing $d = d^* = \beta_1/(\beta_1 + \beta_2)$, which maximizes ε_d, we obtain

$$\varepsilon^* = \varepsilon_{d^*} = \frac{K^* - a}{(K^* - a)\gamma + 2\beta_2}, \quad \mathscr{L}(c) = \hat{\mathscr{L}}(d^*, c). \tag{4.25}$$

Therefore for all $(y, K, z) \in \mathscr{L}(c)$ and for each $\varepsilon < \varepsilon^*$, $\dot{\nu} \leq 0$, and $\dot{\nu} = 0$ only at the equilibrium set $y = 0$, $K =$ constant, $z = 0$. Any solution of (4.9)–(4.11) starting from $\mathscr{L}(c)$ remains inside it. Furthermore, $\nu(y, K, z)$ is a nonincreasing function of time which is bounded from below and hence converges to a finite value ν_∞. Therefore $\dot{\nu}(y(t), K(t), z(t)) \to 0$ as $t \to \infty$, which implies that $y(t) \to 0$, $K(t) \to$ constant and $z(t) \to 0$ as $t \to \infty$.

For any bounded initial conditions $y(0)$, $K(0)$, $z(0)$, we can choose c large enough so that $(y(0), K(0), z(0)) \in \mathscr{L}(c)$. Therefore we conclude that for any bounded initial conditions there exists an $\varepsilon_1 > 0$ such that for each $\varepsilon \in (0, \varepsilon_1)$ the solution $(y(t), K(t), z(t))$ is bounded and $y(t) \to 0$, $K(t) \to$ constant and $z(t) \to 0$ as $t \to \infty$. This shows that the adaptive control law (4.4), (4.5) is robust with respect to the presence of a small parasitic time constant. On the other hand, if ε is not small enough, the stability properties of (4.9)–(4.11) may be lost. In particular, linearization of (4.9)–(4.11) at the equilibrium ($y = 0$, $K = K_0$, $z = 0$) gives the characteristic equation

$$\lambda\left[\lambda^2 + \left(\frac{1}{\varepsilon} - a - K_0\right)\lambda + \frac{1}{\varepsilon}(K_0 - a)\right] = 0,$$

which has an unstable root for $\varepsilon < 1/(a + K_0)$. Since $K(t)$ is nondecreasing, it is clear that instability occurs if $\varepsilon > 1/(a + K(0))$.

As an illustration of the stability properties established above, let $a = 4$, $K^* = 7$, $b = 5$, $y(0) = 1$, $K(0) = 3$, $z(0) = -2$. Taking $c = 4.23$, it can be verified that $(y(0), K(0), z(0)) \in \mathscr{L}(c)$ and $\varepsilon^* = 0.012$. Hence stability is guaranteed for each $\varepsilon < 0.012$. On the other hand, instability occurs for $\varepsilon > 1/(a + K(0)) = 0.143$. Simulation results (Ioannou and Kokotović, 1984a, b) show that at $\varepsilon = 0.05$ the response is stable, while at $\varepsilon = 0.07$ it is unstable. This shows that the bound $\varepsilon^* = 0.012$ estimated by our stability analysis is not overly conservative. Finally, we use the estimates ε^* and $\mathscr{L}(c)$ to observe that increasing the adaptive gain b for a fixed ε reduces the size of $\mathscr{L}(c)$, and the stability properties can no longer be guaranteed if $b \geq O(1/\varepsilon)$. This is seen from the fact that ε^* depends on the product cb and not on c and b individually. Increasing b can be offset by reducing c, which reduces the size of $\mathscr{L}(c)$. Moreover, for sufficiently large b and fixed c we have $\gamma = O(b^{1/2})$, $\beta_2 = O(b)$, and hence $\varepsilon^* = O(1/b)$, which shows that the inequality $\varepsilon < \varepsilon^*$ is violated if $b > O(1/\varepsilon)$.

7.5 Stability Analysis: Nonautonomous Systems

In Section 7.2, we studied autonomous systems where f and g are not functions of ε. In this section, it is shown that dependence of f and g on t and ε can be allowed as long as inequalities similar to (2.20), (2.21), (2.24) and (2.25) hold uniformly in t and ε. Consider the system

$$\dot{x}(t) = f(t, x(t), z(t), \varepsilon), \tag{5.1a}$$

$$\varepsilon\dot{z}(t) = g(t, x(t), z(t), \varepsilon). \tag{5.1b}$$

Assume that the following conditions hold for all $(t, x, z, \varepsilon) \in [t_0, \infty) \times B_x \times B_z \times [0, \varepsilon_1]$.

Assumption 5.1

The origin ($x = 0$, $z = 0$) is the unique equilibrium of (5.1), i.e.

$$0 = f(t, 0, 0, \varepsilon), \tag{5.2a}$$

$$0 = g(t, 0, 0, \varepsilon). \tag{5.2b}$$

Moreover, the equation

$$0 = g(t, x, z, 0) \tag{5.3}$$

has a unique root $z = h(t, x)$ and there exists a class κ function† $p(\cdot)$ such that

$$\|h(t, x)\| \leq p(\|x\|). \tag{5.4}$$

Assumption 5.2: Reduced system

There exists a Lyapunov function $V(t, x)$ satisfying

(i) $V(t, x)$ is positive-definite and decrescent, that is,

$$0 < q_1(\|x\|) \leq V(t, x) \leq q_2(\|x\|),$$

for some class κ functions $q_1(\cdot)$ and $q_2(\cdot)$;

† A continuous function $\alpha: R \to R$ is said to be class κ if (i) $\alpha(\cdot)$ is nondecreasing, (ii) $\alpha(0) = 0$ and (iii) $\alpha(p) > 0$ whenever $p > 0$.

(ii) $$\frac{\partial V}{\partial t} + \frac{\partial V}{\partial x} f(t, x, h(t, x), 0) \le -\alpha_1 \psi^2(x), \quad \alpha_1 > 0, \tag{5.5}$$

where $\psi(\cdot)$ is a continuous scalar function of x that vanishes only at $x = 0$.

Assumption 5.3: Boundary-layer system

There exists a Lyapunov function $W(t, x, z)$ satisfying

(i) $0 < q_3(\|z - h(t, x)\|) \le W(t, x, z) \le q_4(\|z - h(t, x)\|)$ for some class κ functions $q_3(\cdot)$ and $q_4(\cdot)$

(ii) $$\frac{\partial W}{\partial z} g(t, x, z, 0) \le -\alpha_2 \phi^2(z - h(t, x)), \quad \alpha_2 > 0, \tag{5.6}$$

where $\phi(\cdot)$ is a continuous function of an R^m vector z_f which vanishes only at $z_f = 0$.

Assumption 5.4: Interconnection conditions

V and W satisfy the following inequalities:

(i) $$\frac{\partial V}{\partial x}[f(t, x, z, \varepsilon) - f(t, x, h(t, x), 0)] \le \beta_1 \psi(x)\phi(y - h(t, x)) + \varepsilon\gamma_1 \psi^2(\mathrm{x}), \tag{5.7}$$

(ii) $$\frac{\partial W}{\partial z}[g(t, x, z, \varepsilon) - g(t, x, z, 0)] \le \varepsilon\gamma_2' \phi^2(z - h(t, x)) + \varepsilon\beta_2' \psi(x)\phi(z - h(t, x)), \tag{5.8}$$

(iii) $$\frac{\partial W}{\partial t} + \frac{\partial W}{\partial x} f(t, x, z, \varepsilon) \le \gamma_2'' \phi^2(z - h(t, x)) + \beta_2'' \psi(x)\phi(z - h(t, x)). \tag{5.9}$$

For simplicity the constants β_1, β_2', β_2'', γ_1, γ_2' and γ_2'' are assumed to be nonnegative. The term $\varepsilon\gamma_1\psi^2$ in (5.7) is added to allow for more general dependence of f on ε. It drops out when f is independent of ε. Similarly, (5.8) drops out when g is independent of ε. In arriving at the above inequalities, one might need to obtain uniform bounds on ε-dependent terms like $K_1 + \varepsilon K_2$; that is why ε may be restricted to an interval $[0, \varepsilon_1]$.

Consider a Lyapunov function candidate

$$\nu(t, x, z) = (1 - d)V(t, x) + dW(t, x, z). \tag{5.10}$$

From the properties of V and W and inequality (5.4) it follows that $\nu(t, x, z)$ is positive-definite and decresent. Computing $\dot{\nu}$ with respect to (5.1) and using (5.5)–(5.9), we obtain

$$\dot{\nu} \leq -\begin{bmatrix} \psi \\ \phi \end{bmatrix}^{\mathrm{T}} \begin{bmatrix} (1-d)(\alpha_1 - \varepsilon\gamma_1) & -\frac{1}{2}(1-d)\beta_1 - \frac{1}{2}d\beta_2 \\ -\frac{1}{2}(1-d)\beta_1 - \frac{1}{2}d\beta_2 & \dfrac{d}{\varepsilon}(\alpha_2 - \varepsilon\gamma_2) \end{bmatrix} \begin{bmatrix} \psi \\ \phi \end{bmatrix},$$

where $\beta_2 = \beta_2' + \beta_2''$ and $\gamma_2 = \gamma_2' + \gamma_2''$. It can be easily seen that for all

$$\varepsilon < \varepsilon_d = \frac{\alpha_1\alpha_2}{\alpha_1\gamma_2 + \alpha_2\gamma_1 + \dfrac{1}{4(1-d)d}[(1-d)\beta_1 + d\beta_2]^2} \tag{5.12}$$

the right-hand side of (5.11) is a negative-definite quadratic form in ψ and ϕ. Hence there exists a positive constant K (possibly dependent on ε and d) such that

$$\dot{\nu} \leq -K[\psi^2(x) + \phi^2(z - h(t, x))].$$

Since $\phi^2(z_{\mathrm{f}})$ is a continuous positive-definite function of z_{f}, there exists a class κ function $q(\|z_{\mathrm{f}}\|)$ such that

$$\phi^2(z_{\mathrm{f}}) \geq q(\|z_{\mathrm{f}}\|).$$

Therefore

$$\dot{\nu} \leq -K[\psi^2(x) + q(\|z - h(t, x)\|)].$$

Using (5.4), it can be shown that $\psi^2(x) + q(\|z - h(t, x)\|)$ is positive-definite. Therefore $\dot{\nu}$ is negative-definite. Recalling Fig. 7.1, it is seen that ε_d is maximum with the choice $d = \beta_1/(\beta_1 + \beta_2)$, which yields

$$\varepsilon_d\big|_{d=\beta_1/(\beta_1+\beta_2)} = \varepsilon^* = \frac{\alpha_1\alpha_2}{\alpha_1\gamma_2 + \alpha_2\gamma_1 + \beta_1\beta_2}. \tag{5.13}$$

Our conclusions are summarized in Theorem 5.1.

Theorem 5.1

Suppose that Assumptions 5.1–5.4 hold. Then the origin ($x = 0$, $z = 0$) is a uniformly asymptotically stable equilibrium of the singularly perturbed system (5.1) for all $\varepsilon \in (0, \min(\varepsilon^*, \varepsilon_1))$, where ε^* is given by (5.13). Moreover, for every $d \in (0, 1)$, $\nu(t, x, z) = (1-d)V(t, x) + dW(t, x, z)$ is a Lyapunov function for all $\varepsilon \in (0, \min(\varepsilon_d, \varepsilon_1))$, where ε_d is given by (5.12).

Example 5.1

As an illustration of Theorem 5.1, we prove Lemma 4.2 of Chapter 5. We use the same notation as in Chapter 5. Choosing V and W as

$$V(t, x) = x^{\mathrm{T}} P_{\mathrm{s}}(t) x,$$
$$W(t, x, z) = (z + L_0(t)x)^{\mathrm{T}} P_{\mathrm{f}}(t)(z + L_0(t)x),$$

where P_{s} and P_{f} are respectively the solutions of the Lyapunov equations (4.8) and (4.9) of Chapter 5. It can be easily verified (Exercise 7.4) that V and W satisfy Assumptions 5.2–5.4 with $\psi(x) = \|x\|_2$, $\phi(z - h(t, x)) = \|z + L_0(t)x\|_2$, $\alpha_1 = \alpha_2 = 1$, $\gamma_1 = 0$ and β_1, β_2 and γ_2 are upper bounds on the time functions $2\|P_{\mathrm{s}}(t)A_{12}(t)\|_2$, $2\|P_{\mathrm{f}}(t)(\dot{L}_0(t) + L_0(t)A_0(t))\|_2$ and $\|\dot{P}_{\mathrm{f}}(t) + 2P_{\mathrm{f}}(t)L_0(t)A_{12}(t)\|_2$ respectively. Hence the linear system (4.1) of Chapter 5 is uniformly asymptotically stable for all

$$\varepsilon < \varepsilon^* = \frac{1}{\gamma_2 + \beta_1\beta_2},$$

which is the upper bound of Lemma 4.2 of Chapter 5.

7.6 Composite Feedback Control

The composite feedback control method, introduced in Chapter 3 for linear systems, is now extended to nonlinear autonomous systems

$$\dot{x} = f(x, z, u), \quad x \in R^n, \tag{6.1a}$$
$$\varepsilon \dot{z} = g(x, z, u), \quad z \in R^m, \tag{6.1b}$$

where $u \in R^r$ is a control input. Let us assume that the open-loop system (6.1) is a standard singularly perturbed system for every $u \in B_u \subset R^r$, that is to say, the equation

$$0 = g(x, z, u) \tag{6.2}$$

has a unique root $z = h(x, u)$ in $B_x \times B_z \times B_u$. The composite control method seeks the control u as the sum of slow and fast controls,

$$u = u_{\mathrm{s}} + u_{\mathrm{f}}, \tag{6.3}$$

where u_{s} is a feedback function of x,

$$u_{\mathrm{s}} = \Gamma_{\mathrm{s}}(x), \tag{6.4}$$

and u_f is a feedback function of x and z.

$$u_f = \Gamma_f(x, z). \tag{6.5}$$

The fast feedback function $\Gamma_f(x, z)$ is designed to satisfy two crucial requirements. First, when the feedback control (6.3) is applied to (6.1), the closed-loop system should remain a standard singularly perturbed system, i.e. the equation

$$0 = g(x, z, \Gamma_s(x) + \Gamma_f(x, z)) \tag{6.6}$$

should have a unique root $z = H(x)$ in $B_x \times B_z$. This requirement assures that the choice of Γ_f will not destroy this property of the function g in the open-loop system. This is best illustrated in the linear case, where

$$g(x, z, u) = A_{21}x + A_{22}z + B_2 u$$

and

$$u = -G_s x - G_f[z + A_{22}^{-1}(A_{21}x - B_2 G_s x)].$$

It is clear that, when (A_{22}, B_2) is controllable, the matrix $A_{22} - B_2 G_f$ can be singular even when A_{22} is nonsingular. We did not need to explicitly require $A_{22} - B_2 G_f$ to be nonsingular in the linear case, since it is implied by the stronger stability requirement $\operatorname{Re}\lambda(A_{22} - B_2 G_f) < 0$. The second requirement on $\Gamma_f(x, z)$ is that it be "inactive" for $z = h(x, u_s)$, i.e.

$$\Gamma_f(x, h(x, \Gamma_s(x)) = 0. \tag{6.7}$$

The importance of (6.7) can be seen from the closed-loop equation

$$\dot{x} = f(x, z, u_s + u_f), \tag{6.8a}$$

$$\varepsilon\dot{z} = g(x, z, u_s + u_f). \tag{6.8b}$$

The requirement (6.7) guarantees that $z = h(x, \Gamma_s(x))$ is a root of

$$0 = g(x, z, \Gamma_s(x) + \Gamma_f(x, z)). \tag{6.9}$$

By (6.6), equation (6.9) has a unique root $z = H(x)$. Thus, in view of (6.6) and (6.7),

$$H(x) = h(x, \Gamma_s(x)) \tag{6.10}$$

holds as an identity. With (6.7) and (6.10), the reduced model of the closed-loop system (6.8) is given by

$$\dot{x} = f(x, h(x, u_s), u_s), \tag{6.11}$$

which is independent of Γ_f and is the same reduced model obtained from the open-loop system (6.1) when u is taken as u_s. Thus the design of the slow control $u_s = \Gamma_s(x)$ can be carried out independently of the fast design Γ_f.

Once $\Gamma_s(x)$ has been chosen, the boundary layer model of the closed-loop system is defined as

$$\frac{dz}{d\tau} = g(x, z, \Gamma_s(x) + u_f), \tag{6.12}$$

where x is treated as a fixed parameter. The requirement (6.7) is now interpreted as a requirement on the feedback control $u_f = \Gamma_f(x, z)$ not to shift the equilibrium $z = h(x, \Gamma_s(x))$ of the boundary layer system (6.12). The design of u_f must guarantee that $z = h(x, \Gamma_s(x))$ is an asymptotically stable equilibrium of (6.12) uniformly in x. This alone may not be enough, and stronger technical conditions may be needed. A procedure based on Theorem 2.1 to design Γ_s and Γ_f such that the origin ($x = 0$, $z = 0$) becomes an asymptotically stable equilibrium of the closed-loop system is given below.

Stabilization design

Step 1: Design a slow control $u_s = \Gamma_s(x)$ for the reduced system (6.11) such that $x = 0$ is its unique asymptotically stable equilibrium in B_x, and a Lyapunov function $V: R^n \rightarrow R_+$, exists, guaranteeing for all $x \in B_x$

$$\frac{\partial V}{\partial x} f(x, h(x, \Gamma_s(x)), \Gamma_s(x)) \leq -\alpha_1 \psi^2(x), \quad \alpha_1 > 0. \tag{6.13}$$

Step 2: With the knowledge of $u_s = \Gamma_s(x)$, design a fast control $u_f = \Gamma_f(x, z)$ such that

(i) $\Gamma_f(x, h(x, \Gamma_s(x))) = 0$;

(ii) $g(x, z, \Gamma_s(x) + \Gamma_f(x, z)) = 0$ has a unique root $z = h(x, \Gamma_s(x))$ in $B_x \times B_z$;

(iii) $z = h(x, \Gamma_s(x))$ is an asymptotically stable equilibrium of the closed-loop boundary layer system (6.12), and there is a Lyapunov function $W: R^n \times R^m \rightarrow R_+$ such that for all $(x, z) \in B_x \times B_z$

$$\frac{\partial W}{\partial z} g(x, z, \Gamma_s(x) + \Gamma_f(x, z)) \leq -\alpha_2 \phi^2(z - h(x, \Gamma_s(x)),$$

$$\alpha_2 > 0. \tag{6.14}$$

Step 3: Verify the interconnection conditions

$$\text{(i)}\quad \frac{\partial W}{\partial x} f(x, z, \Gamma_s(x) + \Gamma_f(x, z)) \leqslant \gamma\phi^2(z - h(x, \Gamma_s(s))) + \beta_2\psi(x)\phi(z - h(x, \Gamma_s(x))) \qquad (6.15)$$

$$\text{(ii)}\quad \frac{\partial V}{\partial x} [f(x, z, \Gamma_s(x) + \Gamma_f(x, z)) - f(x, h(x, \Gamma_s(x)), \Gamma_s(x))] \leq \beta_1\psi(x)\phi(z - h(x, \Gamma_s(x))). \qquad (6.16)$$

The comparison functions ψ and ϕ are as defined in Section 7.2.

The composite feedback control

$$u_c = \Gamma_s(x) + \Gamma_f(x, z) \qquad (6.17)$$

designed in these three steps has the following stabilization property.

Theorem 6.1

If u_c of (6.17) is designed following the stabilization design, the origin is an asymptotically stable equilibrium of the closed-loop system

$$\dot{x} = f(x, z, u_c),$$
$$\varepsilon\dot{z} = g(x, z, u_c)$$

for all $\varepsilon < \varepsilon_d$, where ε_d is given by (2.29) and $d \in (0, 1)$. Moreover, $\nu(x, z) = (1 - d)V(x) + dW(x, z)$ is a Lyapunov function of this system.

Proof This follows from Theorem 2.1. □

Example 6.1

Consider

$$\dot{x} = xz^3,$$
$$\varepsilon\dot{z} = z + u,$$

and let $B_x = [-1, 1]$ and $B_z = [-\frac{1}{2}, \frac{1}{2}]$. It is desired to design a feedback control law to stabilize the system at the equilibrium point $x = 0$, $z = 0$. Notice that the desired stabilization cannot be achieved by linearization at the equilibrium point because of an uncontrollable pole at the origin. We go through the design procedure as follows.

Step 1: The reduced system is given by

$$\dot{x} = -xu_s^3.$$

We design $u_s = x^{4/3}$. The closed-loop reduced system is $\dot{x} = -x^5$, and with the Lyapunov function $V(x) = \frac{1}{6}x^6$ we get $\psi(x) = |x|^5$ and $\alpha_1 = 1$.

Step 2: The boundary layer system is defined by

$$\frac{dz}{d\tau} = z + u_s + u_f = z + x^{4/3} + u_f.$$

We design $u_f = -3(z + x^{4/3})$. The closed-loop boundary layer system is $dz/d\tau = -2(z + x^{4/3})$. The choice $W = \frac{1}{2}(z + x^{4/3})^2$ yields $\phi = |z + x^{4/3}|$ and $\alpha_2 = 2$.

Step 3: It is straightforward to verify that the interconnection conditions hold with $\beta_1 = \frac{7}{4}$, $\beta_2 = \frac{4}{3}$ and $\gamma = \frac{7}{3}$.

The composite control is formed as

$$u_c = u_s + u_f = x^{4/3} - 3(z + x^{4/3}) = -3z - 2x^{4/3}.$$

Now the choice $d = \frac{21}{37}$ yields $\varepsilon_d = \frac{3}{7}$. Therefore the origin is an asymptotically stable equilibrium of the closed-loop system

$$\begin{aligned} \dot{x} &= xz^3, \\ \varepsilon\dot{z} &= -2x^{4/3} - 2z, \end{aligned} \tag{6.18}$$

for all $\varepsilon < \frac{3}{7}$.

Remark 6.1 In this particular example, the estimate ε_d is conservative since the origin is asymptotically stable for all $\varepsilon > 0$, as can be seen from the Lyapunov function

$$\tilde{\nu} = \frac{3}{2}x^{4/3} + \frac{\varepsilon}{4}z^4, \tag{6.19}$$

whose derivative with respect to (6.18) is given by

$$\dot{\tilde{\nu}} = 2x^{1/3}(xz^3) + z^3(-2x^{4/3} - 2z) = -2z^4$$

and $\dot{\tilde{\nu}}(t) \equiv 0 \Rightarrow z(t) \equiv 0 \Rightarrow x(t) \equiv 0$. Theorem 6.1 guarantees only local stability with an estimate of the domain of attraction, and only for ε less than the given bound.

We conclude this section by observing that the boundary layer system (6.12) is dependent on the slow control $\Gamma_s(x)$. This implies that the design of the fast control $\Gamma_f(x, z)$ will, in general, depend on $\Gamma_s(x)$. An important class of systems for which such dependence can be eliminated is the case when g is linear in z and u, i.e.

$$g(x, z, u) = a_2(x) + A_2(x)z + B_2(x)u. \tag{6.20}$$

In this case, the existence of a unique root of $0 = g$ is equivalent to the nonsingularity of $A_2(x)$, and the boundary layer system (6.12) takes the form

$$\frac{dz}{d\tau} = a_2(x) + A_2(x)z + B_2(x)\Gamma_s(x) + B_2(x)u_f. \tag{6.21}$$

Shifting the equilibrium of (6.21) to the origin via the change of variables

$$z_f = z + A_2^{-1}(x)[a_2(x) + B_2(x)\Gamma_s(x)], \tag{6.22}$$

yields

$$\frac{dz_f}{d\tau} = A_2(x)z_f + B_2(x)u_f, \tag{6.23}$$

which is independent of $\Gamma_s(x)$. The design of the fast feedback control $u_f = \Gamma_f(x, z)$ is greatly simplified. In fact, all the requirements on Γ_f stated in Step 2 of the stabilization design will be satisfied if Γ_f is taken as

$$\begin{aligned}\Gamma_f(x, z) &= -G_f(x)z_f \\ &= -G_f(x)[z + A_2^{-1}(x)(a_2(x) + B_2(x)\Gamma_s(x))],\end{aligned} \tag{6.24}$$

where $G_f(x)$ is designed such that

$$\operatorname{Re}\lambda(A_2(x) - B_2(x)G_f(x)) \leq -\sigma < 0 \quad \forall x \in B_x. \tag{6.25}$$

The existence of $G_f(x)$ satisfying (6.25) is guaranteed if the pair $(A_2(x), B_2(x))$ is controllable uniformly in x, $\forall x \in B_x$, i.e. the corresponding controllability Grammian is bounded from below by a positive-definite matrix. If, in addition, f is linear in z and u, i.e.

$$f(x, z, u) = a_1(x) + A_1(x)z + B_1(x)u, \tag{6.26}$$

the design problem becomes much more tractable, since the reduced system (6.11) will be linear in u_s.

7.7 Near-Optimal Feedback Design

The composite control method proposed in the previous section is now shown to yield near-optimal feedback design for a class of singularly perturbed nonlinear regulators. The composite control stabilizes the desired equilibrium and produces a finite cost which tends to the optimal cost for a slow problem as $\varepsilon \to 0$.

The problem considered is to optimally control the nonlinear system

$$\dot{x} = a_1(x) + A_1(x)z + B_1(x)u = f(x, z, u), \quad x(0) = x_0, \tag{7.1a}$$

$$\varepsilon\dot{z} = a_2(x) + A_2(x)z + B_2(x)u = g(x, z, u), \quad z(0) = z_0, \tag{7.1b}$$

with respect to the performance index

$$J = \int_0^\infty [p(x) + s^{\mathrm{T}}(x)z + z^{\mathrm{T}}Q(x)z + u^{\mathrm{T}}R(x)u]\,\mathrm{d}t$$

$$\triangleq \int_0^\infty L(x, z, u)\,\mathrm{d}t, \tag{7.2}$$

where $x \in R^n$ and $z \in R^m$ are states, and $u \in R^r$ is the control. We make an assumption which, in addition to the differentiability and positivity properties of terms in (7.1), (7.2), also guarantees that the origin is the desired equilibrium.

Assumption 7.1

There exists a closed set $B_x \subset R^n$, containing $x = 0$ as an interior point, such that for all $x \in B_x$ functions a_1, a_2, A_1, A_2, A_2^{-1}, B_1, B_2, p, s, R and Q are continuously differentiable with respect to x; a_1, a_2, p and s are zero only at $x = 0$; Q and R are positive-definite matrices for all $x \in B_x$; matrices $A_1, A_2, B_1, B_2, R, Q, A_2^{-1}, R^{-1}$ and Q^{-1} are uniformly bounded on B_x; the scalar $p + s^{\mathrm{T}}z + z^{\mathrm{T}}Qz$ is a positive-definite function of its arguments, that is, it is positive except for $x = 0$, $z = 0$, where it is zero.

Notice that the boundedness of $A_1, A_2, B_1, B_2, R, Q, A_2^{-1}, R^{-1}$ and Q^{-1} will hold on any bounded set B_x by the continuity of these matrices with respect to x.

We define two, slow and fast, optimal control problems and combine their solutions to form a composite control. The slow problem is defined by taking $u = u_s$, setting $\varepsilon = 0$ in (7.1b) and eliminating z from (7.1a) and

(7.2) using

$$z = -A_2^{-1}(a_2 + B_2 u_s) \triangleq h(x, u_s). \tag{7.3}$$

The resulting slow problem is to optimally control the slow system

$$\dot{x} = a_0(x) + B_0(x)u_s, \quad x(0) = x_0 \tag{7.4}$$

with respect to

$$\begin{aligned} J_s &= \int_0^\infty [p_0(x) + 2s_0^T(x)u_s + u_s^T R_0(x)u_s]\,dt \\ &= \int_0^\infty L(x, h(x, u_s), u_s)\,dt, \end{aligned} \tag{7.5}$$

where

$$\begin{aligned} a_0 &= a_1 - A_1 A_2^{-1} a_2, \\ B_0 &= B_1 - A_1 A_2^{-1} B_2, \\ p_0 &= p - s^T A_2^{-1} u_2 + a_2^T A_2^{-T} Q A_2^{-1} a_2. \\ s_0 &= B_2^T A_2^{-T}(Q A_2^{-1} a_2 - \tfrac{1}{2}s), \\ R_0 &= R + B_2^T A_2^{-T} Q A_2^{-1} B_2. \end{aligned}$$

We note that $x = 0$ is the desired equilibrium of the slow system (7.4), since in view of Assumption 7.1, $a_0(0) = 0$, and $L(x, h(x, u_s), u_s) > 0$, for $x \neq 0$ and $u_s \neq 0$.

Our next assumption imposes some growth conditions on the slow problem.

Assumption 7.2

There exists a scalar positive-definite differentiable function $\psi(x)$, whose derivative $\partial\psi/\partial x$ is bounded on B_x, such that the following inequalities hold for all $x \in B_x$:

$$c_1 \psi^2(x) \le p_0 - s_0^T R_0^{-1} s_0 \le c_2 \psi^2(x), \tag{7.6}$$

$$|a_0 - B_0 R_0^{-1} B_0^T s_0| \le c_3 \psi(x), \tag{7.7}$$

$$|s - 2QA_2^{-1} a_2| \le c_4 \psi(x). \tag{7.8}$$

From optimal control theory (Athans and Falb, 1966), it is well known that if $V(x)$ is a unique positive-definite solution of the Hamilton–Jacobi equation

$$\left.\begin{aligned} &0 = (p_0 - s_0^{\mathrm{T}} R_0^{-1} s_0) + \frac{\partial V}{\partial x}(a_0 - B_0 R_0^{-1} s_0) - \frac{1}{4}\frac{\partial V}{\partial x} B_0 R_0^{-1} B_0^{\mathrm{T}} \left(\frac{\partial V}{\partial x}\right)^{\mathrm{T}}, \\ &V(0) = 0, \end{aligned}\right\} \tag{7.9}$$

then the minimizing control of the slow problem is given by

$$u_s = R_0^{-1}\left(s_0 + \tfrac{1}{2}B_0^{\mathrm{T}}\left(\frac{\partial V}{\partial x}\right)^{\mathrm{T}}\right), \tag{7.10}$$

and the optimal value of J_s is $V(x_0)$.

Assumptions 7.1 and 7.2 are checked on the given data of the problem. Now we make a mild assumption (mild in view of (7.6) and (7.7)) on the solution of the Hamilton–Jacobi equation (7.9).

Assumption 7.3

For all $x \in B_x$ (7.9) has a unique differentiable positive-definite solution $V(x)$ which satisfies

$$\left|\frac{\partial V}{\partial x}\right| \leq c\psi(x). \tag{7.11}$$

(When x is scalar and $B_0 R_0^{-1} B_0^{\mathrm{T}} \geq c_5 > 0 \quad \forall x \in B_x$, Assumption 7.3 follows from Assumptions 7.1 and 7.2.)

Assumption 7.3 allows us to use $V(x)$ as a Lyapunov function for the closed-loop slow system

$$\dot{x} = a_0 - B_0 R_0^{-1}\left(s_0 + \tfrac{1}{2}B_0^{\mathrm{T}}\left(\frac{\partial V}{\partial x}\right)^{\mathrm{T}}\right) \triangleq \bar{a}_0(x) = f(x, h(x, u_s), u_s). \tag{7.12}$$

Let us verify that $V(x)$ satisfies inequality (6.13) of the stabilization design of the previous section. Using the Hamilton–Jacobi equation (7.9), we have

$$\frac{\partial V}{\partial x}\bar{a}_0(x) = -\left[(p_0 - s_0^{\mathrm{T}} R_0^{-1} s_0) + \frac{1}{4}\frac{\partial V}{\partial x} B_0 R_0^{-1} B_0^{\mathrm{T}}\left(\frac{\partial V}{\partial x}\right)^{\mathrm{T}}\right]. \tag{7.13}$$

Inequalities (7.6) and (7.11) imply that $(\partial V/\partial x)\bar{a}_0$ satisfies (7.14):

$$-\alpha_0 \psi^2(x) \leq \frac{\partial V}{\partial x}\bar{a}_0(x) \leq -\alpha_1 \psi^2(x), \quad \alpha_0 > 0, \quad \alpha_1 > 0. \tag{7.14}$$

The right-hand-side inequality shows that $V(x)$ satisfies (6.13); the left-hand-side inequality will be used later on in studying the performance index.

The fast problem is defined by freezing the slow variable x in a stretched time scale τ and shifting the equilibrium of the fast system to the origin. The fast system is given by

$$\frac{dz_f}{d\tau} = A_2(x)z_f + B_2(x)u_f, \quad z_f(0) = z_0 - h(x_0, u_s(x_0)), \tag{7.15}$$

where $z_f = z - h(x, u_s)$. With (7.15) we associate the performance index

$$J_f = \int_0^{\infty} (z_f^T Q(x) z_f + u_f^T R(x) u_f) \, d\tau. \tag{7.16}$$

This problem has the familiar linear–quadratic form for each fixed $x \in B_x$.

Assumption 7.4

The pair $[A_2(x), B_2(x)]$ is controllable uniformly in x for all $x \in B_x$.

Recalling also that $R(x) \geq k_1 I_m$ and $Q(x) \geq k_2 I_m$ for all $x \in B_x$, we obtain, for each $x \in B_x$, the optimal solution of the fast problem

$$u_f(z_f, x) = -R^{-1}(x) B_2^T(x) K(x) z_f, \tag{7.17}$$

where $K(x)$ is the positive-definite solution of the x-dependent Riccati equation

$$0 = KA_2 + A_2^T K + Q - KB_2 R^{-1} B_2^T K. \tag{7.18}$$

The control (7.17) is stabilizing in the sense that

$$\operatorname{Re} \lambda(A_2 - B_2 R^{-1} B_2^T K) \leq -\sigma < 0 \quad \forall x \in B_x. \tag{7.19}$$

It is straightforward to verify that u_f satisfies requirements (i) and (ii) of Step 2 of the stabilization design of Section 7.6. To show that requirement (iii) is also satisfied, consider the Lyapunov function

$$W(x, z) = z_f^T K(x) z_f = [z - h(x, u_s)]^T K(x) [z - h(x, u_s)], \tag{7.20}$$

$$\begin{aligned}
\frac{\partial W}{\partial z} g(x, z, u_s + u_f) &= \frac{\partial W}{\partial z_f} g(x, z, u_s + u_f) \\
&= z_f^T [K(A_2 - B_2 R^{-1} B_2^T K) + (A_2 - B_2 R^{-1} B_2^T K)^T K] z_f \\
&= -z_f^T (Q + KB_2 R^{-1} B_2^T K) z_f \\
&\leq -\alpha_2 \|z_f\|^2, \quad \alpha_2 > 0
\end{aligned} \tag{7.21}$$

Thus inequality (6.14) is satisfied with $\phi = \|z_{\mathrm{f}}\|$.

The composite control is now formed as

$$\begin{aligned} u_{\mathrm{c}}(x, z) &= u_{\mathrm{s}}(x) - R^{-1}B_2^{\mathrm{T}}Kz_{\mathrm{f}} \\ &= -R_0^{-1}\left(s_0 + \tfrac{1}{2}B_0^{\mathrm{T}}\left(\frac{\partial V}{\partial x}\right)^{\mathrm{T}}\right) \\ &\quad -R^{-1}B_2^{\mathrm{T}}K\left[z + A_2^{-1}a_2 - A_2^{-1}B_2R_0^{-1}\left(s_0 + \tfrac{1}{2}B_0^{\mathrm{T}}\left(\frac{\partial V}{\partial x}\right)^{\mathrm{T}}\right)\right]. \end{aligned} \tag{7.22}$$

We shall establish that the composite control (7.22) possesses three important properties. First, it is a stabilizing control. Second, the full system (7.1) controlled by u_{c} results in a bounded value $J_{\mathrm{c}}(x, z)$ of the performance index. Third, we obtain upper and lower bounds on $J_{\mathrm{c}}(x, z)$ in terms of $V(x)$, $\psi(x)$ and $W(x, z)$.

The fact that (7.22) is a stabilizing control will follow from Theorem 6.1 if $V(x)$ and $W(x, z)$ are shown to satisfy the interconnection conditions (6.15) and (6.16). To verify (6.15) we note that

$$\begin{aligned} \|f(x, z, u_{\mathrm{s}} + u_{\mathrm{f}}) - f(x, h(x, u_{\mathrm{s}}), u_{\mathrm{s}})\| &= \|(A_1 - B_1R^{-1}B_2^{\mathrm{T}}K)z_{\mathrm{f}}\| \\ &\le b_1\|z_{\mathrm{f}}\| = b_1\phi. \end{aligned} \tag{7.23}$$

Assumptions 7.3 and (7.23) yield

$$\left\| \frac{\partial V}{\partial x}[f(x, z, u_{\mathrm{s}} + u_{\mathrm{f}}) - f(x, h(x, u_{\mathrm{s}}), u_{\mathrm{s}})\right\| \le \beta_1\psi\phi. \tag{7.24}$$

Furthermore, in view of (7.7), (7.23) and Assumption 7.3, we have

$$\begin{aligned} \|f(x, z, u_{\mathrm{s}} + u_{\mathrm{f}})\| \le \|f(x, h(x, u_{\mathrm{s}}), u_{\mathrm{s}})\| &+ \|f(x, z, u_{\mathrm{s}} + u_{\mathrm{f}}) \\ &- f(x, h(x, u_{\mathrm{s}}), u_{\mathrm{s}})\| \le b_1\phi + b_2\psi. \end{aligned} \tag{7.25}$$

Also, from (7.20)

$$\frac{\partial W}{\partial x} = \sum_{i,j} z_{\mathrm{f}i}z_{\mathrm{f}j}\frac{\partial K_{ij}}{\partial x} + 2z_{\mathrm{f}}^{\mathrm{T}}K\frac{\partial}{\partial x}[A_2^{-1}(a_2 + B_2u_{\mathrm{s}})].$$

If K is independent of x, and B_x is chosen such that $(\partial/\partial x)[A_2^{-1}(a_2 + B_2u_{\mathrm{s}})]$ is bounded, which is always possible at least by taking B_x to be a bounded set, it will follow that $|\partial W/\partial x| \le c_6\phi$. If K is a function of x it is still possible to show that $|\partial W/\partial x|$ is bounded by ϕ by restricting (x, z) to a bounded set $B_x \times B_z$ where z_{f} is bounded. Thus it is always possible to show that $|\partial W/\partial x| \le c_6\phi$, which together with (7.25) imply (7.26):

$$\left\|\frac{\partial W}{\partial x} f(x, z, u_s + u_f)\right\| \le \gamma\phi^2 + \beta_2\psi\phi. \tag{7.26}$$

With inequalities (7.14), (7.21), (7.24) and (7.26), all the conditions of Theorem 6.1 are satisfied and the origin is an asymptotically stable equilibrium of the closed-loop system

$$\dot{x} = f(x, z, u_c), \tag{7.27a}$$

$$\varepsilon\dot{z} = g(x, z, u_c) \tag{7.27b}$$

for all $\varepsilon < \varepsilon_d$, where ε_d is given by (2.29) for some $d \in (0, 1)$. Furthermore, an estimate of the domain of attraction can be obtained by using the Lyapunov function $\nu = (1 - d)V + dW$. Let $D \subset B_x \times B_z$ be such an estimate. From now on, all our discussions are limited to $(x_0, z_0) \in D$.

Asymptotic stability of an equilibrium point at the origin is not sufficient to guarantee that an integral of the type (7.2) will be finite along the trajectories asymptotically converging to this equilibrium (for an example see Exercise 7.8). To qualify as a candidate for near optimality, u_c must produce a bounded cost for all $(x_0, z_0) \in D$. To show that this is the case we need first to verify the inequality

$$|L(x, z, u_c) - L(x, h(x, u_s), u_s)| \le \delta_1\phi^2 + \delta_2\psi\phi, \tag{7.28}$$

which is implied by Assumptions 7.1–7.3 since

$$\begin{aligned} L(x, z, u_c) - L(x, h(x, u_s), u_s) = {} & z_f^T(s - 2QA_2^{-1}a_2) \\ & + 2z_f^T(K + QA_2^{-1})B_2R_0^{-1} \\ & \times \left(s_0 + \tfrac{1}{2}B_0^T\left(\frac{\partial V}{\partial x}\right)^T\right) \\ & + z_f^T(Q + KB_2R^{-1}B_2^TK)z_f. \end{aligned}$$

The boundedness of J_c is established in Theorem 7.1.

Theorem 7.1

Suppose that Assumptions 7.1–7.4 hold and let α_0, α_1, α_2, β_1, β_2, γ, δ_1 and δ_2 be defined by inequalities (7.14), (7.21), (7.24), (7.26) and (7.28). Let d and e be positive numbers such that $d \in (0, 1)$ and $e > \alpha_0/\alpha_1(1 - d)$. Then there exists an $\varepsilon^*(d, e) < \varepsilon_d$ such that for all $\varepsilon \in (0, \varepsilon^*(d, e))$, (7.29) holds:

$$J_c(x_0, z_0) \le e[(1 - d)V(x_0) + dW(x_0)] = e\nu(x_0, z_0) \quad \forall (x_0, z_0) \in D. \tag{7.29}$$

Proof It can be seen that by inequalities (7.14), (7.21), (7.24), (7.26) and (7.28) we have

$$L(x,z,u_c) + e\dot{\nu} \le -e\begin{bmatrix}\psi\\ \phi\end{bmatrix}^{\mathrm{T}} \times \begin{bmatrix}(1-d)\alpha_1' & -\tfrac{1}{2}(1-d)\beta_1' - \tfrac{1}{2}d\beta_2' \\ -\tfrac{1}{2}(1-d)\beta_1' - \tfrac{1}{2}d\beta_2' & d\left(\dfrac{\alpha_2'}{\varepsilon} - \gamma\right)\end{bmatrix}\begin{bmatrix}\psi\\ \phi\end{bmatrix},$$

where

$$\alpha_1' = \alpha_1 - \frac{\alpha_0}{(1-d)e}, \quad \alpha_2' = \alpha_2, \quad \beta_1' = \beta_1 + \frac{\delta_2}{(1-d)e},$$

$$\beta_2' = \beta_2, \quad \gamma' = \gamma + \frac{\delta_1}{de}.$$

Hence, for all

$$\varepsilon \le \varepsilon^*(d,e) = \frac{\alpha_1'\alpha_2'}{\alpha_1'\gamma' + [\beta_1'(1-d) + \beta_2' d]^2/4d(1-d)}, \tag{7.30}$$

$$L(x,z,u_c) + e\dot{\nu} \le 0. \tag{7.31}$$

Integrating (7.31) from 0 to ∞ and using the fact that $\nu(x,z) \to 0$ as $t \to \infty$, the result follows. □

Notice that $\varepsilon^*(d,e) \to \varepsilon_d$ as $e \to \infty$, which means that we can have $\varepsilon^*(d,e)$ arbitrarily close to ε_d and still have bounded J_c.

Having established that J_c is bounded, we move now to compute upper and lower bounds on $J_c - V$. Let us define the constants $b_3 > 0$, $b_4 > 0$ and $b_5 \ge 0$ by means of the following inequalities, which follow from our Assumptions 7.1–7.4:

$$b_3^2\phi^2(z_f) \le W(x,z) \le b_4^2\phi^2(z_f), \tag{7.32}$$

$$\left\|\frac{\partial}{\partial x}\psi(x)\right\| \le b_5. \tag{7.33}$$

Let $(\tilde{x}(\sigma), \tilde{z}(\sigma))$ be the trajectory of the closed-loop system (7.27) starting at (t, x, z). For convenience, in the sequel we shall adopt the notation $f_c(x,z) = f(x,z,u_c)$ and $L_c(x,z) = L(x,z,u_c)$.

To obtain an upper bound on $J_c(x,z) - V(x)$ we consider the function

$$q(x,z) = V(x) - J_c(x,z) + \varepsilon[b_6V(x) + b_7W(x,z) + b_8\psi(x)W^{1/2}(x,z)], \tag{7.34}$$

where b_6, b_7 and b_8 will be specified later. If we can show that $q(x, z) \geq 0$ $\forall (x, z) \in D$, it will follow that $J_c - V \leq \varepsilon(b_6 V + b_7 W + b_8 \psi W^{1/2})$. The function q can be written as

$$q(x, z) = \int_t^\infty \left[-(1 + \varepsilon b_6) \frac{\mathrm{d}V(\tilde{x})}{\mathrm{d}\sigma} - L_c(\tilde{x}, \tilde{z}) - \varepsilon b_7 \frac{\mathrm{d}W(\tilde{x}, \tilde{z})}{\mathrm{d}\sigma} - \varepsilon b_8 \frac{\mathrm{d}}{\mathrm{d}\sigma}(\psi(\tilde{x}) W^{1/2}(x, z)) \right] \mathrm{d}\sigma. \tag{7.35}$$

The time derivative of V can be expressed as

$$\begin{aligned} \frac{\mathrm{d}}{\mathrm{d}\sigma} V(\tilde{x}) &= \frac{\partial V}{\partial x} f_c(\tilde{x}, \tilde{z}) \\ &= \frac{\partial V}{\partial x} \bar{a}_0(\tilde{x}) + \frac{\partial V}{\partial x} [f_c(\tilde{x}, \tilde{z}) - \bar{a}_0(\tilde{x})]. \end{aligned}$$

Using (7.13), which holds as an identity in x, yields

$$\frac{\mathrm{d}}{\mathrm{d}\sigma} V(\tilde{x}) = -L_0(\tilde{x}) + \frac{\partial V}{\partial x} [f_c(\tilde{x}, \tilde{z}) - \bar{a}_0(\tilde{x})], \tag{7.36}$$

where $L_0(x) \triangleq L(x, h(x, u_s(x)), u_s(x))$. Substituting (7.36) into (7.35), we obtain

$$\begin{aligned} q(x, z) &= \int_t^\infty \Big\{ (1 + \varepsilon b_6) L_0(\tilde{x}) - L_c(\tilde{x}, \tilde{z}) - \varepsilon b_7 \frac{\mathrm{d}W}{\mathrm{d}\sigma}(\tilde{x}, \tilde{z}) \\ &\quad - \varepsilon b_8 \frac{\mathrm{d}}{\mathrm{d}\sigma} [\psi(\tilde{x}) (W(\tilde{x}, \tilde{z}))^{1/2}] - (1 + \varepsilon b_6) \frac{\partial V}{\partial x} \\ &\quad \times (f_c(\tilde{x}, \tilde{z}) - \bar{a}_0(\tilde{x})) \Big\} \mathrm{d}\sigma \\ &\triangleq \int_t^\infty Q(\tilde{x}, \tilde{z}) \, \mathrm{d}\sigma. \end{aligned} \tag{7.37}$$

Using inequalities (7.14), (7.21), (7.24), (7.25), (7.26), (7.28), (7.32) and (7.33), it can be shown that

$$Q(x, z) \geq \begin{bmatrix} \psi(x) \\ \phi(z_f) \end{bmatrix}^{\mathrm{T}} \begin{bmatrix} a_{11} & -a_{12} \\ -a_{12} & a_{22} \end{bmatrix} \begin{bmatrix} \psi(x) \\ \phi(z_f) \end{bmatrix}, \tag{7.38}$$

where

$$a_{11} = \varepsilon\left(b_6\alpha_1 - \frac{\beta_2 b_8}{2b_3}\right),$$

$$a_{12} = 0.5\varepsilon\left[\beta_1 b_6 + \beta_2 b_7 + \frac{\gamma b_8}{2b_3} + b_2 b_4 b_5 b_8\right] + 0.5\left[\delta_2 + \beta_1 - \frac{\alpha_2 b_8}{2b_4}\right],$$

$$a_{22} = \alpha_2 b_7 - \delta_1 - \varepsilon b_7 \gamma - \varepsilon b_1 b_4 b_5 b_8.$$

Choosing b_6, b_7 and b_8 to satisfy

$$b_8 = 2b_4(\delta_2 + \beta_1)/\alpha_2, \quad b_6 > \frac{\beta_2 b_8}{2\alpha_1 b_3}, \quad b_7 > \frac{\delta_1}{\alpha_2}, \tag{7.39}$$

it can easily be seen that the right side of (7.38) will be nonnegative for small ε. Indeed, it is nonnegative for all $\varepsilon \in (0, \varepsilon_1)$ where

$$\left.\begin{aligned} \varepsilon_1 &= \frac{\tilde{\alpha}_1\tilde{\alpha}_2}{\tilde{\alpha}_1\gamma + [\tilde{\beta}_1 + b_7\tilde{\beta}_2]^2/4b_7}, \\ \tilde{\alpha}_1 &= \alpha_1 b_6 - \frac{\beta_2 b_8}{2b_3}, \quad \tilde{\alpha}_2 = \alpha_2 - \frac{\delta_1}{b_7}, \\ \tilde{\beta}_1 &= \beta_1 b_6 + \frac{\gamma b_8}{2b_3} + b_2 b_4 b_5 b_8, \\ \tilde{\beta}_2 &= \beta_2, \quad \tilde{\gamma} = \gamma + \frac{b_1 b_4 b_5 b_8}{b_7}. \end{aligned}\right\} \tag{7.40}$$

A lower bound on $J_c - V$ is obtained, similarly, by showing that

$$p(x, z) = -V(x) + J_c(x, z) + \varepsilon[b_6 V(x) + b_7 W(x, z) + b_8 \psi(x)(W(x, z))^{1/2}] \tag{7.41}$$

is nonnegative for all $(x, z) \in D$. It is interesting that this is true for $\varepsilon \in (0, \varepsilon_1)$, with the same ε_1 as in (7.40). Now, taking $t = 0$, $x = x_0$ and $z = z_0$, we arrive at the following theorem.

Theorem 7.2

Suppose that Assumptions 7.1–7.4 hold and let the constants α_0, α_1, α_2, β_1, β_2, γ, δ_1, δ_2 and b_1 to b_5 be defined by inequalities (7.14), (7.21),

(7.24), (7.25), (7.26), (7.28), (7.32) and (7.33). Choose positive numbers d, e, b_6 and b_7 that satisfy

$$0 < d < 1, \quad e > \frac{\alpha_0}{(1-d)\alpha_1}, \quad b_6 > \frac{b_4\beta_2(\beta_1+\delta_2)}{\alpha_1\alpha_2 b_3}, \quad b_7 > \frac{\delta_1}{\alpha_2}.$$

Then, for all $\varepsilon \in (0, \min\{\varepsilon^*(e,d), \varepsilon_1\}$ and for all initial conditions $(x_0, z_0) \in D$, we have

$$|J_c(x_0,z_0) - V(x_0)| \leq \varepsilon[b_6 V(x_0) + b_7 W(x_0,z_0) + b_8\psi(x_0)(W(x_0,z_0))^{1/2}], \tag{7.42}$$

where $\varepsilon^*(e, d)$, ε_1 and b_8 are given by (7.30), (7.40) and (7.39) respectively.

As a consequence of Theorem 7.2 we have

Corollary 7.1

Under Assumptions 7.1–7.4 and for all $(x_0, z_0) \in D$

$$J_c(x_0,z_0) \to V(x_0) \quad \text{as } \varepsilon \to 0. \tag{7.43}$$

Corollary 7.1 shows that u_c is near-optimal in the sense that as $\varepsilon \to 0$ the performance of the system under the composite control tends to the optimal performance at $\varepsilon = 0$.

In calculating the bounds on ε and the performance deviation bound (7.42), one does not have to start from Assumption 7.2, 7.3, and work one's way through verifying the various inequalities as we have just done in our analysis. A more direct and less conservative procedure is to obtain the Lyapunov functions $V(x)$ and $W(x, z)$ and then proceed to verify the inequalities listed in the statement of Theorem 7.2. The fact that this group of inequalities is indeed what is needed to establish the bound (7.42) brings out a fundamental point in our analysis: neither optimality of u_s and u_f nor the linearity in z and u is crucial for the analysis. What is really needed are u_s, u_f and V, W for which that group of inequalities hold. For further treatment of the problem in that general framework the reader is referred to Saberi and Khalil (1985).

Example 7.1

The steps of the composite control design and the calculation of the bounds will be illustrated by a simple example:

$$\dot{x} = -\tfrac{3}{4}x^3 + z, \tag{7.44a}$$

$$\varepsilon\dot{z} = -z + u, \tag{7.44b}$$

$$J = \int_0^\infty [x^6 + \tfrac{3}{4}z^2 + \tfrac{1}{4}u^2]\, dt. \tag{7.45}$$

Assumption 7.1 is clearly satisfied $\forall x \in R$. The slow problem is given by

$$\dot{x} = -\tfrac{3}{4}x^3 + u_s, \tag{7.46}$$

$$J_s = \int_0^\infty (x^6 + u_s^2)\, dt. \tag{7.47}$$

Assumption 7.2 is satisfied with $\psi(x) = |x|^3$. For $\partial\psi/\partial x = 3x^2 \operatorname{sgn} x$ to be bounded on B_x, we have to take B_x as a bounded set. Let $B_x = [-\tfrac{1}{2}, \tfrac{1}{2}]$. The Hamilton–Jacobi equation is

$$x^6 - \tfrac{3}{4}x^3 \frac{\partial V}{\partial x} - \frac{1}{4}\left(\frac{\partial V}{\partial x}\right)^2 = 0, \quad V(0) = 0,$$

whose positive-definite solution satisfies

$$\frac{\partial V}{\partial x} = x^3, \quad V(0) = 0.$$

Thus

$$V(x) = \tfrac{1}{4}x^4, \quad u_s = -\tfrac{1}{2}x^3, \quad \bar{a}_0(x) = -\tfrac{5}{4}x^3.$$

Obviously, Assumption 7.3 is satisfied. Moreover,

$$\frac{\partial V}{\partial x}\bar{a}_0(x) = -\tfrac{5}{4}x^6,$$

which shows that (7.14) holds with $\alpha_0 = \alpha_1 = \tfrac{5}{4}$. The fast problem is given by

$$\frac{dz_f}{d\tau} = -z_f + u_f,$$

$$J_f = \int_0^\infty (\tfrac{3}{4}z_f^2 + \tfrac{1}{4}u_f^2)\, d\tau.$$

where $z_f = z - u_s = z + \tfrac{1}{2}x^3$, which yields

$$-2K + \tfrac{3}{4} - 4K^2 = 0,$$

the positive root of which is $K = \frac{1}{4}$, so that

$$u_{\mathrm{f}} = -z_{\mathrm{f}}.$$

Therefore, $W(x, z) = \frac{1}{4}z_{\mathrm{f}}^2$ and

$$\frac{\partial W}{\partial z}(-z + u_{\mathrm{s}} + u_{\mathrm{f}}) = -z_{\mathrm{f}}^2,$$

so that (7.21) is satisfied with $\alpha_2 = 1$. It is now straightforward to verify (7.24), (7.25), (7.26), (7.28), (7.32) and (7.33) to obtain the constants $\beta_1 = 1$, $b_1 = 1$, $b_2 = \frac{5}{4}$, $\gamma = \frac{3}{16}$, $\beta_2 = \frac{15}{64}$, $\delta_1 = 1$, $\delta_2 = \frac{1}{2}$, $b_3 = b_4 = \frac{1}{2}$ and $b_5 = \frac{3}{4}$. Let us start by obtaining an estimate of the domain of attraction. We need to choose d. Because the fast problem is linear–quadratic, we can arbitrarily extend the domain of attraction in the direction of the z-axis by choosing d sufficiently small. However, we know from Fig. 7.1 that $\varepsilon_d \to 0$ as $d \to 0$, so d should be kept large enough to produce an acceptable bound ε_d. Let us suppose that ε takes values less than one, so that any $\varepsilon_d > 1$ will be acceptable. We choose $d = \frac{1}{4}$, which, using (2.29), yields $\varepsilon_d = 1.13$. Taking into consideration that the domain of attraction should be contained in $B_x \times R$, where $B_x = [-\frac{1}{2}, \frac{1}{2}]$, we take D as

$$D = \{(x, z) : x^4 + \tfrac{1}{3}(z + \tfrac{1}{2}x^3)^2 \le \tfrac{1}{16}\}.$$

So as to apply Theorem 7.2, we start by choosing e. As we mentioned before, the choice of e is not crucial. Typically e will be chosen large enough to yield $\varepsilon^*(d, e)$ very close to ε_d. Taking e large increases the upper bound on J_{c} given by Theorem 7.1, but this is not important since a sharper bound on J_{c} is given by Theorem 7.2. Taking $e = 1000$ and using (7.30), we obtain $\varepsilon^*(d, e) = 1.123$. Now we choose b_6 and b_7, which should satisfy

$$b_6 > 0.281, \quad b_7 > 1.$$

Let us take $b_6 = 1$ and $b_7 = 8$, then using (7.40) we have $\varepsilon_1 = 1.127$. Thus we conclude that for all $\varepsilon \in (0, 1.123)$ and for all $(x_0, z_0) \in D$ the following bound holds:

$$|J_{\mathrm{c}}(x_0, z_0) - V(x_0)| \le \varepsilon[0.25x_0^4 + 2(z_0 + 0.5x_0^3)^2 + 0.75|x_0|^3|z_0 + 0.5x_0^3|].$$

In particular, using the fact that for any $(x, z) \in D$, $|x| \le 0.5$ and $|z + 0.5x^3| \le \frac{1}{4}\sqrt{3}$, we obtain

$$|J_{\mathrm{c}}(x_0, z_0) - V(x_0)| \le 0.431\varepsilon.$$

7.8 Exercises

Exercise 7.1

Consider the second-order system

$$\dot{x} = z^2 - 2x^3,$$

$$\varepsilon\dot{z} = x^3 - \tan z,$$

which has an isolated equilibrium point at the origin.

(a) Show that asymptotic stability of the origin cannot be established by linearization.

(b) Using Theorem 2.1, show that the origin is asymptotically stable for all $\varepsilon \in (0, \varepsilon^*)$. Obtain estimates of ε^* and the domain of attraction.

Hint: Let $B_x = \{x \in R : |x| \leq 1\}$, $B_z = \{z \in R : |z| \leq a\}$, where a is less than, but arbitrarily close to, $\frac{1}{2}\pi$; $V(x) = \frac{1}{4}x^4$, $W(x, z) = \frac{1}{2}(z - \tan^{-1} x^3)^2$.

Exercise 7.2

Consider the tunnel diode circuit shown in Fig. 7.3. The nonlinearity $\xi(v)$ is sketched in Fig. 7.4. Suppose the values of E and R are such that there is a unique equilibrium point (i_0, v_0) as shown in Fig. 7.4.

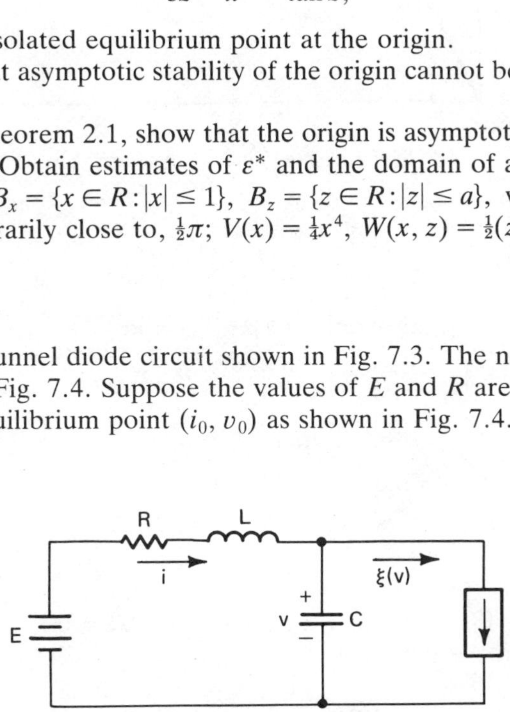

Fig. 7.3. Tunnel diode circuit of Exercise 7.2.

(a) Choosing $x = (v - v_0)/v_0$ and $z = (i - i_0)/i_0$ as dimensionless state variables, $t' = t/CR$ as dimensionless time and $\varepsilon = (L/R)/CR = L/CR^2$, show that the state equation takes the singularly perturbed form

$$\dot{x} = -\eta(x) + az,$$

$$\dot{z} = -\frac{1}{a}x - z,$$

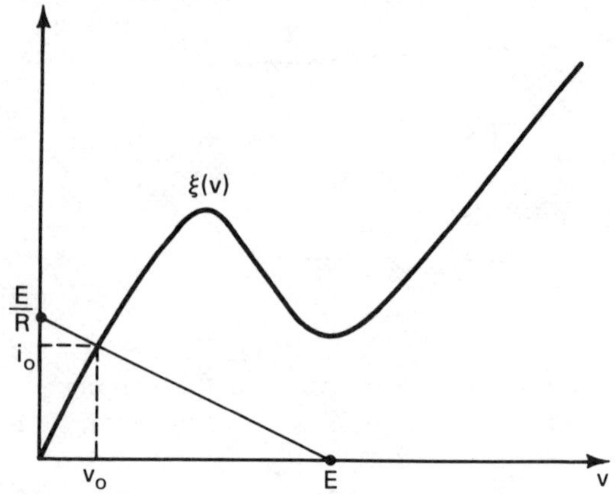

Fig. 7.4. Tunnel diode nonlinearity.

where

$$\eta(x) = \frac{R}{v_0}[\xi(xv_0 + v_0) - \xi(v_0)]$$

and

$$a = \frac{Ri_0}{v_0} = \frac{E - v_0}{v_0}.$$

(b) Show that the reduced system is given by

$$\dot{x} = -\eta(x) - x,$$

and verify using Fig. 7.4 that $x[\eta(x) + x] > 0 \quad \forall x \neq 0$.

(c) Verify that the functions

$$V(x) = \int_0^x [\eta(y) + y]\, dy$$

and

$$W(x, z) = \frac{1}{2}\left(z + \frac{1}{a}x\right)^2$$

satisfy all the assumptions of Theorem 2.1. In particular, show that (2.20), (2.21), (2.24) and (2.25) hold $\forall (x, z) \in R^2$ as equalities with $\alpha_1 = \alpha_2 = 1$, $\beta_1 = a$, $\beta_2 = -1/a$, $\gamma = 1$,

$$\psi(x) = \eta(x) + x$$

and

$$\phi(z - h(x)) = z + \frac{1}{a}x.$$

(d) Show that the derivative of

$$\nu(x, z) = (1 - d)\int_0^x [\eta(y) + y]\,dy + \frac{d}{2}\left(z + \frac{1}{a}x\right)^2$$

along the trajectory of the singularly perturbed system is given by

$$\dot{\nu} = -\begin{bmatrix} \psi \\ \phi \end{bmatrix}^{\mathrm{T}} \begin{bmatrix} 1 - d & -\tfrac{1}{2}(1 - d)a + \dfrac{d}{2a} \\ -\tfrac{1}{2}(1 - d)a + \dfrac{d}{2a} & d\left(\dfrac{1}{\varepsilon} - 1\right) \end{bmatrix} \begin{bmatrix} \psi \\ \phi \end{bmatrix}.$$

(e) Show that the equilibrium is globally asymptotically stable for all $\varepsilon < 1$.

Exercise 7.3

Verify that V and W given in the proof of Corollary 2.3 satisfy all the assumptions of Corollary 2.2.

Exercise 7.4

Verify that V and W of Example 5.1 satisfy Assumptions 5.2–5.4 as claimed.

Exercise 7.5

The stability analysis of Sections 7.2 and 7.5 employs quadratic-type Lyapunov functions. In this exercise, an alternative analysis that employs linear-type Lyapunov functions is developed. Consider the system (5.1) and let Assumption 5.1 hold. Let $V(t, x)$ and $W(t, x, z)$ be Lyapunov functions satisfying part (i) of Assumptions 5.2 and 5.3 as well as the following inequalities:

$$\frac{\partial V}{\partial t} + \frac{\partial V}{\partial x} f(t, x, h(t, x), 0) \le -\alpha_1 \psi(x), \qquad \alpha_1 > 0,$$

$$\frac{\partial W}{\partial z} g(t, x, z, 0) \le -\alpha_2 \phi(z - h(t, x)), \quad \alpha_2 > 0,$$

$$\frac{\partial V}{\partial x}[f(t, x, z, \varepsilon) - f(t, x, h(t, x), 0)] \le \beta_1 \phi(z - h(t, x)) + \varepsilon\gamma_1 \psi(x),$$

$$\frac{\partial W}{\partial z}[g(t,x,z,\varepsilon)-g(t,x,z,0)] \le \varepsilon\gamma_2'\phi(z-h(t,x))+\varepsilon\beta_2'\psi(x),$$

$$\frac{\partial W}{\partial t}+\frac{\partial W}{\partial x}f(t,x,z,\varepsilon) \le \gamma_2''\phi(z-h(t,x))+\beta_2''\psi(x),$$

where $\psi(\cdot)$ and $\phi(\cdot)$ are positive-definite functions. For convenience let β_1, γ_1,β_2', β_2'', γ_2' and γ_2'' be nonnegative.

Let $\nu(t,x,z)=(1-d)V(t,x)+dW(t,x,z)$. Show that for all

$$\varepsilon<\varepsilon^*=\frac{\alpha_1\alpha_2}{\alpha_1\gamma_2+\alpha_2\gamma_1+\beta_1\beta_2},$$

where $\beta_2=\beta_2'+\beta_2''$ and $\gamma_2=\gamma_2'+\gamma_2''$, d can be chosen such that the derivative of ν along the trajectory of (5.1) is negative-definite. Give the range of permissible values of d.

Exercise 7.6

Apply the stability analysis of Exercise 7.5 to Example 2.1. Take $V(x)=(x^2)^{1/2}$ and $W(x,z)=[(x+z)^2]^{1/2}$. Calculate a bound on ε and show your choice of d.

Exercise 7.7

Let all the assumptions of Corollary 2.2 hold with $\psi(x)=\|x\|$ and $\phi(z-h(x))=\|z-h(x)\|$, and take $\alpha\in(0,\alpha_1/2e_2)$. Show that d in $\nu(x,z)=(1-d)V(x)+dW(x,z)$ can be chosen such that $\dot{\nu}\le -2\alpha\nu$ for all

$$\varepsilon<\varepsilon_\alpha=\frac{(\alpha_1-2\alpha e_2)\alpha_2}{(\alpha_1-2\alpha e_2)(\gamma+2\alpha e_4)+\beta_1\beta_2}.$$

Hence α is an estimate of the degree of exponential stability. Estimate the degree of exponential stability of the system

$$\dot{x}=-4x+2z,$$

$$\varepsilon\dot{z}=0.2x-4z$$

when $\varepsilon=0.1$.

Exercise 7.8

Consider the system $\dot{x}=x^2+u$ and the cost $J=\int_0^\infty(x^4+\frac{1}{2}u^2)\,dt$. Let $u=-x^2-x^5$.

(a) Show that the closed-loop system has an asymptotically stable equilibrium point at $x = 0$.

(b) Show that for $x(0) = x_0 \neq 0$, $x(t) = \operatorname{sign}(x_0)\,[4t + x_0^{-4}]^{-1/4}$.

(c) Show that J is infinite.

Exercise 7.9

(a) Consider the linear–quadratic regulator defined by (4.1)–(4.3) of Chapter 3. Suppose that $[M_1 \quad M_2]$ is a square nonsingular matrix, and the slow system pair (A_0, B_0) and the fast system pair (A_{22}, B_2) are each controllable. Verify that Assumptions 7.1–7.4 are satisfied.

(b) Suppose that (A_0, B_0) and (A_{22}, B_2) are only stabilizable. Clearly Assumption 7.4 is not satisfied. Verify, however, that all the inequalities stated in Theorem 7.2 are satisfied. Hence Theorem 7.2 holds.

Exercise 7.10

Apply Theorem 7.2 to Example 4.1 of Chapter 3 to calculate a bound on $|J_c(x_0, z_0) - V(x_0)|$ as in (7.42).

7.9 Notes and References

Stability properties of nonlinear singularly perturbed systems have been extensively studied in the literature. The exposition of Sections 7.2, 7.3 and 7.5 using quadratic-type Lyapunov functions follows the recent treatment of Saberi and Khalil (1984). For further reading consult Klimushchev and Krasovskii (1962), Hoppensteadt (1967), Habets (1974), Chow (1978a) and Grujic (1981). The adaptive control example is patterned on Ioannou and Kokotović (1984a,b). A key point in the composite control development of Section 7.6 is the design of the fast control to be inactive on the slow manifold of the singularly perturbed system under slow control. As a consequence of this design, the slow system is invariant with respect to the fast design. This invariance property, which was observed by Suzuki (1981), has been shown in Chapter 3 for linear systems using matrix manipulations. The presentation of Section 7.6 is based on Saberi and Khalil (1985). The near-optimal regulation of Section 7.7 was first given in Chow and Kokotović (1981). Out treatment is based on Saberi and Khalil (1985), where the restriction of the linear–quadratic form of the regulator problem with respect to the fast states and control inputs is removed.

REFERENCES

Abed, E. H. (1985a). Multiparameter singular perturbation problems: Iterative expansions and asymptotic stability. *Syst. Control Lett.* **5,** 279–282.

Abed, E. H. (1985b). A new parameter estimate in singular perturbations. *Syst. Control Lett.* **6,** 193–198.

Ahmed-Zaid, S., Sauer, P. W., Pai, M. A. and Sarioglu, M. K. (1982). Reduced-order modeling of synchronous machines using singular perturbations. *IEEE Trans. Circuits Systems* **29,** 782–786.

Allemong, J. J. and Kokotović, P. V. (1980). Eigensensitivities in reduced-order modeling, *IEEE Trans. Autom. Control* **25,** 821–822.

Altshuler, D. and Haddad, A. H. (1978). Near-optimal smoothing for singularly perturbed linear systems. *Automatica* **14,** 81–87.

Anderson, B. D. O. and Moore, J. B. (1971). *Linear Optimal Control.* Prentice-Hall, Englewood Cliffs, New Jersey.

Anderson, B. D. O., Bitmead, R. R., Johnson, C. R. Jr., Kokotović, P. V., Kosut, R. L., Mareels, J. M., Praly, L. and Riedle, B. D. (1986). *Stability of Adaptive Systems: Passivity and Averaging Analysis.* Series in Signal Processing, Optimization, and Control. M.I.T. Press, Cambridge, Mass.

Anderson, L. (1978). Decomposition of two-time-scale linear systems. *Proc. JACC*, pp. 153–163.

Anderson, P. M. and Fouad, A. A. (1977). *Power System Control and Stability.* Iowa State University Press.

Andreev, Y. H. (1982). Differential geometry method in control theory—a survey. *Automatika Telemachanika* **10,** 5–46

Ardema, M. D. (1976). Solution of the minimum time-to-climb problem by matched asymptotic expansions. *AIAA J.* **14,** 843–850.

Ardema, M. D. (1979). Linearization of the boundary layer equations for the minimum time to climb problem. *AIAA J. Guidance Control* **2,** 434–436.

Ardema, M. D. (1980). Nonlinear singularly perturbed optimal control problems with singular arcs. *Automatica* **16,** 99–104.

Ardema, M. D. (1983). Singular perturbations in systems and control. *CISM Courses and Lectures*, 180. Springer, New York.

Asatani, K. (1974). Suboptimal control of fixed-end-point minimum energy problem via singular perturbation theory. *J. Math. Anal. Applic.* **45,** 684–697.

Asatani, K. (1976). Near-optimum control of distributed parameter systems via singular perturbation theory. *J. Math. Anal. Applic.* **54,** 799–819.

Asatani, K., Shiotani, M. and Huttoni, Y. (1977). Suboptimal control of nuclear reactors with distributed parameters using singular perturbation theory. *Nucl. Sci. Engng* **6,** 119.

Athans, M. and Falb, P. L. (1966). *Optimal Control: An Introduction to the Theory and its Applications.* McGraw-Hill, New York.

Avramović, B. (1979). Subspace iteration approach to the time scale separation. *Proc. IEEE Conf. on Decision and Control*, pp. 684–697.

Avramović, B. (1980). Time scales, coherency, and weak coupling. Ph.D. thesis, Coordinated Science Laboratory; Report R-895, Univ. Illinois, Urbana.

Avramović, B., Kokotović, P. V., Winkelman, J. R. and Chow, J. H. (1980). Area decomposition of electromechanical models of power systems. *Automatica* **16,** 637–648

Bagirova, N., Vasil'eva, A. B. and Imanaliev, M. I. (1967). The problem of asymptotic solutions of optimal control problems. *Diff. Eqns* **3,** 985–988

Balas, M. J. (1978). Observer stabilization of singularly perturbed systems. *AIAA J. Guidance Control* **1,** 93–95.

Balas, M. J. (1982). Reduced order feedback control of distributed parameter systems via singular perturbation methods. *J. Math. Anal. Applic.* **87,** 281–294

Bell, D. J. and Jacobson, D. H. (1975). *Singular Optimal Control Problems.* Academic Press, New York.

Bensoussan, A. (1981). Singular perturbation results for a class of stochastic control problems. *IEEE Trans. Autom. Control* **26,** 1071–1080.

Bensoussan, A. (1984). On some singular perturbation problems arising in optimal control. *Stochastic Anal. Applic.* **2,** 13–53.

Bensoussan, A. (1986). *Perturbation Methods in Optimal Control.* Dunod and Wiley, New York.

Bensoussan, A., Lions, J. L. and Papanicolaou, G. C. (1978). *Asymptotic Analysis for Periodic Structures.* North-Holland, New York.

Blankenship, G. (1978). On the separation of time scales in stochastic differential equations. *Proc. 7th IFAC Congress, Helsinki*, pp. 937–944.

Blankenship, G. (1979). Asymptotic analysis in mathematical physics and control theory: some problems with common features. *Richerche di Automatica* **10,** 2.

Blankenship, G. (1981). Singularly perturbed difference equations in optimal control problems. *IEEE Trans. Autom. Control.* **26,** 911–917.

Blankenship, G. (Ed.) (1986). *Asymptotic Methods in Modelling and Control.* Springer Verlag, Berlin.

Blankenship, G. and Meyer D. (1977). Linear filtering with wide band noise disturbances. *Proc. 16th IEEE Conf. on Decision and Control*, pp. 580–584.

Blankenship, G. and Papanicolaou, G. C. (1978). Stability and control of stochastic systems with wide-band noise disturbance. *SIAM J. Appl. Math.* **34**, 437–476.

Blankenship, G. and Sachs, S. (1979). Singularly perturbed linear stochastic ordinary differential equations. *SIAM J. Math. Anal.* **10,** 306–320.

Bogoliubov, N. N. and Mitropolsky, Y. A. (1961). *Asymptotic Methods in the Theory of Non-Linear Oscillations*, 2nd edn. Hindustan Publishing, Delhi.

Bratus, A. S. (1977). Asymptotic solutions of some probabilistic optimal control problems. *Appl. Math. Mech.* **41,** 13.

Brauner, C. M. (1978). Optimal control of a perturbed system in enzyme kinetics. *Proc. 7th IFAC Congress, Helsinki*, pp. 945–948.

Brockett, R. W. (1970). *Finite Dimensional Linear Systems.* Wiley, New York.

Bryson, A. E. and Ho, Y. C. (1975). *Applied Optimal Control.* Hemisphere, Washington, DC.

Butuzov, V. F. and Fedoryuk, M. V. (1970). Asymptotic methods in theory of ordinary differential equations. *Progress in Mathematics*, Vol. 8, ed. R. V. Gamkrelidze, pp. 1–82. Plenum Press, New York.

Butuzov, V. F. and Vasil'eva, A. B. (1970). Differential and difference equation systems with a small parameter in the case when unperturbed (degenerated) system is on the spectrum. *Diff. Eqns* **6,** 499–510.

Calise, A. J. (1976). Singular perturbation methods for variational problems in aircraft flight. *IEEE Trans. Autom. Control* **21,** 345–353.

Calise, A. J. (1978). A new boundary layer matching procedure for singularly perturbed systems. *IEEE Trans. Autom. Control* **23,** 434–438.

Calise, A. J. (1979). A singular perturbation analysis of optimal aerodynamic and thrust magnitude control. *IEEE Trans. Autom Control* **24,** 720–730.

Calise, A. J. (1980). A singular perturbation analysis of optimal thrust control with proportional navigation guidance. *AIAA J. Guidance Control* **3,** 312–318.

Calise, A. J. (1981). Singular perturbation theory for on-line optimal flight path control. *AIAA J. Guidance Control* **4,** 398–405.

Calise, A. J. and Moerder, D. D. (1985). Optimal output feedback design of systems with ill-conditioned dynamics. *Automatica* **21,** 271–276.

Campbell, S. L. (1978). Singular perturbation of autonomous linear systems II. *Diff. Eqns* **29,** 362–373.

Campbell, S. L. (1980). *Singular Systems of Differential Equations.* Pitman, New York.

Campbell, S. L. (1981). A more singular singularly perturbed linear system. *IEEE Trans. Autom. Control* **26,** 507–510.

Campbell, S. L. (1981). On an assumption guaranteeing boundary layer convergence of singularly perturbed systems. *Automatica* **17,** 645–646.

Campbell, S. L. (1982). *Singular Systems of Differential Equations II.* Pitman, New York.

Campbell, S. L. and Rose, N. J. (1978). Singular perturbation of autonomous linear systems III. *Houston J. Math.* **4,** 527–539.

Campbell, S. L. and Rose, N. J. (1979). Singular perturbation of autonomous linear systems. *SIAM J. Math. Anal.* **10,** 542–551.

Carr, J. (1981). *Applications of Centre Manifold Theory.* Lecture Notes in Applied Mathematical Sciences, Vol. 35. Springer, New York.

Chakravarty, A. (1984). Optimal selection of cost index for airline fleet hub operation. *AIAA Guidance and Control Conf., Seattle, Washington.*

Chakravarty, A. (1985). Four-dimensional aircraft guidance in the presence of winds. *AIAA J. Guidance Control Dynam.* **8,** 16–22.

Chakravarty, A. and Borrows, J. W. (1984). Time controlled aircraft guidance in uncertain winds and temperatures. *American Control Conf., San Diego, California.*

Chakravarty, A. and Vagners, J. (1981). Application of singular perturbation theory to onboard aircraft trajectory optimization. *AIAA Paper* 81-0019.

Chakravarty, A. and Vagners, J. (1982). Development of 4D guidance laws using singular perturbation methodology. *America Control Conf., Arlington, Virginia.*

Chakravarty, A. and Vagners, J. (1983). 4D aircraft flight path management in real time. *American Control Conf., San Francisco, California.*

Chang, K. W. (1969). Remarks on a certain hypothesis in singular perturbations. *Proc. Am. Math. Soc.* **23,** 41–45.

Chang, K. W. (1972). Singular perturbations of a general boundary problem. *SIAM J. Math. Anal.* **3,** pp. 520–526.

Chang, K. W. and Coppel, W. A. (1969). Singular perturbations of initial value problems over a finite interval. *Arch. Rat. Mech. Anal.* **32,** 268–280.

Chen, C. T. (1984). *Linear System Theory and Design.* Holt, Rinehart and Winston, New York.

Chow, J. H. (1977a). Singular perturbation of nonlinear regulators and systems with oscillatory modes. Ph.D. thesis, Coordinated Science Laboratory; Report R-801, Univ. Illinois, Urbana.

Chow, J. H. (1977b). Preservation of controllability in linear time invariant perturbed systems. *Int. J. Control* **25,** 697–704.

Chow, J. H. (1978a). Asymptotic stability of a class of nonlinear singularly perturbed systems. *J. Franklin Inst.* **306,** 275–278.

Chow, J. H. (1978b). Pole-placement design of multiple controllers via weak and strong controllability. *Int. J. Systems Sci.* **9,** 129–135.

Chow, J. H. (1979). A class of singularly perturbed nonlinear, fixed endpoint control problems. *J. Optim. Theory Applic.* **29,** 231–251.

Chow, J. H. (Ed.) (1982). Time Scale Modeling of Dynamic Networks. *Lecture Notes in Control and Information Sciences*, **47**. Springer, New York.

Chow, J. H., Allemong, J. J. and Kokotović, P. V. (1978). Singular perturbation analysis of systems with sustained high frequency oscillations. *Automatica* **14,** 271–279.

Chow, J. H., Cullum, J. and Willoughby, R. A. (1984). A sparity-based technique for identifying slow-coherent areas in large power systems. *IEEE Trans. Power Appar. Systems* **103,** 463–471.

Chow, J. H. and Kokotović, P. V. (1976a). A decomposition of near-optimum regulators for systems with slow and fast modes. *IEEE Trans. Autom. Control* **21,** 701–705.

Chow, J. H. and Kokotović, P. V. (1976b). Eigenvalue placement in two-time-scale systems. *Proc. IFAC Symp. on Large Scale Systems*, *Udine*, *Italy*, pp. 321–326.

Chow, J. H. and Kokotović, P. V. (1978a). Near-optimal feedback stabilization of a class of nonlinear singularly perturbed systems. *SIAM J. Control Optim.* **16,** 756–770.

Chow, J. H. and Kokotović, P. V. (1978b). Two-time-scale feedback design of a class of nonlinear systems. *IEEE Trans. Autom. Control* **23,** 438–443.

Chow, J. H. and Kokotović, P. V. (1981). A two-stage Lyapunov–Bellman feedback design of a class of nonlinear systems. *IEEE Trans. Autom. Control* **26,** 656–663.

Chow, J. H. and Kokotović, P. V. (1983). Sparsity and time scales. *Proc. 1983 American Control Conf.*, *San Francisco*, Vol. 2, pp. 656–661.

Chow, J. H. and Kokotović, P. V. (1985). Time-scale modeling of sparse dynamic networks. *IEEE Trans. Autom. Control* **30,** 714–722.

Chow, J. H. Kokotović, P. V. and Hwang Y. K. (1983). Aggregate modeling of dynamic networks with sparse interconnections. *Proc. 22nd Decision and Control Conf. San Antonio*, pp. 223–229.

Cobb, D. (1981). Feedback and pole-placement in descriptor-variable systems. *Int. J. Control* **33,** 1135–1146.

Cobb, D. (1983). Descriptor-variable systems and optimal state regulation. *IEEE Trans. Autom. Control* **28,** 601–611.

Cobb, D. (1984). Controllability, observability and duality in singular systems. *IEEE Trans. Autom. Control* **2,** 1076–1082.

Cobb, D. (1984). Slow and fast stability in singular systems. *Proc. 23rd Decision and Control Conf., Las Vegas, Nevada*, pp. 280–282.

Coderch, M., Willsky, A. S., Sastry, S. S. and Castanon, D. A. (1983). Hierarchical aggregation of linear systems with multiple time scales. *IEEE Trans. Autom. Control* **28,** 1017–1030.

Collins, W. B. (1973). Singular perturbations of linear time-optimal control. *Recent Mathematical Developments in Control.* ed. D. J. Bell, pp. 123–136. Academic Press, New York.

Coppel, W. A. (1965). *Stability and Asymptotic Behavior of Differential Equations*. D. C. Heath, Boston.

Coppel, W. A. (1967). Dichotomies and reducibility. *J. Diff. Eqns.* **3,** 500–521.

Cori, R. and Maffezzoni, C. (1984). Practical optimal control of a drum boiler power plant. *Automatica* **20,** 163–173.

Dauphin-Tanguy, G. and Borne, P. (1985). Order reduction of multi-time scale systems using bond graphs, the reciprocal system and the singular perturbation method. *J. Franklin Inst.* **319,** 157–171.

Delebecque, F. (1983). A reduction process for perturbed Markov chains. *SIAM J. Appl. Math.* **43,** 325–350.

Delebecque, F. and Quadrat, J. P. (1978). Contribution of stochastic control, singular perturbation averaging and team theories to an example of large scale systems: management of hydropower production. *IEEE Trans. Autom. Control* **23,** 209–222.

Delebecque, F. and Quadrat, J. P. (1981). Optimal control of Markov chains admitting strong and weak interactions. *Automatica* **17,** 281–296.

Delebecque, F., Quadrat, J. P. and Kokotović, P. V. (1984). A unified view of aggregation and coherency in networks and Markov chains. *Int. J. Control* **40,** 939–952.

Demello, F. P. and Concordia, C. (1969). Concepts of synchronous machine stability as affected by excitation control. *IEEE Trans. Power Appar. Systems* **88,** 316–329.

Desoer, C. A. (1970). Singular perturbation and bounded input bounded state stability. *Electron. Lett.* **6,** 16–17.

Desoer, C. A. (1977). Distributed networks with small parasitic elements: input–output stability. *IEEE Trans. Circuits Syst.* **24,** 1–8.

Desoer, C. A. and Lin, C. A. (1985). Tracking and disturbance rejection of MIMO nonlinear systems with PI controller. *IEEE Trans Autom. Control* **30**, 861–867.

Desoer, C. A. and Shensa, M. J. (1970). Network with very small and very large parasitics: natural frequencies and stability. *Proc. IEEE* **58,** 1933–1938.

Dieudonné, J. (1982). *Treatise on Analysis,* Vol. III. Academic Press, New York and London.

Dmitriev, M. G. (1978). On a class of singularly perturbed problems of optimal control. *J. Appl. Math. Mech.* **42,** 238–242.

Dontchev, A. L. (1983). Perturbations. Approximations and Sensitivity Analysis of Optimal Control Systems. *Lecture Notes in Control and Information Sciences,* **52**. Springer, New York.

Dontchev, A. L. and Veliov, V. M. (1983). Singular perturbation in Mayer's problem for linear systems. *SIAM J. Control Optim.* **21,** 566–581.

Dontchev, A. L. and Veliov, V. M. (1985). Singular perturbations in linear control systems with weakly coupled stable and unstable fast sub systems. *J. Math. Analysis Applic.* **110,** 1–30.

Dragan, V. and Halanay, A. (1982). Suboptimal stabilization of linear systems with several time scales. *Int. J. Control* **36,** 109–126.

Eckhaus, W. (1973). *Matched Asymptotic Expansions and Singular Perturbations*. North-Holland/American Elsevier, New York.

Eckhaus, W. (1977). Formal approximation and singular perturbations. *SIAM Rev.* **19,** 593–633.

Eitelberg, E. (1985). A transformation of nonlinear dynamical systems with single singular singularly perturbed differential equation. *Int. J. Control* **41,** 271–276.

El-Ansary, M. and Khalil, H. (1982). Reduced-order modeling of nonlinear singularly perturbed systems driven by wide-band noise. *Proc. 21st IEEE Conf. on Decision and Control, Orlando, Florida.*

El-Ansary, M. and Khalil, H. K. (1986). On the interplay of singular perturbations and wide-b and stochastic fluctuations. *SIAM J. Control Optim.* **24**, 83–94.

Elliott, J. R. (1977). NASA's advanced control law program for the F-8 digital fly-by-wire aircraft. *IEEE Trans. Autom. Control* **22,** 753–757.

Etkin, B. (1972), *Dynamics of Atmospheric Flight.* Wiley, New York.

Farber, N. and Shinar, J. (1980). Approximate solution of singularly perturbed nonlinear pursuit-evasion games. *J. Optim. Theory Applic.* **32,** 39–73.

Fenichel, N. (1979). Geometric singular perturbation theory for ordinary differential equations. *J. Diff. Eqns.* **31,** 53–98.

Ficola, A., Marino, R. and Nicosia, S. (1983). A singular perturbation approach to the dynamic control of elastic robots. *Proc. 21st Allerton Conf. Comm., Control, Comput., University of Illinois*, pp. 335–342.

Fossard, A., Berthelot, J. M. and Magni, J. F. (1983). On coherency-based decomposition algorithms. *Automatica* **19,** 247–253.

Fossard, A. G. and Magni, J. S. (1980). Frequential analysis of singularly perturbed systems with state or output control. *J. Large Scale Systems* **1,** 223–228.

Francis, B. A. (1979). The optimal linear–quadratic time-invariant regulator with cheap control. *IEEE Trans. Autom. Control* **24,** 616–621.

Francis, B. A. (1982). Convergence in the boundary layer for singularly perturbed equations. *Automatica* **18,** 57–62.

Francis, B. A. and Glover, K. (1978). Bounded peaking in the optimal linear regulator with cheap control. *IEEE Trans. Autom. Control* **23,** 608–617.

Freedman, M. I. (1977). Perturbation analysis of an optimal control problem involving bang-bang-controls. *J. Diff. Eqns.* **25,** 11–29.

Freedman, M. I. and Granoff, B. (1976). Formal asymptotic solution of a singularly perturbed nonlinear optimal control problem. *J. Optim. Theory Applic.* **19,** 301–325.

Freedman, M. I. and Kaplan, J. L. (1976). Singular perturbations of two point boundary value problems arising in optimal control. *SIAM J. Control Optim.* **14,** 189–215.

Gaitsgori, V. G. (1979). Perturbation method in optimal control problems. *J. Systems Sci.* **5,** 91–102.

Gaitsgori, V. G. (1980). On the optimization of weakly controlled stochastic systems. *Sov. Math. Dokl.* **21,** 408–410.

Gaitsgori, V. G. and Pervozvanskii A. A. (1975). Aggregation of states in a Markov chain with weak interactions. *Kibernetika* **3,** 91–98. (In Russian.)

Gajić, Z. (1986). Numerical fixed-point solution for near-optimum regulators of linear quadratic Gaussian control problems for singularly perturbed systems. *Int. J. Control* **43,** 373–387

Gajić, Z. and Khalil, H. (1986). Multimodel strategies under random disturbances and imperfect partial observations. *Automatica* **22**, 121–125.

Gardner, B. F. and Cruz, J. B. (1978). Well-posedness of singularly perturbed Nash games. *J. Franklin Inst.* **306,** 355–374.

Gicev, T. R. and Dontchev, A. L. (1979). Convergence of the solutions of the singularly perturbed time optimal problem. *Appl. Math. Mech.* **43,** 466–474.

Glizer, V. J. (1976). On a connection of singular perturbations with the penalty function method. *Sov. Math. Dokl.* **17,** 1503–1505.

Glizer, V. J. (1977). On the continuity of the regulator problem with respect to singular perturbations. *Appl. Math. Mech.* **41,** 573–576.

Glizer, V. J. (1978). Asymptotic solution of a singularly perturbed Cauchy problem in optimal control. *Diff. Eqns.* **14,** 601–612.

Glizer, V. J. (1979). Singular perturbations and generalized functions *Sov. Math. Dokl.* **20,** 1360–1364.

Glizer, V. J. and Dmitriev, M. G. (1975). Singular perturbations in a linear optimal control problem with quadratic functional. *Sov. Math. Dokl.* **16,** 1555–1558.

Grasman, J. (1982). On a class of optimal control problems with an almost cost-free solution. *IEEE Trans. Autom. Control* **27,** 441–445.

Grishin, S. A. and Utkin V. I. (1980). On redefinition of discontinuous systems. *Differential Equations* **16,** 227–235.

Grujic, L. T. (1979). Singular perturbations, large scale systems and asymptotic stability of invariant sets. *Int. J. Systems Science* **12,** 1323–1341.

Grujic, L. T. (1981). Uniform asymptotic stability of non-linear singularly perturbed large-scale systems. *Int. J. Control* **33,** 481–504.

Habets, P. (1974). Stabilité asymptotique pour des problèmes de perturbations singulières. In *Bressanone* pp. 3–18. Edizioni Cremonese, Rome.

Haddad, A. H. and Kokotović, P. V. (1971). Note on singular perturbation of linear state regulators. *IEEE Trans. Autom. Control* **16,** 279–281.

Haddad, A. H. (1976). Linear filtering of singularly perturbed systems. *IEEE Trans. Autom. Control* **31,** 515–519.

Haddad, A. H. and Kokotović, P. V. (1977). Stochastic control of linear singularly perturbed systems. *IEEE Trans. Autom. Control* **22,** 815–821.

Hadlock, C. R. (1970). Singular perturbations of a class of two point boundary value problems arising in optimal control. Ph.D. thesis, Coordinated Science Laboratory; Report R-481, Univ. Illinois, Urbana.

Hadlock, C. A. (1973). Existence and dependence on a parameter of solutions of a nonlinear two-point boundary value problem. *J. Diff. Eqns.* **14,** 498–517.

Halanay, A. and Mirica, S. (1979). The time optimal feedback control for singularly perturbed linear systems. *Rev. Roum. Mat. Pures et Appl.* **24,** 585–596.

Hale, J. K. (1980). *Ordinary Differential Equations.* Krieger Publishing Company.

Harris, W. A. (1960). Singular perturbations of two-point boundary problems for systems of ordinary differential equations. *Arch. Rat. Mech. Anal.* **5,** 212–225.

Hopkins, W. E. and Blankenship, G. L. (1981). Perturbation analysis of a system of quasi-variational inequalities for optimal stochastic scheduling. *IEEE Trans. Autom. Control* **26,** 1054–1070.

Hoppensteadt, F. (1966). Singular perturbation on the infinite interval. *Trans. Am. Math. Soc.* **123,** 521–535.

Hoppensteadt, F. (1967). Stability in systems with parameters. *J. Math. Anal. Applic.* **18,** 129–134.

Hoppensteadt, F. (1971). Properties of solutions of ordinary differential equations with small parameters. *Commun. Pure Appl. Math.* **34,** 807–840.

Hoppensteadt, F. (1974). Asymptotic stability in singular perturbation problems, II. *J. Diff. Eqns.* **15,** 510–521.

Hoppensteadt, F. and Miranker, W. (1977). Multitime methods for systems of difference equations. *Stud. Appl. Math.* **56,** 273–289.

Howes, F. A. (1976). Effective characterization of the asymptotic behaviour of solutions of singularly perturbed boundary value problems. *SIAM J. Appl. Math.* **30,** 296–306.

Ioannou, P. (1981). Robustness of absolute stability. *Int. J. Control* **34,** 1027–1033.

Ioannou, P. A. (1982). Robustness of model reference adaptive schemes with respect to modeling errors. Ph.D. thesis, Coordinated Science Laboratory; Report R-955, Univ. Illinois, Urbana.

Ioannou, P. A. (1984). Robust direct adaptive controller. *Proc. 23rd IEEE Conf. on Decision and Control, Las Vegas, Nevada*, pp. 1015–1019.

Ioannou, P. and Kokotović, P. V. (1982). An asymptotic error analysis of identifiers and adaptive observers in the presence of parasitics. *IEEE Trans. Autom. Control* **27,** 921–927.

Ioannou, P. A. and Kokotović, P. V. (1983). *Adaptive Systems with Reduced Models.* Lecture Notes in Control and Information Sciences, Vol. 47. Springer, New York.

Ioannou, P. A. and Kokotović, P. V. (1984a). Robust redesign of adaptive control. *IEEE Trans. Autom. Control* **29,** 202–211.

Ioannou, P. A. and Kokotović, P. V. (1984b). Instability analysis and improvement of robustness of adaptive control: *Automatica* **20,** 583–594.

Ioannou, P. A. and Kokotović, P. V. (1985). Decentralized adaptive control of interconnected systems with reduced-order models. *Automatica* **21,** 401–412.

Jameson, A. and O'Malley, R. E. (1975). Cheap control of the time-invariant regulator. *Appl. Math. Optim.* **1,** 337–354.

Jamshidi, M. (1974). Three-stage near-optimum design of nonlinear control processes. *Proc. IEEE* **121,** 886–892.

Javid, S. H. (1977). The time-optimal control of singularly perturbed systems. Ph.D. Thesis, Coordinated Science Laboratory; Report R-794, Univ. Illinois, Urbana.

Javid, S. H. (1978a). The time optimal control of a class of non-linear singularly perturbed systems. *Int. J. Control* **27,** 831–836.

Javid, S. H. (1978b). Uniform asymptotic stability of linear time varying singularly perturbed systems. *J. Franklin Inst.* **305,** 27–37.

Javid, S. H. (1980). Observing the slow states of a singularly perturbed system. *IEEE Trans. Autom. Control* **25,** 277–280.

Javid, S. H. (1982). Stabilization of time varying singularly perturbed systems by observer based slow state feedback. *IEEE Trans. Autom. Control* **27,** 702–704.

Javid, S. H. and Kokotović, P. V. (1977). A decomposition of time scales for iterative computation of time optimal controls. *J. Optim. Theory Applic.* **21,** 459–468.

Kailath, T. (1980). *Linear systems.* Prentice-Hall, Englewood Cliffs, NJ.

Kalman, R. E. (1960), Contributions to the theory of optimal control. *Bol. Soc. Mat. Mex.* **5,** 102–119.

Kelley, H. J. (1970a). Boundary layer approximations to powered-flight attitude transients. *J. Spacecraft Rockets* **7,** 879.

Kelley, H. J. (1970b). Singular perturbations for a Mayer variational problem. *AIAA J.* **8,** 1177–1178.

Kelley, H. J. (1971a). Flight path optimization with multiple time scales. *J. Aircraft* **8,** 238.

Kelley, H. J. (1971b). Reduced-order modeling in aircraft mission analysis. *AIAA J.* **9,** 349.

Kelley, H. J. (1973). Aircraft manoeuver optimization by reduced-order approximations. *Control and Dynamic Systems*, ed. C. T. Leon des, pp 131–178. Academic Press, New York.

Kelley, H. J. and Edelbaum, T. N. (1970). Energy climbs, energy turns and asymptotic expansions. *J. Aircraft* **7,** 93–95.

Khalil, H. K. (1978a). Multimodeling and multiparameter singular perturbation in control and game theory. Ph.D. thesis, Coordinated Science Laboratory; Report T-65, Univ. Illinois, Urbana.

Khalil, H. K. (1978b). Control of linear singularly perturbed systems with colored noise disturbances. *Automatica* **14,** 153–156.

Khalil, H. K. (1979). Stabilization of multiparameter singularly perturbed systems. *IEEE Trans. Autom. Control* **24,** 790–791.

Khalil, H. K. (1980). Multimodel design of a Nash strategy. *J. Optim. Theory Applic.* **31,** 553–564.

Khalil, H. K. (1981a). Asymptotic stability of a class of nonlinear multiparameter singularly perturbed systems. *Automatica* **17,** 797–804.

Khalil, H. K. (1981b) On the robustness of output feedback control methods to modeling errors. *IEEE Trans. Autom. Control* **28,** 524–528.

Khalil, H. K. (1984a). A further note on the robustness of output feedback control to modeling errors. *IEEE Trans. Automat. Control* **29,** 861–862.

Khalil, H. K. (1984b). Time-scale-decomposition of linear implicit singularly perturbed systems. *IEEE Trans. Autom. Control* **29,** 1054–1056.

Khalil, H. K. (1984c). Feedback control of implicit singularly perturbed systems. *Proc. 23rd IEEE Conf. on Decision and Control, Las Vegas, Nevada*, pp. 1219–1223.

Khalil, H. K. (1985). Output feedback control of linear two time scale systems. *Proc. 1985 American Control Conf., Boston*, pp. 1397–1400.

Khalil, H. K. (1986). Output feedback stabilization of singularly perturbed systems. *Proc. 25th IEEE Conf. on Decision and Control, Athens, Greece.*

Khalil, H. K. and Gajic, Z. (1984). Near-optimal regulators for stochastic linear singularly perturbed systems. *IEEE Trans. Autom. Control* **29,** 531–541.

Khalil, H. K., Haddad, A. and Blankenship, G. (1978). Parameter scaling and well-posedness of stochastic singularly perturbed control systems. *Proc. 12th Asilomar Conferences, Pacific Grove, CA* pp. 407–411.

Khalil, H. K. and Kokotović, P. V. (1978). Control strategies for decision makers using different models of the same system. *IEEE Trans. Autom. Control* **23,** 289–298.

Khalil, H. K. and Kokotović, P. V. (1979a). D-stability and multiparameter singular perturbations. *SIAM J. Control Optim.* **17,** 56–65.

Khalil, H. K. and Kokotović, P. V. (1979b). Control of linear systems with multiparameter singular perturbations. *Automatica* **15,** 197–207.

Khalil, H. K. and Kokotović, P. V. (1979c). Feedback and well-posedness of singularly perturbed Nash games. *IEEE Trans. Autom. Control* **24,** 699–708.

Khalil, H. K. and Medanić, J. V. (1980). Closed-loop Stackelberg strategies for singularly perturbed linear quadratic problems. *IEEE Trans. Autom. Control* **25,** 66–71.

Khorasani, K. and Kokotović, P. V. (1985). Feedback linearization of a flexible manipulator near its rigid body manifold. *Syst. Control Lett.* **6,** 187–192.

Khorasani, K. and Kokotović, P. V. (1986). A corrective feedback design for nonlinear systems with fast actuators. *IEEE Trans. Autom. Control* **31**, 67–69

Khorasani, K. and Pai, M. A. (1985). Asymptotic stability improvements of multiparameter nonlinear singularly perturbed systems. *IEEE Trans. Autom. Control* **30,** 802–804.

Khorasani, K. and Mai, M. A. (1985b) Asymptotic stability of nonlinear singularly perturbed systems using higher order corrections. *Automatica* **21**, 717–727.

Klimushchev, A. I. and Krasovskii, N. N. (1962). Uniform asymptotic stability of systems of differential equations with a small parameter in the derivative terms *J. Appl. Math. Mech.* **25,** 1011–1025.

Koda, M. (1982). Sensitivity analysis of singularly perturbed systems. *Int. J. Systems Sci.* **13,** 909–919.

Kokotović, P. V. (1975). A Riccati equation for block-diagonalization of ill-conditioned systems. *IEEE Trans. Autom. Control* **20,** 812–814.

Kokotović, P. V. (1981). Subsystems, time-scales and multimodeling. *Automatica* **17,** 789–795.

Kokotović, P. V. (1984). Applications of singular perturbation techniques to control problems. *SIAM Rev.* **26,** 501–550.

Kokotović, P. V. (1985). Recent trends in feedback design: an overview. *Automatica* **21,** 225–236.

Kokotović, P. V., Allemong, J. J., Winkelman, J. R. and Chow, J. H. (1980). Singular perturbation and iterative separation of time scales. *Automatica* **16,** 23–33.

Kokotović, P. V., Avramović, B., Chow, J. H. and Winkelman J. R. (1982). Coherency-based decomposition and aggregation *Automatica* **17,** 47–56.

Kokotović, P. V. and Haddad, A.H. (1975a). Controllability and time-optimal control of systems with slow and fast modes. *IEEE Trans. Autom. Control* **20,** 111–113.

Kokotović, P. V. and Haddad, A. H. (1975b). Singular perturbations of a class of time-optimal controls. *IEEE Trans. Autom. Control* **20,** 163–164.

Kokotović, P. V. and Khalil, H. K. (Eds.) (1986). *Singular Perturbations in Systems and Control* (Reprint). IEEE Press, New York.

Kokotović, P. V., O'Malley R. E. and Sannuti P. (1976). Singular perturbations and order reduction in control theory—an overview. *Automatica* **12,** 123–132.

Kokotović, P. V. and Sannuti, P. (1968). Singular perturbation method for reducing the model order in optimal control design. *IEEE Trans. Autom. Control* **13,** 377–384.

Kokotović, P. V. and Yackel, R. A. (1972). Singular perturbation of linear regulators: basic theorems. *IEEE Trans. Autom. Control* **17,** 29–37.

Kopel, N. (1979). A geometric approach to boundary layer problems exhibiting resonance. *SIAM J. Appl. Math.* **37,** 436–458.

Kouvaritakis, B. (1978). The optimal root loci of linear multivariable systems. *Int. J. Control* **28,** 33–62.

Kouvaritakis, B. and Edmunds, J. M. (1979). A multivariable root loci: a unified approach to finite and infinite zeros. *Int. J. Control* **29,** 393–428.

Krtolica, R. (1984). A singular perturbation model of reliability in systems control. *Automatica* **2,** 51–57.

Kung, C. F. (1976). Singular perturbation of an infinite interval linear state regulator problem in optimal control. *J. Math. Anal. Appl.* **55,** 365–374.

Kurina, G. A. (1977). Asymptotic solutions of a classical singularly perturbed optimal control problem. *Sov. Math. Dokl.* **18,** 722–726.

Kuruoghu, N., Clough, D. E. and Ramirez, W. F. (1981). Distributed parameter estimation for systems with fast and slow dynamics. *Chem. Engng. Sci.* **3,** 1357.

Kushner, H. J. (1982). A cautionary note on the use of singular perturbation methods for 'small noise' models. *Stochastics* **6,** 117–120.

Kushner, H. J. (1984). *Approximations and Weak Convergence Methods for Random Processes with Application to Stochastic System Theory*. M.I.T. Press.

Kwakernaak, H. and Sivan, S. (1972). *Linear Optimal Control Systems.* Wiley, New York.

Ladde, G. S. and Siljak, D. D. (1983). Multiparameter singular perturbations of linear systems with multiple time scales. *Automatica* **19,** 385–394.

Lagerstrom, P. A. and Casten R. G. (1972). Basic concepts underlying singular perturbation techniques. *SIAM Rev.* **14,** 63–120.

Lakin, W. D. and Van der Driessche, P. (1977). Time-scales in population biology. *SIAM J. Appl. Math.* **32,** 694–705.

Lehtomaki, N. A., Castanon, D. A., Levy, B. C., Stein G., Sandell, N. R. and Athans, M. (1984). Robustness and modeling error chacterization. *IEEE Trans. Autom. Control* **29,** 212–220.

Levin, J. (1957). The asymptotic behavior of the stable initial manifold of a system of nonlinear differential equations. *Trans. Am. Math. Soc.* **85,** 357–368.

Levin, J. J. and Levinson, N. (1954). Singular perturbations on non-linear systems of differential equations and an associated boundary layer equation. *J. Rat. Mech. Anal.* **3,** 274–280.

Levinson, N. (1950). Perturbations of discontinuous solutions of non-linear systems of differential equations. *Acta Math.* **82,** 71–106.

Lions, J. L. (1983). Perturbations singulières dans les problemes aux limites et en controle optimal. Lecture Notes in Mathematics, Vol. 323. Springer, New York.

Litkouhi, B. and Khalil, H. (1984). Infinite-time regulators for singularly perturbed difference equations. *Int. J. Control* **39,** 587–598.

Litkouhi, B. and Khalil, H. (1985). Multirate and composite control of two-time-scale discrete-time systems. *IEEE Trans. Autom. Control* **30,** 645–651.

Lomov, S. A. (1981). *Introduction to the General Theory of Singular Perturbations.* Nauka, Moscow. (In Russian.)

Lukyanov, A. G. and Utkin, V. I. (1981). Methods for reduction of dynamic system equations to a regular form, *Automn Remote Control* **4,** 5–13.

Luse, D. W. (1984). Basic results for multiple frequency scale systems. *Proc. 1984 American Control Conference, San Diego, CA*, pp. 1366–1367.

Luse, D. W. (1985). A continuation method for pole-placement for singularly perturbed systems. *American Control Conf., Boston, MA*, pp. 1392–1396.

Luse, D. W. and Khalil, H. K. (1985). Frequency domain results for systems with slow and fast dynamics. *JEEE Trans. Autom. Control* **30**, 1171–1179.

Mahmoud, M. S. (1982). Structural properties of discrete systems with slow and fast modes. *J. Large Scale Systems* **3,** 227–236.

Marino, R. and Kokotović, P. V. (1986). A geometric approach to composite control of two-time-scale systems. *Proc. 25th IEEE Conf. on Decision and Control, Athens, Greece.*

Mehra, R. K., Washburn, R. B., Sajon, S. and Corell, J. V. (1979). A study of the application of singular perturbation theory. *NASA* CR3167.

Miller, R. K. and Michel, A. N. (1982). *Ordinary Differential Equations.* Academic Press. New York.

Mitropolsky, Y. A. and Lykova, O. B. (1973). *Integral Manifolds in Nonlinear Mechanics.* Nauka, Moscow.

Moiseev, N. N. and Chernousko, F. L. (1981). Asymptotic methods in the theory of optimal control. *IEEE Trans. Autom. Control* **26,** 993–1000.

Narendra, K. S., Lin, Y. H. and Valavani, V. S. (1980). Stable adaptive controller design, part II: proof of stability. *IEEE Trans. Autom. Control* **25,** 440–448.

O'Malley, R. E. (1971). Boundary layer methods for nonlinear initial value problems. *SIAM Rev.* **13,** 425–434.

O'Malley, R. E. (1972a). The singularly perturbed linear state regulator problem. *SIAM J. Control* **10,** 399–413.

O'Malley, R. E. (1972b). Singular perturbation of the time invariant linear state regulator problem. *J. Diff. Eqns* **12,** 117–128.

O'Malley, R. E. (1974a). Boundary layer methods for certain nonlinear singularly perturbed optimal control problems. *J. Math. Anal. Appl.* **45,** 468–484.

O'Malley, R. E. (1974b). The singularly perturbed linear state regulator problem, II. *SIAM J. Control* **13,** 327–337.

O'Malley, R. E. (1974c). *Introduction to Singular Perturbations.* Academic Press, New York.

O'Malley, R. E. (1975). On two methods of solution for a singularly perturbed linear state regulator problem. *SIAM Rev.* **17,** 16–37.

O'Malley, R. E. (1976). A more direct solution of the nearly singular linear regulator problem. *SIAM J. Control Optim.* **14,** 1063–1077.

O'Malley, R. E. (1978a). Singular perturbations and optimal control. *Mathematical Control Theory.* Lecture Notes in Mathematics, Vol. 680. Springer, New York.

O'Malley, R. E. (1978b). On singular singularly-perturbed initial value problems. *Applic. Anal.* **8,** 71–81.

O'Malley, R. E. (1979). A singular singularly-perturbed linear boundary value problem. *SIAM J. Math. Anal.* **10,** 695–708.

O'Malley, R. E. (1982). Book Reviews. *Bull. Am. Math. Soc.* (New Series) **7,** 414–420.

O'Malley, R. E. (1983). Slow/fast decoupling—analytical and numerical aspects. *CISM Courses and Lectures*, 280, ed. M. Ardema, pp. 143–159. Springer, New York.

O'Malley, R. E. and Anderson, R. L. (1978). Singular perturbations and slow mode approximation for large-scale linear systems. *Proc. IFAC/IRIA Workshop on Singular Perturbations in Control, France*, pp. 113–121.

O'Malley, R. E. and Anderson, R. L. (1982). Time-scale decoupling and order reduction for linear time-varying systems. *Optim. Control Meth. Applic.* **3,** 135–154.

O'Malley, R. E. and Flaherty, J. E. (1977). Singular singular perturbation problems. *Singular Perturbations and Boundary Layer Theory.* Lecture Notes in Mathematics, Vol. 594, pp. 422–436. Springer, New York.

O'Malley, R. E. and Flaherty, J. E. (1980). Analytical and numerical methods for nonlinear singular singularly perturbed initial value problems. *SIAM J. Appl. Math.* **38,** 225–248.

O'Malley, R. E. and Jameson, A. (1975). Singular perturbations and singular arcs—part I. *IEEE Trans. Autom. Control* **20,** 218–226.

O'Malley, R. E. and Jameson A. (1977). Singular perturbations and singular arcs—part II. *IEEE Trans. Autom. Control* **22,** 328–337.

O'Malley, R. E. and Kung, C. F. (1974). The matrix Riccati approach to a singularly perturbed regulator problem. *J. Diff. Eqns.* **17,** 413–427.

O'Reilly, J. (1979a). Two time scale feedback stabilization of linear time varying singularly perturbed systems. *J. Franklin Inst.* **308,** 465–474.

O'Reilly, J. (1979b). Full order observers for a class of singularly perturbed linear time varying systems. *Int. J. Control* **30,** 745–756.

O'Reilly, J. (1980). Dynamical feedback control for a class of singularly perturbed linear systems using a full order observer. *Int. J. Control* **31,** 1–10.

O'Reilly, J. (1983a). Partial cheap control of the time-invariant regulator. *Int. J. Control* **37,** 909–927.

O'Reilly, J. (1983b). *Observers for Linear Systems.* Academic Press, London.

O'Reilly, J. (1985). The robustness of linear feedback control systems to unmodelled high frequency dynamics. *Proc. IEE 'Control 85' Conf. Cambridge*, 405–408.

O'Reilly, J. (1986). The robustness of linear feedback control systems to unmodelled high frequency dynamics. *Int. J. Control* **43.**

Othman, H. A., Khraishi, N. M. and Mahmoud, M. S. (1985). Discrete regulators with time scale separation. *IEEE Trans. Autom. Control* **30**, 293–297.

Özgüner, U. (1979). Near-optimal control of composite systems: the multi-time scale approach. *IEEE Trans. Autom. Control* **24,** 652–655.

Pai, M. A. (1981). *Power System Stability Analysis by the Direct Method of Lyapunov.* North-Holland, Amsterdam.

Pai, M. A., Sauer, P. W. and Khorasani, K. (1984). Singular perturbations and large-scale power system stability. *Proc. 23rd Decision and Control Conf., Las Vegas, Nevada*, pp. 173–178.

Papoulis, A. (1965). *Probability, Random Variables and Stochastic Processes.* McGraw-Hill, New York.

Peponides, G. M. (1982). Nonexplicit singular perturbations and interconnected systems. Ph.D. Thesis, Coordinated Science Laboratory; Report R-960, Univ. Illinois, Urbana.

Peponides, G. and Kokotović, P. V. (1983). Weak connections, time scales and aggregation of nonlinear systems. *IEEE Trans. Autom. Control* **28,** 729–735.

Peponides, G., Kokotović, P. V. and Chow, J. H. (1982). Singular perturbations and time scales in nonlinear models of power systems. *IEEE Trans. Circuits Systems* **29,** 758–767.

Pervozvanskii, A. A. (1979a). *Decomposition, Aggregation and Suboptimization.* Nauka, Moscow. (In Russian.)

Pervozvanskii, A. A. (1979b). Perturbation method for LQ problems: duality in degenerate cases. *Proc. 2nd Warsaw Workshop on Multilevel Control.*

Pervozvanskii, A. A. (1980). On aggregation of linear control systems. *Automn Remote Control* **8,** 88–95.

Pervozvanskii, A. A. (1981). Degeneracy in LQ and LQG problems of optimal control; possibilities to simplify the synthesis. *Proc. 8th IFAC Congress, Kyoto.*

Pervozvanskii, A. A. and Gaitsgori, V. G. (1978). Suboptimization, decomposition and aggregation. *7th IFAC World Congress, Helsinki.*

Phillips, R. G. (1980a). Reduced order modelling and control of two time scale discrete systems. *Int. J. Control* **31,** 765–780.

Phillips, R. G. (1980b). Decomposition of time scales in linear systems and Markovian decision processes. Ph.D. thesis, Coordinated Science Laboratory; Report R-902, Univ. Illinois, Urbana.

Phillips, R. G. (1981). A two stage design of linear feedback controls. *IEEE Trans. Autom. Control* **26,** 1220–1222.

Phillips, R. G. (1983). The equivalence of time-scale decomposition techniques used in the analysis and design of linear systems. *Int. J. Control* **37,** 1239–1257.

Phillips, R. G. and Kokotović, P. V. (1981). A singular perturbation approach to modelling and control of Markov chains. *IEEE Trans. Automat. Control* **26,** 1087–1094.

Pliss, V. A. (1966a). *Nonlocal Problems of the Theory of Oscillations.* Academic Press, New York.

Pliss, V. A. (1966b). On the theory of invariant surfaces. *Differentialniye Uraynaniya* **2,** 1139–1150.

Porter, B. (1974). Singular perturbation methods in the design of stabilizing feedback controllers for multivariable linear systems. *Int. J. Control* **20,** 689–692.

Porter, B. and Shenton, A. T. (1975). Singular perturbation analysis of the transfer function matrices of a class of multivariable linear systems. *Int. J. Control* **21,** 655–660.

Porter, B. (1977). Singular perturbation methods in the design of state feedback controllers for multivariable linear systems. *Int. J. Control* **26,** 583–587.

Price, D. B. (1979). Comments on linear filtering of singularly perturbed systems. *IEEE Trans. Autom. Control* **24,** 675–677.

Razevig, V. D. (1978). Reduction of stochastic differential equations with small parameters and stochastic integrals. *Int. J. Control* **28,** 707–720.

Reddy, P. B. and Sannuti, P. (1975). Optimal control of a coupled-core nuclear reactor by a singular perturbation method. *IEEE Trans. Autom. Control* **20,** 766–769.

Riedle, B. and Kokotović, P. V. (1985). A stability–instability boundary for disturbance-free slow adaptation and unmodeled dynamics. *IEEE Trans. Autom. Control* **30,** 1027–1030

Riedle, B. and Kokotović, P. V. (1986). Integral manifolds of slow adaptation. *IEEE Trans. Autom. Control* **31,** 316–324

Saberi, A. (1983). Stability and control of nonlinear singularly perturbed systems with application to high-gain feedback. Ph.D. thesis, Michigan State University.

Saberi, A. and Khalil, H. (1984). Quadratic-type Lyapunov functions for singularly perturbed systems. *IEEE Trans. Autom. Control* **29,** 542–550.

Saberi, A. and Khalil, H. (1985). Stabilization and regulation of nonlinear singularly perturbed systems—composite control. *IEEE Trans. Autom. Control* **30,** 739–747.

Saberi, A. and Khalil, H. K. (1985b). An initial value theorem for nonlinear singularly perturbed systems. *Syst. Control Lett.* **4,** 301–305

Saberi, A. and Khalil, H. K. (1985c). Decentralized stabilization of interconnected systems using feedback. *Int. J. Control* **41,** 1461–1475.

Saberi, A. and Khalil, H. K. (1986). Adoptive stabilization of SISO systems with unknown high-frequency gains. *Proc.* 1986 *American Control Conf., Seattle.*

Saberi, A. and Sannuti, P. (1985). Time-scale decomposition of a class of linear and nonlinear cheap control problems. *1985 American Control Conf., Boston, MA*, pp. 1414–1421.

Saberi, A. and Sannuti, P. (1986). Cheap and singular controls for linear quadratic regulators. *IEEE Trans. Autom. Control*, to appear.
Saksena, V. R. and Basar, T. (1982). A multimodel approach to stochastic team problems *Automatica* **18,** 713–720.
Saksena, V. R. and Cruz, J. B. (1981a). Stabilization of singularly perturbed linear time-invariant systems using low order observer. *IEEE Trans. Autom. Control* **28,** 510–513.
Saksena, V. R. and Cruz, J. B. (1981b). Nash strategies in decentralized control of multi-parameter singularly perturbed large-scale systems. *J. Large Scale Systems* **2,** 219–234.
Saksena, V. R. and Cruz, J. B. (1982). A multimodel approach to stochastic Nash games. *Automatica* **18,** 295–305.
Saksena, V. R. and Cruz, J. B. (1984). Robust Nash strategies for a class of non-linear singularly perturbed problems. *Int. J. Control* **39,** 293–310.
Saksena, V. R. and Cruz, J. B. (1985). Optimal and near-optimal incentive strategies in the hierarchical control of Markov-chains. *Automatica* **21,** 181–191.
Saksena, V. R. and Kokotović, P. V. (1981). Singular perturbation of the Popov–Kalman–Yakubovich lemma. *Syst. Control Lett.* **1,** 65–68.
Saksena, V. R., O'Reilly, J. and Kokotović, P. V. (1984). Singular perturbations and time-scale methods in control theory: survey 1976–1983. *Automatica* **20,** 273–293.
Salman, M. A. and Cruz, J. B. (1979). Well posedness of linear closed Stackelberg strategies for singularly perturbed systems. *J. Franklin Inst.* **308,** 25–37.
Salman, M. A. and Cruz, J. B. (1983). Optimal coordination of multimode interconnected systems with slow and fast modes. *J. Large Scale Systems* **5,** 207–219.
Sandell, N. R. (1979). Robust stability of systems with applications to singular perturbation. *Automatica* **15,** 467–470.
Sannuti, P. (1968). Singular perturbation method in the theory of optimal control. Ph.D. thesis, Coordinated Science Laboratory; Report R-379, Univ. Illinois, Urbana.
Sannuti, P. (1969). Singular perturbation method for near-optimum design of high order nonlinear systems. *Automatica* **5,** 773–779.
Sannuti, P. (1974a). A note on obtaining reduced order optimal control problems by singular perturbations. *IEEE Trans. Autom. Control* **19,** 256
Sannuti, P. (1974b). Asymptotic solution of singularly perturbed optimal control problems. *Automatica* **10,** 183–194.
Sannuti, P. (1975). Asymptotic expansions of singularly perturbed quasi-linear optimal systems. *SIAM J. Control* **13,** 572–592.
Sannuti, P. (1977). On the controllability of singularly perturbed systems. *IEEE Trans. Autom. Control* **22,** 622–624.
Sannuti, P. (1978). On the controllability of some singularly perturbed nonlinear systems. *J. Math. Anal. Applic.* **64,** 579–591.
Sannuti, P. (1981). Singular perturbations in the state space approach of linear electrical networks. *Circuit Theory Applic.* **9,** 47–57.
Sannuti, P. (1983). Direct singular perturbation analysis of high-gain and cheap control problems. *Automatica* **19,** 41–51.
Sannuti, P. (1984). Determination of multivariable root-loci. *Proc. 18th Conference Information Science, Princeton.*
Sannuti, P. and Kokotović, P. V. (1969a). Near optimum design of linear systems by a singular perturbation method. *IEEE Trans. Autom. Control* **14,** 15–22.
Sannuti, P. and Kokotović, P. V. (1969b). Singular perturbation method for near-optimum design of high order nonlinear systems. *Automatica* **5,** 773–779.
Sannuti, P. and Wason, H. (1983). Singular perturbation analysis of cheap control problems. *Proc. 22nd Decision and Control Conf., San Antonio, TX*, pp. 231–236.

Sannuti, P. and Wason, H. S. (1985). Multi-time scale decomposition in cheap control problems—singular control. *IEEE Trans. Autom. Control* **30,** 633–644.

Sastry, S. S. and Desoer, C. A. (1981). Jump behaviour of circuits and systems. *IEEE Trans. Circuits Systems* **28,** 1109–1124.

Sastry, S. S. and Desoer, C. A. (1983). Asymptotic unbounded root loci-formulas and computation, *IEEE Trans. Autom Control* **28,** 557–568.

Sastry, S. S. and Hijab, O. (1981). Bifurcation in the presence of small noise. *Syst. Control Lett.* **1,** 159–167.

Sauer, P. W., Ahmed-Zaid, S. and Pai, M. A. (1984a). Systematic inclusion of stator transients in reduced order synchronous machine models. *IEEE Trans. Power Appar. Systems* **103,** 1348–1355.

Sauer, P. W., Ahmed-Zaid, S. and Pai, M. A. (1984b). Nonlinear decoupling in a class of quadratic two time-scale systems with application to synchronous machine models. *Proc. 23rd IEEE Conf. on Decision and Control, Las Vegas, Nevada*, pp. 481–484.

Schuss, Z. (1980). Singular perturbation methods in stochastic differential equations of mathematical physics. *SIAM Rev.* **22,** 119–155.

Sebald, A. V. and Haddad, A. H. (1978). State estimation for singularly perturbed systems with uncertain perturbation parameter. *IEEE Trans. Autom. Control* **23,** 464–469.

Shaked, U. (1976). Design techniques for high-feedback gain stability. *Int. J. Control* **24,** 137–144.

Shaked, U. (1978). The asymptotic behaviour of the root loci of multivariable optimal regulators. *IEEE Trans. Autom. Control* **23,** 425–430.

Shimitzu, K. and Matsubara, M. (1985). Singular perturbation for the dynamic interaction measure. *IEEE Trans. Autom. Control* **30,** 790–792.

Shinar, J. (1981). Solution techniques for realistic pursuit–evasion games. *Advances in Control and Dynamic Systems*, Vol. 17, ed. C. T. Leondes, pp. 63–124. Academic Press, New York.

Shinar, J. (1983). On applications of singular perturbation techniques in nonlinear optimal control. *Automatica* **19,** 203–211.

Shinar, J. and Farber, N. (1984). Horizontal variable-speed interception game solved by forced singular perturbation technique. *J. Optim. Theory Control* **42,** 603–636.

Siddiqee, M. W. (1968). Transient stability of a.c. generator by Lyapunov's direct method. *Int. J. Control* **8,** 131–144.

Siljak, D. D. (1972). Singular perturbation of absolute stability. *IEEE Trans. Autom. Control* **17,** 720.

Silva-Madriz, R. and Sastry, S. S. (1984). Input–output description of linear systems with multiple time-scales. *Int. J. Control* **40,** 699–721.

Silva-Madriz, R. (1985). Feedback systems and multiple time scales. *Int. J. Control* **43,** 587–600.

Singh, R.-N. P. (1982). The linear–quadratic–Gaussian problem for singularly perturbed systems. *Int. J. Systems Sci.* **13,** 93–100.

Skoog, R. A. and Lau, C. G. Y. (1972). Instability of slowly varying systems. *IEEE Trans. Autom. Control* **17,** 86–92.

Slater, G. L. (1984). Perturbation analysis of optimal integral controls. *Trans. ASME: J. Dyn. Systems Meas. Control* **106,** 114–116.

Soliman, M. A. and Ray, W. H. (1979). Nonlinear filtering for distributed parameter systems with a small parameter. *Int. J. Control* **30,** 757–772.

Sololev, V. A. (1984). Integral manifolds and decomposition of singularly perturbed systems. *Syst. Control Lett.* **5,** 1169–1179.

Spong, M. W., Khorasani, K. and Kokotović, P. V. (1985). A slow manifold approach to feedback control of nonlinear flexible systems. *1985 American Control Conf., Boston, MA*. pp. 1386–1391.

Sridhar, B. and Gupta, N. K. (1980). Missile guidance laws based on singular perturbation methodology. *AIAA J. Guidance Control* **3,** 158–165.

Suzuki, M. (1981). Composite controls for singularly perturbed systems. *IEEE Trans. Autom. Control* **26,** 505–507.

Suzuki, M. and Miura, M. (1976). Stabilizing feedback controllers for singularly perturbed linear constant systems. *IEEE Trans. Autom. Control* **21,** 123–124.

Sycros, G. P. and Sannuti, P. (1984). Near-optimum regulator design of singularly perturbed systems via Chandrasekhar equations. *Int. J. Control* **39,** 1083–1102.

Sycros, G. P. and Sannuti, P. (1986). Singular perturbation modelling and design techniques applied to jet engine control. *Optim. Control Applic. Methods* **7,** 1–17.

Teneketzis, D. and Sandell, N. R. (1977). Linear regulator design for stochastic systems by a multiple time scale method. *IEEE Trans. Autom. Control* **22,** 615–621.

Tikhonov, A. (1948). On the dependence of the solutions of differential equations on a small parameter. *Mat. Sb.* **22,** 193–204. (In Russian.)

Tikhonov, A. N. (1952). Systems of differential equations containing a small parameter multiplying the derivative. *Mat. Sb.* **31,** 575–586. (In Russian.)

Tasi, E. P. (1978). Perturbed stochastic linear regulator problems. *SIAM J. Control* **16,** 396–410.

Utkin, V. I. (1977a) *Sliding Modes and Their Application to Variable Structure Systems*. Mir, Moscow. (In English.)

Utkin, V. I. (1977b). Variable structure systems with sliding modes: a survey. *IEEE Trans. Autom. Control* **22,** 212–222.

Utkin, V. I. (1983). Variable structure systems: state of the art and perspectives. *Automn Remote Control* **9,** 5–25.

Van Harten, A. (1984). Singularly perturbed systems of diffusion type and feedback control. *Automatica* **20,** 79–91.

Vasil'eva, A. B. (1963). Asymptotic behavior of solutions to certain problems involving nonlinear differential equations containing a small parameter multiplying the highest derivatives. *Russian Math. Surveys* **18,** 13–81.

Vasil'eva, A. B. (1975). Singularly perturbed systems containing indeterminancy in the case of degeneracy. *Sov. Math. Dokl.* **16,** 1121–1125.

Vasil'eva, A. B. (1976). Singularly perturbed systems with an indeterminacy in their degenerate equations. *J. Diff. Eqns* **12,** 1227–1235.

Vasil'eva, A. B. and Anikeeva, V. A. (1976). Asymptotic expansions of solutions of nonlinear problems with singular boundary conditions. *J. Diff. Eqns* **12,** 1235–1244.

Vasil'eva, A. B. and Butuzov, V. F. (1973). *Asymptotic Expansions of Solutions of Singularly Perturbed Differential Equations*. Nauka, Moscow. (In Russian.)

Vasil'eva, A. B. and Butuzov, V. F. (1978). *Singularly Perturbed Systems in Critical Cases*, Moscow University Press. (In Russian.)

Vasil'eva, A. B. and Dmitriev, M. (1978). Singular perturbations and some optimal control problems. *Proc. 7th IFAC World Congress*, Paper 23.6.

Vasil'eva, A. B. and Dmitriev, M. G. (1980). Determination of the structure of generalized solutions of nonlinear optimal control problems. *Sov. Math. Dokl.* **21,** 104–109.

Vasil'eva, A. B. and Dmitriev, M. G. (1982). Singular perturbations in optimal control. *Progress in Science and Technology*: *Mathematical Analysis*, Vol. 20, pp. 3–77. Viniti, Moscow.

Vasil'eva, A. B. and Faminskaya, M. V. (1977). A boundary-value problem for singularly perturbed differential and difference systems when the unperturbed system is on a spectrum. *J. Diff. Eqns* **13,** 738–742.

Vasil'eva, A. B. and Faminskaya, M. V. (1981). An investigation of a nonlinear optimal control problem by the methods of singular perturbation theory. *Sov. Math. Dokl.* **21,** 104–108.

Vidyasagar, M. (1978). *Nonlinear Systems Analysis.* Prentice-Hall, Englewood Cliffs, NJ.

Vidyasagar, M. (1984). The graph metric for unstable plants and robustness estimates for feedback stability. *IEEE Trans. Autom. Control* **29,** 403–418.

Vidyasagar, M. (1985). Robust stabilization of singularly perturbed systems. *Syst. Control Lett.* **5,** 413–418.

Vishik, M. I. and Liusternik, L. A. (1958). On the asymptotic behavior of the solutions of boundary problems for quasi-linear differential equations. *Dokl. Akad. Nauk SSSR* **121,** 778–781. (In Russian.)

Wasow, W. (1965). *Asymptotic Expansions for Ordinary Differential Equations.* Wiley-Interscience, New York.

Weston, A., Cliff, G. and Kelley, H. (1985). On board near-optimal climb-dash energy management. *AIAA J. Guidance Control Dynamics* **8,** 320–324.

Wilde, R. R. (1972). A boundary layer method for optimal control of singularly perturbed systems. Ph.D. thesis, Coordinated Science Laboratory; Report R-547, Univ. Illinois, Urbana.

Wilde, R. R. (1973). Optimal open and closed-loop control of singularly perturbed linear systems. *IEEE Trans. Autom. Control* **18,** 616–625.

Wilde, R. R. and Kokotović, P. V. (1972a). Stability of singularly perturbed systems and networks with parasitics. *IEEE Trans. Autom. Control* **17,** 245–246.

Wilde, R. R. and Kokotović, P. V. (1972b). A dichotomy in linear control theory. *IEEE Trans. Automat. Control* **17,** 382–383.

Willems, J. L. (1969). The computation of finite stability regions by means of open Lyapunov surfaces. *Int. J. Control* **10,** 537–544.

Willems, J. C. (1981). Almost invariant subspaces; an approach to high gain feedback design, part I, almost controlled invariant subspaces. *IEEE Trans. Autom. Control* **26,** 235–252.

Willems, J. C. (1982). Almost invariant subspaces: an approach to high gain feedback design, part II, almost conditionally invariant subspaces. *IEEE Trans. Autom. Control* **27,** 1071–1085.

Winkelman, J. R., Chow, J. H., Allemong, J. H. and Kokotović, P. V. (1980). Multi-time-scale analysis of power systems. *Automatica* **16,** 35–43.

Womble, M. E., Potter, J. E. and Speyer, J. L. (1976). Approximations to Riccati equations having slow and fast modes. *IEEE Trans. Autom. Control* **21,** 846–855.

Yackel, R. A. (1971). Singular perturbation of the linear state regulator. Ph.D. thesis, Coordinated Science Laboratory; Report R-532, Univ. Illinois, Urbana.

Yackel, R. A. and Kokotović, P. V. (1973). A boundary layer method for the matrix Riccati equation. *IEEE Trans. Autom. Control* **18,** 17–24.

Young, K. D. (1977). Analysis and synthesis of high gain and variable structure feedback systems. Ph.D. thesis, Coordinated Science Laboratory; Report R-800, Univ. Illinois, Urbana.

Young, K. D. (1978). Multiple time-scales in single-input single-output high-gain feedback systems. *J. Franklin Inst.* **306,** 293–301.

Young, K. D. (1982a). Near insensitivity of linear feedback systems. *J. Franklin Inst.* **314,** 129–142.

Young, K. D. (1982b). Disturbance decoupling by high-gain feedback. *IEEE Trans. Automat., Control* **27,** 970–971.
Young, K. D. (1985). On near optimal decentralized control. *Automatica* **21,** 607–610.
Young, K. D. and Kokotović, P. V. (1982). Analysis of feedback loop interactions with actuator and sensor parasitics. *Automatica* **18,** 577–582.
Young, K. D., Kokotović, P. V. and Utkin, V. I. (1977). A singular perturbation analysis of high-gain feedback systems. *IEEE Trans. Autom. Control* **22,** 931–938.
Zien, L. (1973). An upper bound for the singular parameter in a stable, singularly perturbed system. *J. Franklin Inst.* **295,** 373–381.

References Added in Proof

Kokotović, P. and Riedle, B. (1986). Stability bounds for slow adaptation: integral manifold approach. *Proc.* 1986 *American Control Conf., Seattle.*
Visser, H. G. and Shinar, J. (1986). First-order corrections in optimal feedback control of singularly perturbed nonlinear systems. *I.E.E.E. Trans Automatic Control* **31**, 387–393.

APPENDIX A
APPROXIMATION OF SINGULARLY PERTURBED SYSTEMS DRIVEN BY WHITE NOISE

Consider the linear time-invariant singularly perturbed system

$$\dot{x}(t) = A_{11}(\varepsilon)x(t) + A_{12}(\varepsilon)z(t) + \mu B_1(\varepsilon)w(t), \quad x(0) = x_0(\varepsilon), \tag{A 1a}$$

$$\varepsilon\dot{z}(t) = A_{21}(\varepsilon)x(t) + A_{22}(\varepsilon)z(t) + \mu B_2(\varepsilon)w(t), \quad z(0) = z_0(\varepsilon), \tag{A 1b}$$

where $x \in R^n$, $y \in R^m$, $w \in R^r$ and $0 < \mu = \mu(\varepsilon) < C$. The system matrices are analytic functions of ε at $\varepsilon = 0$, i.e.

$$A_{ij}(\varepsilon) = \sum_{r=0}^{\infty} \frac{\varepsilon^r}{r!} A_{ij}^{(r)}, \quad B_i(\varepsilon) = \sum_{r=0}^{\infty} \frac{\varepsilon^r}{r!} B_i^{(r)}.$$

The input $w(t)$ is zero-mean stationary Gaussian white noise with intensity matrix $V > 0$. The initial conditions $x_0(\varepsilon)$ and $z_0(\varepsilon)$ are jointly Gaussian random vectors independent of w with moments that are analytic at $\varepsilon = 0$. T_0 guarantee asymptotic stability of (A 1) for sufficiently small ε, it is assumed that

$$\begin{aligned} &\operatorname{Re}\lambda[A_{11}(0) - A_{12}(0)A_{22}^{-1}(0)A_{21}(0)] < 0, \\ &\operatorname{Re}\lambda[A_{22}(0)] < 0. \end{aligned} \tag{A 2}$$

The purpose of this Appendix is to study approximations $u(t)$ and $v(t)$ of $x(t)$ and $z(t)$ respectively for small ε, where $u(t)$ and $v(t)$ are defined by

$$\dot{u}(t) = \tilde{A}_{11}(\varepsilon)u(t) + \tilde{A}_{12}(\varepsilon)v(t) + \mu\tilde{B}_1(\varepsilon)w(t), \quad u(0) = u_0(\varepsilon), \tag{A 3a}$$

$$\varepsilon\dot{v}(t) = \tilde{A}_{21}(\varepsilon)u(t) + \tilde{A}_{22}(\varepsilon)v(t) + \mu\tilde{B}_2(\varepsilon)w(t), \quad v(0) = v_0(\varepsilon). \tag{A 3b}$$

The matrices $\tilde{A}_{ij}(\varepsilon)$ and $\tilde{B}_i(\varepsilon)$ are analytic functions of ε which are $O(\varepsilon^N)$ close to the corresponding matrices A_{ij} and B_i, where N is a positive integer, i.e.

$$\left.\begin{aligned} \Delta A_{ij} &\overset{\Delta}{=} A_{ij} - \tilde{A}_{ij} = \sum_{r=N}^{\infty} \frac{\varepsilon^r}{r!} (\Delta A_{ij})^{(r)} = O(\varepsilon^N), \\ \Delta B_i &\overset{\Delta}{=} B_i - \tilde{B}_i = \sum_{r=N}^{\infty} \frac{\varepsilon^r}{r!} (\Delta B_i)^{(r)} = O(\varepsilon^N). \end{aligned}\right\} \tag{A 4}$$

The initial conditions $u_0(\varepsilon)$ and $v_0(\varepsilon)$ are jointly Gaussian with moments that are analytic at $\varepsilon = 0$. It is assumed that

$$\mathrm{E}\left\{\begin{pmatrix} u_0(\varepsilon) - x_0(\varepsilon) \\ v_0(\varepsilon) - z_0(\varepsilon) \end{pmatrix}\begin{pmatrix} u_0(\varepsilon) - x_0(\varepsilon) \\ v_0(\varepsilon) - z_0(\varepsilon) \end{pmatrix}^{\mathrm{T}}\right\} = \sum_{r=2N}^{\infty} \frac{\varepsilon^r}{r!}\Gamma^{(r)} = O(\varepsilon^{2N}). \tag{A 5}$$

In other words, the approximation (A 3) is obtained by making $O(\varepsilon^N)$ perturbations in the matrix coefficients and initial conditions of (A 1), where the perturbations in initial conditions are taken in a mean-square sense. In order to validate approximating $x(t)$ and $z(t)$ by $u(t)$ and $v(t)$, we study the means and variances of the approximation errors $e_x(t) = x(t) - u(t)$ and $e_z(t) \triangleq z(t) - v(t)$ for small ε. The mean of a stochastic process will be denoted by a bar over the variables, e.g. $\bar{x}(t)$ is the mean of $x(t)$. We begin by stating a lemma that will be used repeatedly in our study. Lemma A.1 follows from the decoupling transformation of Chapter 2 in an obvious way.

Lemma A.1

Consider the singularly perturbed system

$$\dot{\xi}(t) = F_{11}(\varepsilon)\xi(t) + F_{12}(\varepsilon)\eta(t) + f_1(t, \varepsilon), \quad \xi(0) = C_1(\varepsilon),$$
$$\varepsilon\dot{\eta}(t) = F_{21}(\varepsilon)\xi(t) + F_{22}(\varepsilon)\eta(t) + f_2(t, \varepsilon), \quad \eta(0) = C_2(\varepsilon),$$

where

$$\begin{aligned} &F_{ij}(\varepsilon) \text{ are analytic at } \varepsilon && = 0, \\ &\mathrm{Re}\,\lambda[F_{11}(0) - F_{12}(0)F_{22}^{-1}(0)F_{21}(0)] && < 0, \\ &\mathrm{Re}\,\lambda[F_{22}(0)] && < 0, \\ &\|f_i(t, \varepsilon)\| && \leq K_i\varepsilon^\beta \quad \forall (t, \varepsilon) \in [0, \infty) \times (0, \varepsilon_0), \\ &\|C_i(\varepsilon)\| && \leq \bar{C}_i\varepsilon^\beta, \quad \varepsilon \in (0, \varepsilon_0], \end{aligned}$$

where β is a nonnegative number. Then there exists an $\varepsilon^* > 0$ such that for all $\varepsilon \in (0, \varepsilon^*)$, $\xi(t)$ and $\eta(t)$ exist for all $t \geq 0$ and satisfy

$$\xi(t, \varepsilon) = O(\varepsilon^\beta), \quad \eta(t, \varepsilon) = O(\varepsilon^\beta),$$

where $O(\varepsilon^\beta)$ holds uniformly for all $t \geq 0$.

The means $\bar{x}(t)$ and $\bar{z}(t)$ are governed by

$$\dot{\bar{x}}(t) = A_{11}(\varepsilon)\bar{x}(t) + A_{12}(\varepsilon)\bar{y}(t), \quad \bar{x}(0) = \bar{x}_0(\varepsilon), \tag{A 6a}$$
$$\varepsilon\dot{\bar{z}}(t) = A_{21}(\varepsilon)\bar{x}(t) + A_{22}(\varepsilon)\bar{z}(t), \quad \bar{z}(0) = \bar{z}_0(\varepsilon). \tag{A 6b}$$

The stability criterion (A 2) and Lemma A.1 guarantee that, for sufficiently small ε, $\bar{x}(t)$ and $\bar{z}(t)$ exist and are uniformly bounded in ε and t for all $t \geq 0$. Employing (A 4) and (A 5), it is easy to see that $\bar{e}_x(t)$ and $\bar{e}_z(t)$ satisfy

$$\bar{e}_x(t) = \tilde{A}_{11}(\varepsilon)\bar{e}_x(t) + \tilde{A}_{12}(\varepsilon)\bar{e}_z(t) + O(\varepsilon^N), \quad \bar{e}_x(0) = O(\varepsilon^N), \tag{A 7a}$$
$$\varepsilon\bar{e}_z(t) = \tilde{A}_{21}(\varepsilon)\bar{e}_x(t) + \tilde{A}_{22}(\varepsilon)\bar{e}_z(t) + O(\varepsilon^N), \quad \bar{e}_z(0) = O(\varepsilon^N). \tag{A 7b}$$

Again, application of Lemma A.1 and the use of (A 2) yields

$$\bar{e}_x(t) = O(\varepsilon^N), \quad \bar{e}_z(t) = O(\varepsilon^N) \quad \forall t \geq 0. \tag{A 8}$$

To study the variances of $e_x(t)$ and $e_z(t)$, we consider the following system of differential equations driven by white noise:

$$\begin{bmatrix} \dot{x}(t) \\ \dot{e}_x(t) \\ \dot{z}(t) \\ \dot{e}_z(t) \end{bmatrix} = \begin{bmatrix} A_{11} & 0 & A_{12} & 0 \\ \Delta A_{11} & \tilde{A}_{11} & \Delta A_{12} & \tilde{A}_{12} \\ \dfrac{A_{21}}{\varepsilon} & 0 & \dfrac{A_{22}}{\varepsilon} & 0 \\ \dfrac{\Delta A_{21}}{\varepsilon} & \dfrac{\tilde{A}_{21}}{\varepsilon} & \dfrac{\Delta A_{22}}{\varepsilon} & \dfrac{\tilde{A}_{22}}{\varepsilon} \end{bmatrix} \begin{bmatrix} x(t) \\ e_x(t) \\ z(t) \\ e_z(t) \end{bmatrix} + \mu \begin{bmatrix} B_1 \\ \Delta B_1 \\ \dfrac{B_2}{\varepsilon} \\ \dfrac{\Delta B_2}{\varepsilon} \end{bmatrix} w(t).$$

The above equation is rewritten as

$$\dot{y}(t) = \mathscr{A} y(t) + \mu \mathscr{B} w(t),$$

with obvious definitions for y, $\mathscr{A}$ and $\mathscr{B}$. Let Q^0 be the variance matrix of the initial condition $y(0)$. The matrix Q^0 is partitioned, compatibly with the partitioning of y, as

$$Q^0 = \begin{bmatrix} Q^0_{11} & Q^0_{12} & Q^0_{13} & Q^0_{14} \\ Q^{0\mathrm{T}}_{12} & Q^0_{22} & Q^0_{23} & Q^0_{24} \\ Q^{0\mathrm{T}}_{13} & Q^{0\mathrm{T}}_{23} & Q^0_{33} & Q^0_{34} \\ Q^{0\mathrm{T}}_{14} & Q^{0\mathrm{T}}_{24} & Q^{0\mathrm{T}}_{34} & Q^0_{44} \end{bmatrix}.$$

The initial condition closeness assumption (A 5) and the use of the Schwartz inequality imply

$$\left.\begin{aligned} Q^0_{ij} &= O(\varepsilon^N) \quad \text{for } ij = 12, 14, 23 \text{ and } 34, \\ Q^0_{ij} &= O(\varepsilon^{2N}) \quad \text{for } ij = 22, 24, \text{ and } 44. \end{aligned}\right\} \tag{A 9}$$

Let $Q(t)$ be the variance matrix of $y(t)$; then $Q(t)$ satisfies the Lyapunov matrix differential equation

$$\dot{Q} = \mathscr{A} Q + Q \mathscr{A}^{\mathrm{T}} + \mu^2 \mathscr{B} V \mathscr{B}^{\mathrm{T}}, \quad Q(0) = Q^0. \tag{A 10}$$

Because of the presence of $1/\varepsilon$ terms in $\mathscr{A}$ and $\mathscr{B}$, Q is sought in the form

$$Q = \begin{bmatrix} Q_{11} & Q_{12} & Q_{13} & Q_{14} \\ Q^{\mathrm{T}}_{12} & Q_{22} & Q_{23} & Q_{24} \\ Q^{\mathrm{T}}_{13} & Q^{\mathrm{T}}_{23} & \dfrac{1}{\varepsilon} Q_{33} & \dfrac{1}{\varepsilon} Q_{34} \\ Q^{\mathrm{T}}_{14} & Q^{\mathrm{T}}_{24} & \dfrac{1}{\varepsilon} Q^{\mathrm{T}}_{34} & \dfrac{1}{\varepsilon} Q_{44} \end{bmatrix} \tag{A 11}$$

Substituting (A 11) in (A 10) and partitioning the Lyapunov equation, we obtain the following ten equations:

$$\dot{Q}_{11} = A_{11}Q_{11} + Q_{11}A_{11}^{T} + A_{12}Q_{13}^{T} + Q_{13}A_{12}^{T} + \mu^2 B_1 V B_1^{T},$$
$$Q_{11}(0) = Q_{11}^{0}, \tag{A 12a}$$

$$\dot{Q}_{12} = A_{11}Q_{12} + A_{12}Q_{23}^{T} + Q_{11}(\Delta A_{11})^{T} + Q_{12}\tilde{A}_{11}^{T} + Q_{13}(\Delta A_{12})^{T}$$
$$+ Q_{14}\tilde{A}_{12}^{T} + \mu^2 B_1 V(\Delta B_1)^{T}, \quad Q_{12}(0) = Q_{12}^{0}, \tag{A 12b}$$

$$\varepsilon\dot{Q}_{13} = \varepsilon A_{11}Q_{13} + A_{12}Q_{33} + Q_{11}A_{21}^{T} + Q_{13}A_{22}^{T} + \mu^2 B_1 V B_2^{T},$$
$$Q_{13}(0) = Q_{13}^{0}, \tag{A 12c}$$

$$\varepsilon\dot{Q}_{14} = \varepsilon A_{11}Q_{14} + A_{12}Q_{34} + Q_{11}(\Delta A_{21})^{T} + Q_{12}\tilde{A}_{21}^{T} + Q_{13}(\Delta A_{22})^{T}$$
$$+ Q_{14}\tilde{A}_{22}^{T} + \mu^2 B_1 V(\Delta B_2)^{T}, \quad Q_{14}^{(0)} = Q_{14}^{0}, \tag{A 12d}$$

$$\dot{Q}_{22} = \Delta A_{11}Q_{12} + \tilde{A}_{11}Q_{22} + \Delta A_{12}Q_{23}^{T} + \tilde{A}_{12}Q_{24}^{T} + Q_{12}^{T}(\Delta A_{11})^{T}$$
$$+ Q_{22}\tilde{A}_{11}^{T} + Q_{23}(\Delta A_{12})^{T} + Q_{24}\tilde{A}_{12}^{T}$$
$$+ \mu^2 \Delta B_1 V(\Delta B_1)^{T}, \quad Q_{22}(0) = Q_{22}^{0}, \tag{A 12e}$$

$$\varepsilon\dot{Q}_{23} = \varepsilon\Delta A_{11}Q_{13} + \varepsilon\tilde{A}_{11}Q_{23} + \Delta A_{12}Q_{33} + \tilde{A}_{12}Q_{34}^{T} + Q_{12}^{T}A_{21}^{T}$$
$$+ Q_{23}A_{22}^{T} + \mu^2 \Delta B_1 V B_2^{T}, \quad Q_{23}(0) = Q_{23}^{0}, \tag{A 12f}$$

$$\varepsilon\dot{Q}_{24} = \varepsilon\Delta A_{11}Q_{14} + \varepsilon\tilde{A}_{11}Q_{24} + \Delta A_{12}Q_{34} + \tilde{A}_{12}Q_{44} + Q_{12}^{T}(\Delta A_{21})^{T}$$
$$+ Q_{22}\tilde{A}_{21}^{T} + Q_{23}(\Delta A_{22})^{T} + Q_{24}\tilde{A}_{22}^{T}$$
$$+ \mu^2 \Delta B_1 V(\Delta B_2)^{T}, \quad Q_{24}(0) = Q_{24}^{0}, \tag{A 12g}$$

$$\varepsilon\dot{Q}_{33} = A_{22}Q_{33} + Q_{33}A_{22}^{T} + \varepsilon A_{21}Q_{13} + \varepsilon Q_{13}^{T}A_{21}^{T}$$
$$+ \mu^2 B_2 V B_2^{T}, \quad Q_{33}(0) = \varepsilon Q_{33}^{0}, \tag{A 12h}$$

$$\varepsilon\dot{Q}_{34} = \varepsilon A_{21}Q_{14} + A_{22}Q_{34} + \varepsilon Q_{13}^{T}(\Delta A_{21})^{T} + \varepsilon Q_{23}^{T}\tilde{A}_{21}^{T} + Q_{33}(\Delta A_{22})^{T}$$
$$+ Q_{34}\tilde{A}_{22}^{T} + \mu^2 B_2 V(\Delta B_2)^{T}, \quad Q_{34}(0) = \varepsilon Q_{34}^{0}, \tag{A 12i}$$

$$\varepsilon\dot{Q}_{44} = \varepsilon\Delta A_{21}Q_{14} + \varepsilon\tilde{A}_{21}Q_{24} + \Delta A_{22}Q_{34} + \tilde{A}_{22}Q_{44} + \varepsilon Q_{14}^{T}(\Delta A_{21})^{T}$$
$$+ \varepsilon Q_{24}^{T}\tilde{A}_{21}^{T} + Q_{34}^{T}(\Delta A_{22})^{T} + Q_{44}\tilde{A}_{22}^{T}$$
$$+ \mu^2 \Delta B_2 V(\Delta B_2)^{T}, \quad Q_{44}(0) = \varepsilon Q_{44}^{0}. \tag{A 12j}$$

Thus, by seeking Q in the form (A 11) the study of the variances of e_x and e_z has been replaced by the study of the solution of the linear time-invariant singularly perturbed system (A 12). We divide (A 12a)–(A 12j) into three groups of equations and apply Lemma A.1 successively. Consider first (A 12a, c, h). These three equations form a singularly perturbed system in Q_{11}, Q_{13} and Q_{33}. This system of matrix equations can be rearranged in the form of vector equations. Employing Lemma A.1 and the stability condition (A 2) shows that, for sufficiently small ε, Q_{11}, Q_{13} and Q_{33} exist for all $t \geq 0$ and are $O(1)$. Next, consider (A 12b, d, f, i). These equations form a singularly perturbed system in Q_{12}, Q_{14}, Q_{23} and Q_{34} with driving terms dependent on Q_{11}, Q_{13} and Q_{33}. Equations (A 4) and (A 9) imply that the driving terms and initial conditions of these four equations are $O(\varepsilon^N)$. Thus by applying Lemma A.1 we conclude that for sufficiently small ε,

$$Q_{ij}(t, \varepsilon) = O(\varepsilon^N) \quad \text{for } ij = 12, 14, 23 \text{ and } 34. \tag{A 13}$$

Finally, consider (A 12e, g, j), which form a singularly perturbed system in Q_{22}, Q_{24} and Q_{44}. Equations (A 4), (A 9) and (A 13) show that the driving terms and initial conditions of these three equations are $O(\varepsilon^{2N})$. Thus Lemma A.1 implies

$$Q_{ij}(t, \varepsilon) = O(\varepsilon^{2N}) \quad \text{for } ij = 22, 24 \text{ and } 44. \tag{A 14}$$

Combining (A 8) with (A 14) and recalling the form of the variance matrix (A 11) completes the proof of the following theorem.

Theorem A.1

Consider (A 1) and (A 3) and assume that (A 2), (A 4) and (A 5) are satisfied. Then there exists an $\varepsilon^* > 0$ such that, for all $\varepsilon \in (0, \varepsilon^*)$, (A 15) holds for all $t \geq 0$:

$$\left.\begin{aligned} x(t) - u(t) &= O(\varepsilon^N), \\ z(t) - v(t) &= O(\varepsilon^{N-1/2}). \end{aligned}\right\} \tag{A 15}$$

Theorem A.1 establishes that $(u(t), v(t))$ is a valid approximation of $(x(t), z(t))$ as $\varepsilon \to 0$. It is interesting to observe that the mean-square order of approximation of the slow variable x is the same as the order of perturbation of parameters and initial conditions, while for the fast variable z the order of approximation is less by $\frac{1}{2}$ owing to the effect of the $O(1/\varepsilon)$ variance of z.

Frequently, the solution of (A 1) is of interest only at the steady state, i.e. as $t \to \infty$. In the steady-state situation the initial condition closeness assumption (A 5) is not required since the part of the solution due to initial conditions decays to zero as $t \to \infty$, in view of the asymptotic stability of the system. Furthermore, sharper bounds on the approximation errors can be obtained for the case $\mu < 1$. Theorem A.2 summarizes our conclusions for the steady-state case.

Theorem A.2

Consider (A 1) and (A 3) and assume that (A 2) and (A 4) are satisfied. Then there exists an $\varepsilon^* > 0$ such that, for all $\varepsilon \in (0, \varepsilon^*)$ (A 16) holds as $t \to \infty$:

$$\left.\begin{aligned} x(t) - u(t) &= \mu O(\varepsilon^N), \\ z(t) - v(t) &= \mu O(\varepsilon^{N-1/2}). \end{aligned}\right\} \tag{A 16}$$

Proof Consider $y^T(t) = (x^T(t), e_x^T(t), z^T(t), e_z^T(t))$ as before. At steady state, the variance of $y(t)$ satisfies an algebraic Lyapunov equation that is the equilibrium of (A 10). Seeking Q in the form (A 11), it follows that the Q_{ij} satisfy algebraic equations that are the equilibrium of (A 12). Writing $Q_{ij} = \mu^2 \bar{Q}_{ij}$ and dividing the equations by μ^2 shows that the $\bar{Q}_{ij}$ satisfy the equilibrium of (A 12) with $\mu = 1$. The expression (A 17) follows as a limiting case of (A 13) and (A 14) when $t \to \infty$.

$$\left.\begin{aligned} \bar{Q}_{ij}(\varepsilon) &= O(\varepsilon^N) \quad \text{for } ij = 12, 14, 23 \text{ and } 34, \\ \bar{Q}_{ij}(\varepsilon) &= O(\varepsilon^{2N}) \quad \text{for } ij = 22, 24, \text{ and } 44. \end{aligned}\right\} \tag{A 17}$$

Equation (A 17) implies (A 16). □

We conclude this Appendix by studying the effect of employing the approximation (A 3) on evaluating a quadratic form in x and z given by

$$\sigma = \lim_{t\to\infty} \mathrm{E}\left\{\begin{pmatrix} x(t) \\ z(t) \end{pmatrix}^{\mathrm{T}} (C_1(\varepsilon) \quad C_2(\varepsilon))^{\mathrm{T}}(C_1(\varepsilon) \quad C_2(\varepsilon))\begin{pmatrix} x(t) \\ z(t) \end{pmatrix}\right\}, \tag{A18}$$

where $C_1(\varepsilon)$ and $C_2(\varepsilon)$ are analytic at $\varepsilon = 0$. Such quadratic forms appear in the steady-state LQG control problem (Section 4.4). We examine the following approximation of σ by $\tilde{\sigma}$:

$$\tilde{\sigma} = \lim_{t\to\infty} \mathrm{E}\left\{\begin{pmatrix} u(t) \\ v(t) \end{pmatrix}^{\mathrm{T}} (\tilde{C}_1(\varepsilon) \quad \tilde{C}_2(\varepsilon))^{\mathrm{T}}(\tilde{C}_1(\varepsilon) \quad \tilde{C}_2(\varepsilon))\begin{pmatrix} u(t) \\ v(t) \end{pmatrix}\right\} \tag{A 19}$$

where

$$\Delta C_i(\varepsilon) = C_i(\varepsilon) - \tilde{C}_i(\varepsilon) = \sum_{r=N}^{\infty} \frac{\varepsilon^r}{r!}(\Delta C_i)^{(r)} = O(\varepsilon^N). \tag{A 20}$$

Theorem A.3 shows the order of magnitude of the error in approximating σ by $\tilde{\sigma}$.

Theorem A.3

Consider (A 1), (A 3) and the scalar quadratic forms (A 18) and (A 19). Assume that (A 2) and (A 4) are satisfied. Then there exists an $\varepsilon^* > 0$ such that, for all $\varepsilon \in (0, \varepsilon^*)$,

$$\frac{\Delta\sigma}{\sigma} \triangleq \frac{\tilde{\sigma} - \sigma}{\sigma} = O(\varepsilon^N). \tag{A 21}$$

Proof σ and $\tilde{\sigma}$ are given by

$$\sigma = \mu^2(\mathrm{tr}\,[C_1^{\mathrm{T}} C_1 \bar{Q}_{11} + 2C_1^{\mathrm{T}} C_2 \bar{Q}_{13}^{\mathrm{T}}] + \mathrm{tr}\,[\varepsilon^{-1} C_2^{\mathrm{T}} C_2 \bar{Q}_{33}]),$$

$$\tilde{\sigma} = \mu^2(\mathrm{tr}\,[\tilde{C}_1^{\mathrm{T}}\tilde{C}_1(\bar{Q}_{11} - 2\bar{Q}_{12} + \bar{Q}_{22}) + 2\tilde{C}_1^{\mathrm{T}}\tilde{C}_2(\bar{Q}_{13}^{\mathrm{T}} - \bar{Q}_{14}^{\mathrm{T}} - \bar{Q}_{23}^{\mathrm{T}} + \bar{Q}_{24}^{\mathrm{T}})]$$
$$+ \mathrm{tr}\,[\varepsilon^{-1}\tilde{C}_2^{\mathrm{T}}\tilde{C}_2(\bar{Q}_{33} - 2\bar{Q}_{34} + \bar{Q}_{44})]),$$

where the $\bar{Q}_{ij}$ are as defined in the proof of Theorem A.2. Using (A 17) and (A 20) we get

$$\Delta\sigma \triangleq \tilde{\sigma} - \sigma = -2\mu^2\varepsilon^{N-1}\,\mathrm{tr}\,[(\Delta C_2)^{(N)\mathrm{T}} C_2(0)\bar{Q}_{33}(0)$$
$$+ C_2^{\mathrm{T}}(0)C_2(0)\bar{Q}_{34}^{(N)}] + O(\varepsilon^N), \tag{A 22}$$

where $\bar{Q}_{34}^{(N)}$ is the Nth-order term in the Maclaurin series of $\bar{Q}_{34}(\varepsilon)$. It can be verified that $\bar{Q}_{33}(0)$ and $\bar{Q}_{34}^{(N)}$ are respectively the unique solutions of the algebraic Lyapunov equations

$$0 = A_{22}(0)\bar{Q}_{33}(0) + \bar{Q}_{33}(0)A_{22}^{\mathrm{T}}(0) + B_2(0)VB_2^{\mathrm{T}}(0), \tag{A 23}$$

$$0 = A_{22}(0)\bar{Q}_{34}^{(N)} + \bar{Q}_{34}^{(N)}A_{22}^{\mathrm{T}}(0) + \bar{Q}_{33}(0)(\Delta A_{22})^{(N)\mathrm{T}}$$
$$+ B_2(0)V(\Delta B_2)^{(N)\mathrm{T}}. \tag{A 24}$$

On the other hand

$$\sigma = \mu^2(\varepsilon^{-1}\,\mathrm{tr}\,[C_2^{\mathrm{T}}(0)C_2(0)\bar{Q}_{33}(0)] + O(1)). \tag{A 25}$$

If $\mathrm{tr}\,[C_2^{\mathrm{T}}(0)C_2(0)\bar{Q}_{33}(0)] \neq 0$, dividing (A 22) by (A 25) yields (A 21). If $\mathrm{tr}\,[C_2^{\mathrm{T}}(0)C_2(0)\bar{Q}_{33}(0)] = 0$ such division will only show that $\Delta\sigma/\sigma = O(\varepsilon^{N-1})$; further analysis is needed in this case to establish (A 21). Suppose that

$$\mathrm{tr}\,[C_2^{\mathrm{T}}(0)C_2(0)\bar{Q}_{33}(0)] = 0, \tag{A 26}$$

and recall that $\bar{Q}_{33}(0)$ satisfies (A 23); hence it is given by

$$\bar{Q}_{33}(0) = \int_0^\infty \mathrm{e}^{A_{22}(0)t} B_2(0) V B_2^{\mathrm{T}}(0) \mathrm{e}^{A_{22}^{\mathrm{T}}(0)t}\, \mathrm{d}t. \tag{A 27}$$

Substituting (A 27) in (A 26) yields

$$\int_0^\infty C_2(0)\mathrm{e}^{A_{22}(0)t} B_2(0) V B_2^{\mathrm{T}}(0) \mathrm{e}^{A_{22}^{\mathrm{T}}(0)t} C_2^{\mathrm{T}}(0)\, \mathrm{d}t = 0,$$

which implies

$$C_2(0)\mathrm{e}^{A_{22}(0)t} B_2(0) \equiv 0, \quad t \geq 0. \tag{A 28}$$

Hence

$$C_2(0)\bar{Q}_{33}(0) = 0. \tag{A 29}$$

Since, by (A 24),

$$\bar{Q}_{34}^{(N)} = \int_0^\infty \mathrm{e}^{A_{22}(0)t}[\bar{Q}_{33}(0)(\Delta A_{22})^{(N)\mathrm{T}} + B_2(0)V(\Delta B_2)^{(N)\mathrm{T}}]\mathrm{e}^{A_{22}^{\mathrm{T}}(0)t}\, \mathrm{d}t,$$

using (A 27) and (A 28) it can be shown that

$$C_2(0)\bar{Q}_{34}^{(N)} = 0. \tag{A 30}$$

Equations (A 29) and (A 30) imply that $\Delta\sigma = O(\varepsilon^N)$, which completes the proof of Theorem A.3. □

The above analysis shows that $\Delta\sigma$ is $O(1)$ if and only if (A 28) is satisfied, which is equivalent to

$$C_2(0)[pI - A_{22}(0)]^{-1} B_2(0) \equiv 0 \quad \forall p.$$

That is to say, $\Delta\sigma$ is $O(1)$ if and only if the fast transfer matrix $G_{\mathrm{f}}(p, \varepsilon)$ from the input w to the output $y_{\mathrm{c}} = C_1 x + C_2 z$ is identically zero at $\varepsilon = 0$.

APPENDIX B

We verify the matrix identity $Q_1(0) - H_1(0)Q_2(0) = Q_s$, which is used in the proof of Theorem 3.1 of Chapter 4.

The matrix identity (B 1) follows by dualizing the coefficients of x in (4.4) and (4.23) of Chapter 3:

$$(P_s C_1^T + P_m C_2^T)V^{-1} = A_{12}A_{22}^{-1}Q_f + Q_s(I - C_2 A_{22}^{-1}Q_f). \qquad \text{(B 1)}$$

Now

$$\begin{aligned} Q_1(0) - H_1(0)C_2(0) &= Q_1(0) - (A_{12} - Q_1(0)C_2)(A_{22} - Q_f C_2)^{-1}Q_f \\ &= Q_1(0)[I + C_2(A_{22} - Q_f C_2)^{-1}Q_f] \\ &\quad -A_{12}(A_{22} - Q_f C_2)^{-1}Q_f. \end{aligned}$$

But

$$(A_{22} - Q_f C_2)^{-1}Q_f = A_{22}^{-1}Q_f[I + C_2(A_{22} - Q_f C_2)^{-1}Q_f].$$

Hence

$$\begin{aligned} Q_1(0) - H_1(0)Q_2(0) &= [(P_s C_1^T + P_m C_2^T)V^{-1} - A_{12}A_{22}^{-1}Q_f] \\ &\quad \times [I + C_2(A_{22} - Q_f C_2)^{-1}Q_f]. \end{aligned}$$

Using (B 1), we obtain

$$Q_1(0) - H_1(0)Q_2(0) = Q_s(I - C_2A_{22}^{-1}Q_f)[I + C_2(A_{22} - Q_f C_2)^{-1}Q_f].$$

Observing that

$$(I - C_2A_{22}^{-1}Q_f)[I + C_2(A_{22} - Q_f C_2)^{-1}Q_f] = I$$

shows that

$$Q_1(0) - H_1(0)Q_2(0) = Q_s.$$

INDEX